Role of Genomics in the Management of Hypertension

Role of Genomics in the Management of Hypertension

Editors

Paolo Mulatero
Silvia Monticone

MDPI • Basel • Beijing • Wuhan • Barcelona • Belgrade • Manchester • Tokyo • Cluj • Tianjin

Editors
Paolo Mulatero
University of Turin
Italy

Silvia Monticone
University of Turin
Italy

Editorial Office
MDPI
St. Alban-Anlage 66
4052 Basel, Switzerland

This is a reprint of articles from the Special Issue published online in the open access journal *International Journal of Molecular Sciences* (ISSN 1422-0067) (available at: https://www.mdpi.com/journal/ijms/special_issues/hypertension).

For citation purposes, cite each article independently as indicated on the article page online and as indicated below:

LastName, A.A.; LastName, B.B.; LastName, C.C. Article Title. *Journal Name* **Year**, *Article Number*, Page Range.

ISBN 978-3-03936-627-9 (Hbk)
ISBN 978-3-03936-628-6 (PDF)

Contents

About the Editors

Paolo Mulatero serves as a full professor of Internal Medicine at the University of Torino, Italy, where he is responsible for the Laboratory of Genetics and Molecular Biology of Arterial Hypertension and he is also head of the Hypertension Unit. After he completed his medical degree in 1991 at the University of Torino, he completed his residency in Internal Medicine at the same university in 1997. Between 1992 and 1993, he worked as a research fellow at the Blood Pressure Unit in Glasgow for studies on glucocorticoid receptors. Between 1995 and 1997, he worked as a research fellow at the Department of Pathologie Vasculaire et Endocrinologie Rénale of the Collège de France INSERM U36, Paris, France (Directed by Prof. P. Corvol), where he studied the 11βhydroxylase and aldosterone synthase genes and their roles in essential and secondary hypertension. Prof Mulatero has over 25 years of clinical and basic research experience in the field of arterial hypertension, particularly primary aldosteronism. In that regard, his research significantly contributed to demonstrating that primary aldosteronism is the most frequent form of secondary hypertension and to our understanding of the genetic determinants of both sporadic and familial forms. Paolo Mulatero is currently a member of the Endocrine Society, European Society of Hypertension, Italian Society of Hypertension, chairman of the ESH Endocrine Hypertension Working Group, and member of the Executive Committee of the Italian Society of Hypertension. He has delivered over 40 invited lectures worldwide and authored 190 scientific publications with an IF > 1000 and over 10.000 citations.

Silvia Monticone earned her Medical Doctor Degree in 2007 from the University of Torino. In 2013, she completed a residency in Internal Medicine and, in 2016, she earned a Ph.D. at the same university. Dr. Monticone has a strong interdisciplinary skillset with research interests spanning both basic science and clinical research topics in the field of arterial hypertension and, in particular, primary aldosteronism. In 2011, she spent one year as a research fellow at Georgia Health Sciences University in the Department of Physiology, where she studied, in the laboratory directed by Prof. W.E. Rainey), the molecular mechanisms responsible for hyperaldosteronism in carriers of KCNJ5 mutations. She has authored over 70 scientific publications and she actively participates in national and international conferences. Dr. Monticone is a member of the Italian Hypertension Society since 2008 and, in 2012, she won the Italian Hypertension Society grant. She currently works as a research scientist at the University of Torino, Italy.

Preface to "Role of Genomics in the Management of Hypertension"

Arterial hypertension, affecting about 1 billion people worldwide, is the strongest modifiable risk factor for cardiovascular disease and related disability, leading to 9.4 million deaths each year. Notwithstanding major advances in understanding the etiology of hypertension, it remains largely underdiagnosed and often undertreated in the general population. The pathophysiology of hypertension involves the complex interplay between environmental (such as high sodium intake, excess alcohol consumption, and mental stress) and genetic factors, which, in turn, result in the disruption of factors involved in blood pressure control.

Monogenic forms of hypertension are rare conditions, due to loss or gain of function mutations in genes ultimately involved in salt and water homeostasis. The identification of the molecular basis of Mendelian forms of hypertension has provided important insights into the pathophysiology of blood pressure regulation and enabled effective etiological therapy.

Twin and family studies have demonstrated that 30–50% of the individual risk comes from genetic factors and a family history of hypertension increases the risk of developing high blood pressure levels by four times. However, genome-wide association studies demonstrated that hypertension-associated variants can only explain 2–3% of blood pressure variance and the effect size for each identified SNP was about 1 mmHg for SBP and 0.5 mmHg for DBP.

Additionally, genome science is providing new tools for understanding variability in drug response and can be applied to patients affected by arterial hypertension, with the ultimate goal of improving drug efficacy and reducing toxicity.

This book focuses on recent advances in genetic and genomic alterations in patients affected by arterial hypertension and their potential clinical and therapeutic implications.

Paolo Mulatero, Silvia Monticone
Editors

 International Journal of
Molecular Sciences

Article

Role of Cryptochrome-1 and Cryptochrome-2 in Aldosterone-Producing Adenomas and Adrenocortical Cells

Martina Tetti [1], Isabella Castellano [2], Francesca Veneziano [2], Corrado Magnino [1], Franco Veglio [1], Paolo Mulatero [1] and Silvia Monticone [1,*]

[1] Division of Internal Medicine and Hypertension, Department of Medical Sciences, University of Torino, 10126 Torino, Italy; tetti.martina@gmail.com (M.T.); cmagnino@libero.it (C.M.); franco.veglio@unito.it (F.V.); paolo.mulatero@unito.it (P.M.)
[2] Division of Pathology, Department of Medical Sciences, University of Torino,10126 Torino, Italy; isabella.castellano@unito.it (I.C.); francesca.veneziano11@gmail.com (F.V.)
* Correspondence: silvia.monticone@unito.it; Tel.: +39-011-633-6959

Received: 7 April 2018; Accepted: 31 May 2018; Published: 5 June 2018

Abstract: Mice lacking the core-clock components, cryptochrome-1 (CRY1) and cryptochrome-2 (CRY2) display a phenotype of hyperaldosteronism, due to the upregulation of type VI 3β-hydroxyl-steroid dehydrogenase (*Hsd3b6*), the murine counterpart to the human type I 3β-hydroxyl-steroid dehydrogenase (*HSD3B1*) gene. In the present study, we evaluated the role of *CRY1* and *CRY2* genes, and their potential interplay with *HSD3B* isoforms in adrenal pathophysiology in man. Forty-six sporadic aldosterone-producing adenomas (APAs) and 20 paired adrenal samples were included, with the human adrenocortical cells HAC15 used as the in vitro model. In our cohort of sporadic APAs, *CRY1* expression was 1.7-fold [0.75–2.26] higher ($p = 0.016$), while *CRY2* showed a 20% lower expression [0.80, 0.52–1.08] ($p = 0.04$) in APAs when compared with the corresponding adjacent adrenal cortex. Type II 3β-hydroxyl-steroid dehydrogenase (*HSD3B2*) was 317-fold [200–573] more expressed than *HSD3B1*, and is the main *HSD3B* isoform in APAs. Both dehydrogenases were more expressed in APAs when compared with the adjacent cortex (5.7-fold and 3.5-fold, respectively, $p < 0.001$ and $p = 0.001$) and *HSD3B1* was significantly more expressed in APAs composed mainly of zona glomerulosa-like cells. Treatment with angiotensin II (AngII) resulted in a significant upregulation of *CRY1* (1.7 ± 0.25-fold, $p < 0.001$) at 6 h, and downregulation of *CRY2* at 12 h (0.6 ± 0.1-fold, $p < 0.001$), through activation of the AngII type 1 receptor. Independent silencing of *CRY1* and *CRY2* genes in HAC15 cells resulted in a mild upregulation of *HSD3B2* without affecting *HSD3B1* expression. In conclusion, our results support the hypothesis that *CRY1* and *CRY2*, being AngII-regulated genes, and showing a differential expression in APAs when compared with the adjacent adrenal cortex, might be involved in adrenal cell function, and in the regulation of aldosterone production.

Keywords: aldosterone-producing adenoma; *CRY1*; *CRY2*; *HSD3B1*; *HSD3B2*

---

1. Introduction

Primary aldosteronism (PA), affecting 6% of the general hypertensive population [1], and up to 20% of patients referred to hypertension units [2,3], is widely recognized as the leading cause of endocrine hypertension. Aldosterone-producing adenoma (APA) and bilateral adrenal hyperplasia (BAH) are the most frequent underlying causes of PA, while unilateral adrenal hyperplasia (UAH) is less common. The last few years witnessed major advances in the understanding of the molecular determinants leading to autonomous aldosterone overproduction in both sporadic and familial

PA. In particular, the introduction of next-generation sequencing allowed the identification of somatic mutations in four genes differently involved in Ca^{2+} homeostasis (*KCNJ5*, *ATP1A1*, *ATP2B3*, and *CACNA1D*), unraveling the genetic basis of approximately 50% of sporadic APAs [4–7]. Similarly, new insight was gained from mice lacking the core-clock components, cryptochrome-1 (CRY1) and cryptochrome-2 (CRY2) (*Cry*-null mice) [8]. Mammals, as well as many other organisms including plants, adapt most of their physiologic processes to a 24-h time cycle, generated by an internal molecular oscillator referred to as the circadian clock [9]. At the cellular level, circadian oscillations are generated by a series of genes, whose proteic products form a transcriptional autoregulatory feedback loop, where clock circadian regulator (CLOCK) and aryl hydrocarbon receptor nuclear translocator-like protein 1 (ARNTL, also known as BMAL1) act as positive regulators, while period (PER) and CRY act as negative regulators [10]. *Cry*-null mice displayed salt-sensitive hypertension due to chronic and autonomous aldosterone overproduction by the adrenal glands, as a consequence of the massive upregulation of type VI 3β-hydroxyl-steroid dehydrogenase (*Hsd3b6*), the murine counterpart to the human type I 3β-hydroxyl-steroid dehydrogenase (*HSD3B1*) gene [8]. HSD3B catalyzes the conversion of pregnenolone to progesterone, an enzymatic reaction required for aldosterone biosynthesis [11]; two different *HSD3B* isoforms are expressed in man—*HSD3B1* is mainly expressed in the placenta, while *HSD3B2* localizes primarily in adrenals and gonads [12]. Immunohistochemistry studies in normal human adrenals showed that *HSD3B2* is the predominant isoform, expressed through the zona glomerulosa and the zona fasciculata (ZF), while *HSD3B1* displays faint immunoreactivity, predominantly in the outermost layer zona glomerulosa (ZG) [8,13,14]. Moreover, in APA samples, *HSD3B1* expression was significantly correlated with the expression of the rate-limiting enzyme for aldosterone production—aldosterone synthase (CYP11B2) [15]. Despite much knowledge being gained from the *Cry*-null animal model, the significance of CRY1 and CRY2 in human adrenal function and aldosterone production is still unknown. So far, few reports have investigated the roles of *HSD3B1* and *HSD3B2* in sporadic PA. Therefore, in this study we aimed to (I) evaluate the expressions of *HSD3B1* and *HSD3B2* in a large cohort of 46 adrenal glands, removed from patients in whom a final diagnosis of unilateral PA was achieved; and (II) investigate the expression of *CRY1* and *CRY2* in unilateral sporadic PA, and their roles in aldosterone production in the HAC15 human adrenocortical cell model.

2. Results

2.1. Expression of CRY1, CRY2, HSD3B1, and HSD3B2 in Adrenal Tissues

The expression levels of *CRY1*, *CRY2*, *HSD3B1*, and *HSD3B2* were determined by real-time PCR in a cohort of 46 sporadic APAs, and 20 paired adjacent adrenal tissues. Within the same sample, the median expression of *CRY1* was 2.1-fold [1.45–2.87] higher than that of *CRY2*, consistently in both APA and UAH (Figure 1A). In our cohort, the expression of both *CRY* genes was neither associated with the cellular composition of the APAs (*CRY1* expression in ZG-like APAs: 1.46 [0.45–2.59], *CRY1* expression in ZF-like APAs: 0.96 [0.49–1.58], *p*-value 0.291; *CRY2* expression in ZG-like APAs: 1.24 [0.57–2.05], *CRY2* expression in ZF-like APAs: 0.84 [0.62–1.28], *p*-value 0.170) nor with the mutational status (*CRY1* expression in wild-type APAs: 1.39 [0.58–2.7], *CRY1* expression in *KCNJ5* mutant APAs: 0.89 [0.42–1.70], *CRY1* expression in *ATP1A1-ATP2B3* mutant APAs: 0.89 [0.49–1.09], *CRY1* expression in *CACNA1D* mutant APAs: 1.4 [0.64–2.61], *p*-value = 0.417; *CRY2* expression in wild-type APAs: 1.08 [0.61–1.98], *CRY2* expression in *KCNJ5* mutant APAs: 0.98 [0.54–1.41], *CRY2* expression in *ATP1A1-ATP2B3* mutant APAs: 0.89 [0.62–1.06], *CRY2* expression in *CACNA1D* mutant APAs: 1.61 [0.80–2.87], *p*-value = 0.170). While the median expression of *CRY1* was 1.7-fold [0.75–2.26] higher in APA tissues when compared with that in the adjacent adrenal cortex (*p* = 0.016), *CRY2* showed a 20% lower expression [0.80, 0.52–1.08] in the nodule when compared with the corresponding surrounding tissue (*p* = 0.04) (Figure 1B). Representative immunohistochemistry staining of frozen tissue sections showing the expression of *CRY1* and *CRY2* in APA and adjacent adrenal cortex is illustrated in Figure 2A–F.

In both APA and UAH samples, *HSD3B2* was the main isoform, with an overall median expression 317-fold [200–573] higher than that of *HSD3B1* ($p < 0.001$) (Figure 1C). *HSD3B1* transcription was significantly more abundant (median fold change 5.2, $p < 0.001$) in APAs that were composed mainly of ZG-like cells when compared with APAs that had a ZF-like morphology (Figure 1D). A tendency towards a higher *HSD3B2* expression in APAs composed mainly of ZG-like cells was observed, but the difference did not reach statistical significance (median fold change 1.8, $p = 0.051$) (Figure 1E). In addition, the median *HSD3B1/HSD3B2* relative ratio was 1.9-fold higher in APA samples composed mainly of ZG-like cells ($p = 0.003$) when compared with APAs composed mainly of ZF-like cells (Figure 1F). No differences in the expression of *HSD3B1* or *HSD3B2*, according to the mutational status (*HSD3B1* expression in wild-type APAs: 1.37 [0.42–4.5], *HSD3B1* expression in *KCNJ5* mutant APAs: 0.51 [0.30–1.97], *HSD3B1* expression in *ATP1A1-ATP2B3* mutant APAs: 0.66 [0.27–1.63], *HSD3B1* expression in *CACNA1D* mutant APAs: 2.27 [1.40–3.12], p-value = 0.212; *HSD3B2* expression in wild-type APAs: 1.0 [0.75–2.37], *HSD3B2* expression in *KCNJ5* mutant APAs: 1.01 [0.68–2.18], *HSD3B2* expression in *ATP1A1-ATP2B3* mutant APAs: 0.69 [0.46–0.93], *HSD3B2* expression in *CACNA1D* mutant APAs: 2.82 [1.16–3.19], p-value = 0.147) or the final diagnosis (*HSD3B1* expression in APAs: 1.37 [0.30–3.00], *HSD3B1* expression in UAHs: 0.66 [0.42–3.61], p-value 0.899; *HSD3B2* expression in APAs: 1.17 [0.66–2.63], *HSD3B2* expression in UAHs: 0.99 [0.71–1.63], p-value 0.523), were observed. Notably, both *HSD3B1* and *HSD3B2* were significantly more expressed in the main nodule when compared with adjacent adrenal tissue (5.7- and 3.5-fold, respectively, $p < 0.001$ and $p = 0.001$) (Figure 1G). Representative immunohistochemistry staining of frozen tissue sections showing the expression of *HSD3B1* and *HSD3B2* in APAs, according to the cellular composition, is illustrated in Figure 3A–D.

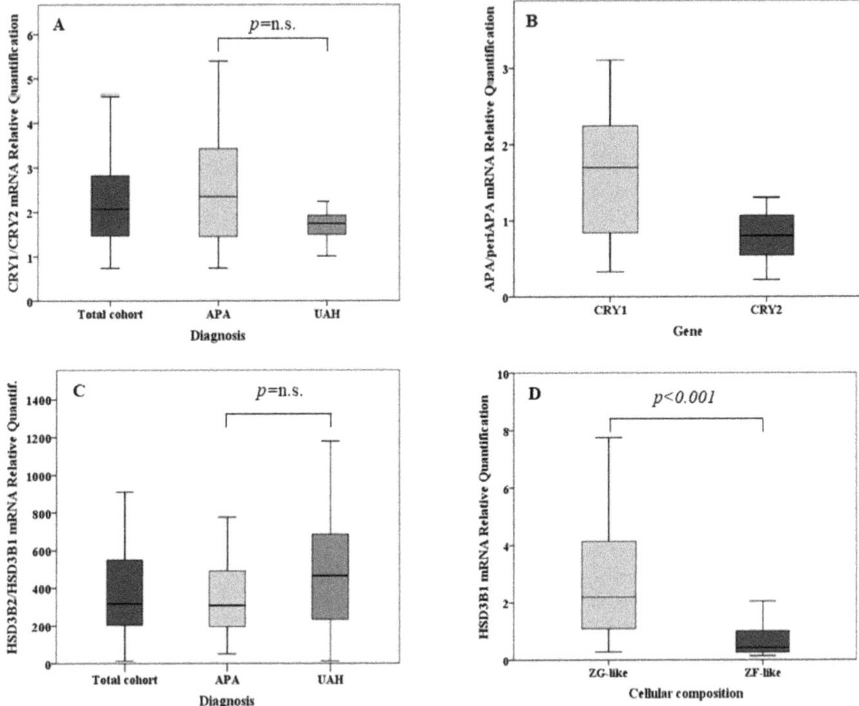

Figure 1. *Cont.*

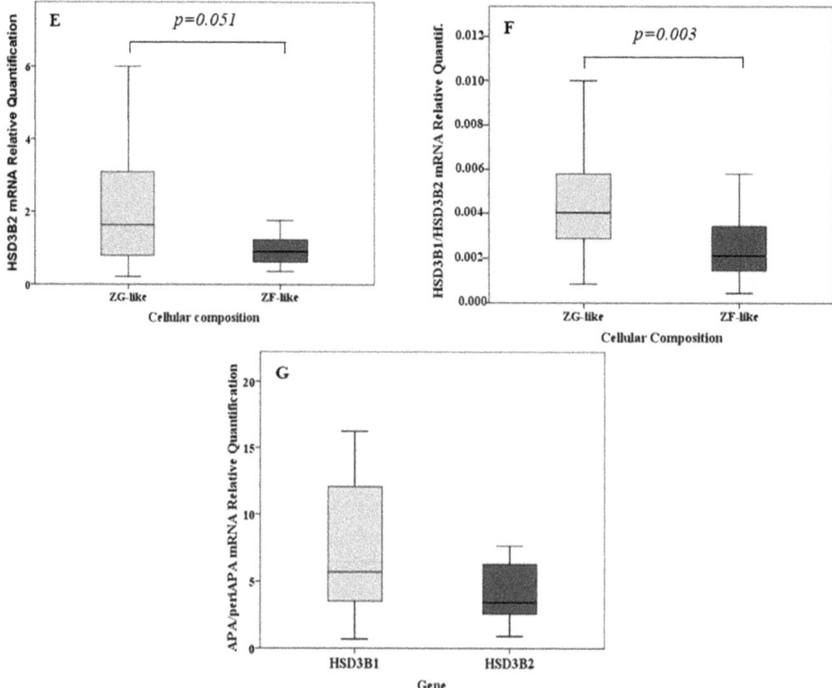

Figure 1. (**A**) Relative quantification of cryptochrome-1 (*CRY1*) mRNA over cryptochrome-2 (*CRY2*) messenger RNA (mRNA) in the total cohort, in aldosterone-producing adenoma (APA) samples (*n* = 35) and in unilateral adrenal hyperplasia (UAH) samples (*n* = 11). (**B**) Relative quantification of *CRY1* and *CRY2* mRNA in APA samples over that in the corresponding adjacent adrenal cortex (*n* = 20). (**C**) Relative quantification of type II 3β-hydroxyl-steroid dehydrogenase (*HSD3B2*) mRNA over type I 3β-hydroxyl-steroid dehydrogenase (*HSD3B1*) mRNA in the total cohort, in APA samples (*n* = 35) and in UAH samples (*n* = 11). (**D**) *HSD3B1* mRNA expression according to the cellular composition in the total cohort of adrenal samples. (**E**) *HSD3B2* mRNA expression according to the cellular composition in the total cohort of adrenal samples. (**F**) Relative quantification of *HSD3B1* mRNA over *HSD3B2* mRNA according to the cellular composition. (**G**) Relative quantification of *HSD3B1* and *HSD3B2* mRNA in APA samples over that in the corresponding adjacent adrenal cortex. For each box plot, the horizontal line represents the median, and the box and bar indicate the 25th to 75th and 5th to 95th percentiles, respectively.

2.2. Regulation of CRY1, CRY2, HSD3B1, and HSD3B2 Expression in HAC15 Cells

To investigate the potential roles of *CRY1* and *CRY2* genes in adrenal cell function and aldosterone production, we used HAC15 adrenocortical cells as an in vitro model. *CRY1* and *CRY2* genes were transcribed in HAC15 cells to a level comparable to that of a pooled set of APA samples, while *HSD3B2* was 35-fold (25–61, *p* < 0.001) more expressed than *HSD3B1*.

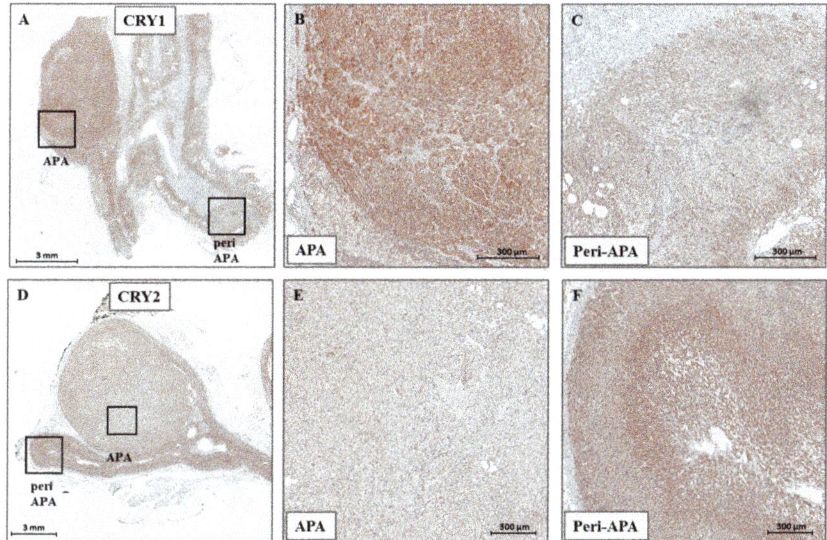

Figure 2. (**A–C**) Representative immunohistochemistry staining for *CRY1* in APAs. (**D–F**) Representative immunohistochemistry staining for *CRY2* in APAs. Magnifications in (**B,C**), and in (**E,F**) correspond to the boxed sections in (**A,D**), respectively.

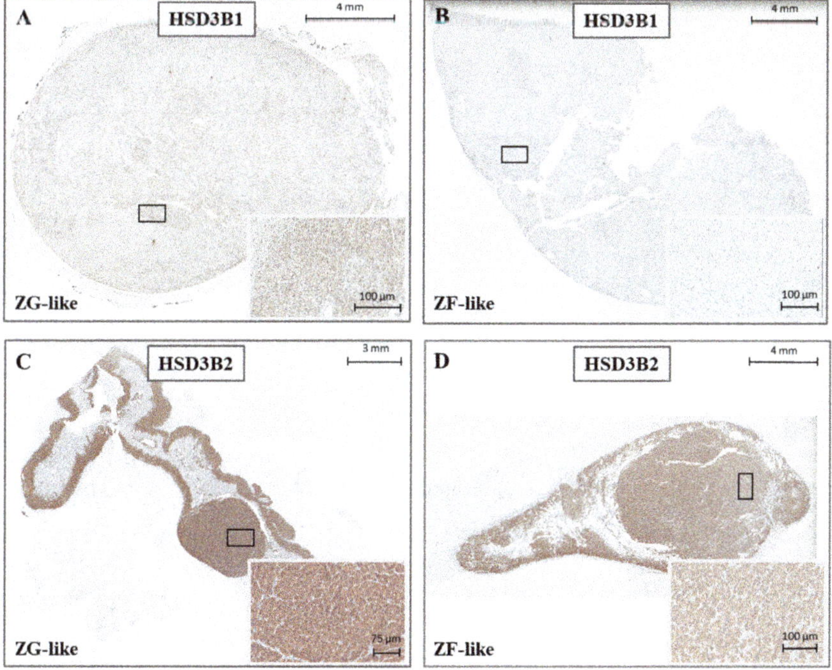

Figure 3. Representative immunohistochemistry staining for *HSD3B1* (**A,B**) and *HSD3B2* (**C,D**) in APAs, according to the cellular composition.

HAC15 cells, which were previously reported to express the AngII type 1 receptor [16], were stimulated with AngII ($\pm$1 μM irbesartan) or forskolin for 6, 12, and 24 h, and were then harvested for RNA extraction and gene-expression studies.

As expected, treatment with AngII (100 nM) resulted in a significant increase in *CYP11B2* expression at 12 h (68 $\pm$ 20-fold over basal, $p < 0.001$).

Treatment with AngII significantly increased the expression of *CRY1* mRNA within 6 h (1.7 $\pm$ 0.25-fold, $p < 0.001$). Following a peak in expression, the levels of *CRY1* mRNA returned to basal levels after 12 h of AngII treatment (Figure 4A). With respect to *CRY2* expression, stimulation with AngII resulted in a significant downregulation (0.6 $\pm$ 0.1-fold, $p < 0.001$) at 12 h (Figure 4B), followed by a return to basal levels at 24 h.

Treatment with forskolin, which mimics adrenocorticotropin (ACTH)-mediated elevation of intracellular cyclic adenosine monophosphate (cAMP), resulted in a downregulation of *CRY1* at 6, 12, and 24 h, and a downregulation of *CRY2* and at 12 and 24 h (Figure 4A,B).

Additionally, AngII and forskolin treatment positively regulated the transcription of both *HSD3B1* and *HSD3B2*. Following a 6-h stimulation with AngII, we observed that *HSD3B1* was 3.2 $\pm$ 2.4-fold ($p = 0.035$) more expressed when compared with basal conditions, while the maximum upregulation of *HSD3B2* was observed at 12 h (3.7 $\pm$ 0.4-fold, $p = 0.002$) (Figure 4C,D). Similarly, forskolin treatment induced a significant upregulation of both *HSD3B1*, with a peak at 6 h, and *HSD3B2*, with a peak at 12 h, (2.1 $\pm$ 1.2-fold and 5.1 $\pm$ 2.1-fold, $p = 0.03$ and $p = 0.001$, respectively).

Consistently, after 6 h of AngII stimulation, we detected a 1.5 $\pm$ 0.2-fold upregulation of *PER1*, that acts as a negative regulator of the core clock together with *CRY*, followed by a 42% reduction at 12 h, when compared with basal levels (Figure 4E).

Pre-treatment with irbersartan (1 μM) reverted the effects of AngII on *PER1*, *CRY1*, and *CRY2* expression (Figure 4F–H), indicating that the observed effects on gene expression were mediated by the activation of the AngII type 1 receptor.

2.3. Effect of CRY1 and CRY2 Silencing in HAC15 Cells

Our observation of the regulation of *CRY* genes by AngII, together with the experimental evidence available from Cry-null mice [8], prompted us to investigate the effect of *CRY* silencing on gene expression in HAC15 cells.

Silencing *CRY* genes by transfection of siRNA resulted in a 62% reduction in *CRY1* mRNA levels, and a 70% reduction in *CRY2* mRNA levels, measured by real-time PCR (Figure 5A,B). Notably, silencing *CRY1* induced a significant upregulation of *CRY2* (1.3 $\pm$ 0.2-fold, $p = 0.005$) (Figure 5B), which resulted in less efficient *CRY2* silencing when the double *CRY1* and *CRY2* siRNA assay was performed, for this reason simultaneous silencing of both genes was not allowed.

The expression of mRNA-encoding key enzymes involved in the production of aldosterone was examined. Transfection with *CRY1* siRNA resulted in a significant upregulation of *HSD3B2* expression (1.30 $\pm$ 0.23-fold, $p = 0.009$) (Figure 5F), and a trend toward the upregulation of *HSD3B1* (1.20 $\pm$ 0.5-fold, $p = $ not significant) (Figure 5E), while the transfection with *CRY2* siRNA did not affect the expression of either *HSD3B1* or *HSD3B2*. Similarly, the expression of *CYP11B2*, and its main transcriptional factor *NR4A2* were not significantly modified at the evaluated timepoint (42 h post-transfection) (Figure 5C,D).

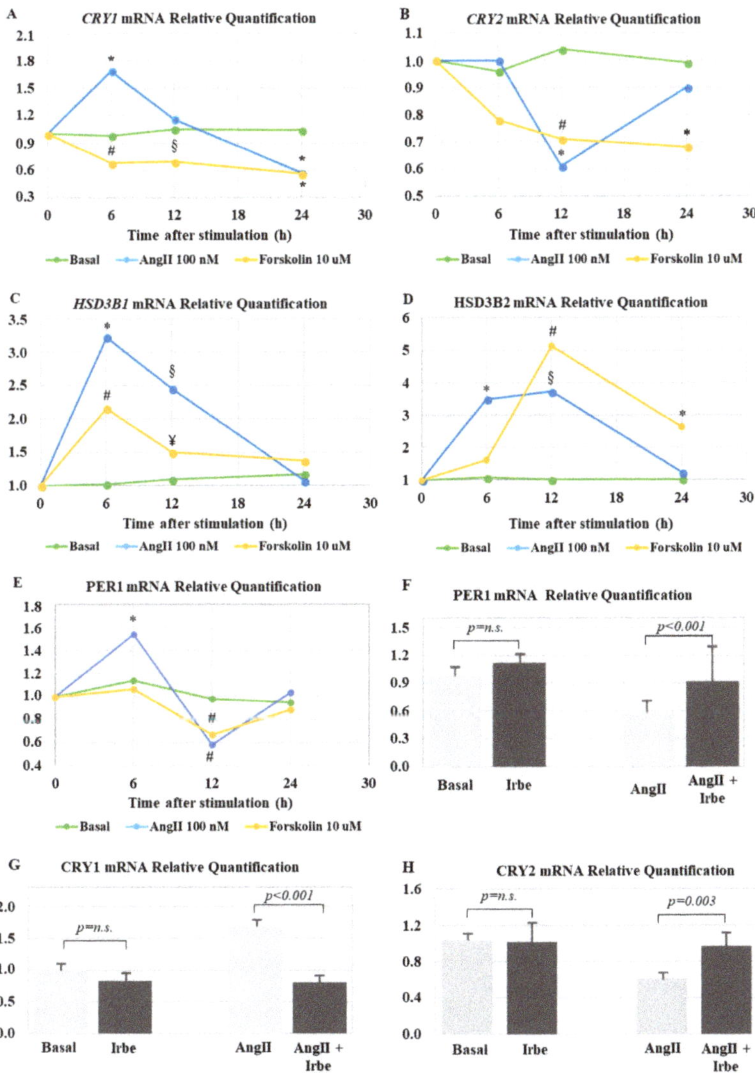

Figure 4. (**A**) Real-time PCR analysis of *CRY1* gene expression. * p-Value < 0.001, # p-value = 0.007, and § p-value = 0.017 when compared with basal. (**B**) Real-time PCR analysis of *CRY2* gene expression. * p-value < 0.001 and # p-value = 0.003 when compared with basal. (**C**) Real-time PCR analysis of *HSD3B1* gene expression. * p-value = 0.022, # p-value = 0.03, § p-value < 0.001, and ¥ p-value = 0.023 when compared with basal. (**D**) Real-time PCR analysis of *HSD3B2* gene expression. * p-value < 0.001, # p-value = 0.001, and § p-value = 0.009 when compared with basal. (**E**) Real-time PCR analysis of period (*PER1*) gene expression. * p-value = 0.001 and # p-value < 0.001 when compared with basal. (**A–E**) Each point expresses the mean fold change over basal expression in at least three independent experiments. (**F**) Real-time PCR analysis of *PER1* gene expression at 6 h, after stimulation with 100 nM AngII ± 1 μM irbesartan. (**G**) Real-time PCR analysis of *CRY1* gene expression at 6 h, after stimulation with 100 nM AngII ± 1 μM irbesartan. (**H**) Real-time PCR analysis of *CRY2* gene expression at 12 h, after stimulation with 100 nM AngII ± 1 μM irbesartan. (**F–H**) Each bar represents the mean ± SD of relative fold change of gene expression in three independent experiments.

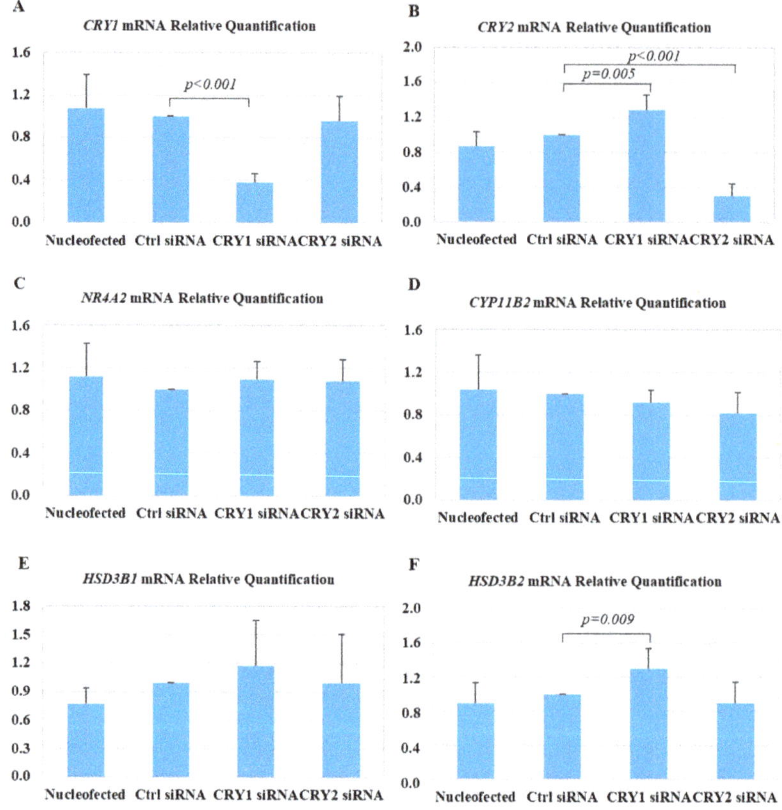

Figure 5. Effect of silencing *CRY1* and *CRY2* on gene expression in HAC15 adrenocortical cells. Real-time PCR analysis of *CRY1* (**A**), *CRY2* (**B**), *HSD3B1* (**C**), *HSD3B2* (**D**), *NR4A2* (**E**), and *CYP11B2* (**F**) gene expression. Each bar expresses the mean ± SD fold change over the expression in cells transfected with a control small interfering RNA (siRNA; Ctrl siRNA) of at least five independent experiments. No significant differences were detected between the cells transfected with Ctrl siRNA, and electroporated cells (Nucleofected).

3. Discussion

Over the last few years, significant knowledge about the molecular mechanisms that regulate aldosterone overproduction was gained from both next-generation sequencing studies [17], and murine models of primary aldosteronism [18]. The *Cry*-null mice, lacking the core-clock components CRY1 and CRY2 [8], displayed hyperaldosteronism and salt-sensitive hypertension, most likely sustained by the upregulation of the type VI 3β-hydroxyl-steroid dehydrogenase (*Hsd3b6*), corresponding to the human type I 3β-hydroxyl-steroid dehydrogenase (*HSD3B1*) gene.

Immunohistochemistry studies in normal human adrenals showed that HSD3B2 was the predominant isoform, while HSD3B1 localized mainly in the outermost layer zona glomerulosa [8,13,14]. In adrenal pathology, HSD3B1 appeared to be strongly expressed in the hyperplastic zona glomerulosa cells of BAH samples, while its expression was low in a series of eight APAs, composed predominantly of zona fasciculata-like cells [13]. Based on these results, it was hypothesized that HSD3B1 overexpression might represent the molecular mechanism responsible for autonomous aldosterone overproduction in BAH [19].

So far, the role and clinical significance of *CRY1* and *CRY2* genes in the regulation of aldosterone production and APA development, together with their potential interplay with *HSD3B* isoforms, were not explored in humans.

In this study, we demonstrated, for the first time, that *CRY1* is overexpressed, while *CRY2* is downregulated in APA tissue, when compared with the paired adjacent adrenal cortex, which represents the optimal control tissue, given the multiplicity of factors that influence the transcription of the core-clock genes [9]. In agreement with previous reports [15], we observed that *HSD3B2*, being over 300-fold more expressed than *HSD3B1*, is the principal isoform in APAs. A previous study showed that HSD3B1 (evaluated as H-score) was more expressed in APAs carrying somatic mutations in the *KCNJ5* gene [15], while in our cohort we did not detect any significant association between the expression of *HSD3B1* (evaluated by real-time PCR) and the mutational status of the samples. On the contrary, we observed that both *HSD3B1* expression and the relative *HSD3B1*/*HSD3B2* ratio were significantly more elevated in APAs composed mainly of zona glomerulosa-like cells (while APAs carrying a mutation in *KCNJ5* are composed mainly of zona fasciculata-like cells [20]).

Additionally, this study demonstrated, for the first time, that the expression of both *CRY1* and *CRY2* genes is modulated by AngII through activation of the AT1R. Similarly, the negative regulator *PER1* showed an AngII-dependent regulation. It was previously reported that stimulation with AngII for three hours induced the negative regulator of the core-clock protein PER1 in H295R adrenocortical cells [21]. Additionally, overexpressing *PER1* in H295R cells was able to induce CYP11B1 and CYP11B2 promoter activity [21]. A role for the circadian-clock protein PER1 in the regulation of aldosterone production was recently reported in both in vitro and in vivo studies. *Per1* knock-out mice displayed lower aldosterone levels when compared with wild-type animals, and also a lower expression of *Hsd3b6* in adrenal gland tissue [22]. Silencing *PER1* in H295R cells was able to decrease the expression of *HSD3B1* isoform by 58%, supporting the hypothesis that *PER1* is involved in the modulation of serum aldosterone levels [22].

In the presented study, we showed that AngII stimulation triggers the expression of both *HSD3B1* and *HSD3B2* in HAC15 cells; our results differ from those reported by Ota T. et al. [23], showing that AngII can induce the expression of *HSD3B1*, but not *HSD3B2* in H295R cells.

To further investigate the potential role of *CRY1* and *CRY2* in the regulation of *HD3B* isoforms, we transfected HAC15 cells with *CRY1* and *CRY2* siRNA. Contrary to what was expected from the *Cry*-null and the *Per* knock-out murine models, silencing *CRY* genes did not modify the expression of *HSD3B1*; however, we observed a mild upregulation of *HSD3B2* in HAC15 cells transfected with *CRY1* siRNA. However, as previously described [24], silencing *CRY1* resulted in a significant upregulation of *CRY2*, which did not allow us to perform an efficient double silencing, and could, therefore, have affected the results, representing a limitation of the presented study.

4. Materials and Methods

4.1. Patients Selection

A total of 46 adrenal adenomas and 20 paired adjacent adrenal samples were included in the presented study. The adrenal glands were removed from patients affected by unilateral PA, diagnosed in our tertiary referral hypertension centre (Division of Internal Medicine 4—Hypertension Unit, at the University of Torino, Italy). The diagnostic work-up for PA was performed according to the recommendations of the Endocrine Society clinical practice guideline [25]. After withdrawal of interfering medications, the ratio of aldosterone to plasma renin activity was used as a screening test for PA. To confirm the diagnosis, either an intravenous (i.v.) saline load test or a captopril challenge test (when acute plasma volume expansion was contraindicated) was performed. All patients with confirmed PA underwent adrenal computed tomography (CT) scanning and adrenal vein sampling (AVS), as previously described [26]. All patients showing lateralization upon AVS underwent unilateral laparoscopic adrenalectomy. The diagnosis of unilateral PA was confirmed based on clinical benefit and a complete biochemical outcome after adrenalectomy, as defined according to a recent consensus (Primary Aldosteronism Surgical Outcome, PASO) [27]. Clinical and biochemical parameters (before and after adrenalectomy) of the included patients are summarized in Table S1. Normal adrenal glands were obtained from normotensive patients who underwent unilateral nephrectomy for renal carcinoma. For all samples, any adrenal gland showing involvement in the tumor lesion was excluded upon histological examination. All patients gave their written informed consent for the use of samples and clinical data, and the protocol of the study was approved by our local ethics committee, (Comitato etico interaziendale A.O.U. Città della Salute e della Scienza di Torino), Project ID CEI/28, date of approval 14 May 2007).

4.2. DNA Sequencing for KCNJ5, ATP1A1, ATP2B3, and CACNA1D

DNA fragments from *KCNJ5*, *ATP1A1*, *ATP2B3*, and *CACNA1D* were amplified by PCR as previously reported [28,29]. The validity of novel mutations was confirmed by sequencing both strands of an independently amplified PCR fragment. Of the presented cohort of adrenal samples, 36 were included in the study by Fernandes-Rosa et al. [29]. Of the included samples, 16 adrenal nodules carried a mutation in the *KCNJ5* gene, six in the *ATP1A1* or *ATP2B3* genes, and five in the *CACNA1D* gene, while 19 samples had no mutations in any of these genes.

4.3. Pathological Analysis

Histological examination was performed by an experienced pathologist (I.C.). All adrenal glands included in the study were embedded in paraffin, cut into 3-μm-thick slices, and stained with hematoxylin and eosin (H&E). After accurate macroscopical and microscopical analysis, the final diagnosis of APA was established when a single nodule was present, while the final diagnosis of unilateral adrenal hyperplasia (UAH) was established in the presence of several nodules of varying sizes (with or without a dominant one). In the case of UAH, the dominant nodule was used for gene-expression studies, provided that it was the one identified with *CYP11B2* expression upon immunohistochemistry analysis.

After examination for the known features of ZF (large, lipid-laden clear cells with round to oval vesicular nuclei), ZG (small, compact cells with a high nuclear/cytoplasmic ratio, and a moderate amount of lipids), and zona reticularis (lipid-sparse cytoplasm, and compact cells) cells [20,30], the tumors were categorized as ZF-like when the percentage of large vacuolated cells was greater than 50%, and ZG-type when the percentage of ZF-like cells was ≤50% and the ZG-like cells were the prevalent cell type. Of the analyzed samples, 36 were previously included in the study by Monticone et al. [31].

The final histopathological diagnosis was APA in 35 cases, and multinodular hyperplasia in 11 cases. Twenty-five samples (23 APA samples and two UAH samples) were composed mainly of ZG-like cells, and 21 samples (19 APA samples and 3 UAH samples) were composed mainly of ZF-like cells.

4.4. Immunohistochemistry Analyses

Immunohistochemistry analysis was performed using the following primary antibodies: CYP11B2 (CYP11B2-41-17) [32], HSD3B1 (Abnova), HSD3B2 (KALKG619) [13], Cry1 (Abgent), and Cry2 (Abcam), as detailed in Table S2.

Formalin-fixed paraffin-embedded tumor samples were cut into sequential 2-μm-thick sections, and deparaffinized and stained at the Pathology Department using a fully automated Ventana BenchMark ULTRA stainer (Ventana, Tucson, AZ, USA), according to the manufacturers' instructions. Binding of peroxidase-coupled antibodies was detected using the ultraView Universal DAB Detection Kit as a substrate, and the sections were counterstained with hematoxylin.

4.5. Cell Culture and Transfection

HAC15 adrenocortical cells were cultured as previously reported [33]. For experiments, cells were plated at a density of 4×10^5 cells/well in a 12-well plate for 48 h. After overnight incubation in low-serum medium (DMEM/F-12 containing 0.1% cosmic calf serum, and antibiotics), cells were stimulated with 100 nM AngII (reference value in normotensive individuals 24 ± 17 pM [34]) (Sigma #A9525) ± 1 μM irbesartan (Sigma #I2286) or forskolin (10 μM, Sigma #F6886) for 6, 12, and 24 h, and then harvested for RNA extraction, and gene-expression studies.

CRY1 and *CRY2* gene silencing was performed using the Amaxa technology (Program X-005). One million cells were electroporated in 100 μL of Nucleofector solution R, using 2 μL of a 100-μmol/L solution of Silencer Select predesigned small interfering RNA (siRNA) (Thermo Fisher Scientific, Waltham, MA, USA). After electroporation, cells were plated in a six-well plate, and recovered for 24 h. The medium was then changed to experimental low-serum medium (0.1% cosmic calf serum), and the cells were starved overnight. The following morning, cells were harvested for RNA extraction, and gene-expression studies.

4.6. RNA Extraction, and Gene-Expression Studies

RNA isolation from adrenal tissue and cultured cells, and subsequent reverse transcription were both performed as previously reported [33]. Real-time PCR was performed in triplicate using TaqMan gene-expression assays (Thermo Scientific) for *CRY1*, *CRY2*, *HSD3B1*, *HSD3B2*, *PER1*, *NR4A2*, and *CYP11B2*. Gene-expression levels were analyzed using the $2^{-\Delta\Delta Ct}$ relative quantification method, using 18S RNA or *GAPDH* as endogenous reference genes.

4.7. Statistical Analyses

IBM SPSS statistics v. 24 was used for the statistical analysis. Data were expressed as mean $\pm$ standard deviation or median $[25°–75°]$. Differences between variables were evaluated using one-way ANOVA followed by Bonferroni's or Dunnett's post-hoc tests, when appropriate, and paired (to compare the expression between the adrenal nodule and the corresponding adjacent adrenal cortex) or unpaired t-tests or Mann–Whitney tests. A probability of less than 0.05 was considered as statistically significant.

5. Conclusions

Our results supported the hypothesis that *CRY1* and *CRY2*, being AngII-regulated genes, and showing a differential expression in APAs when compared with the adjacent adrenal cortex, might be involved in adrenal cell function, and in the regulation of aldosterone production. However, silencing *CRY1* and *CRY2* expression in HAC15 adrenocortical cells resulted only in a modest upregulation of the *HSD3B2* gene, which was not consistent with the experimental observations in the *Cry*-null animal model. Species differences should be considered when studying the role of these genes in adrenal function, and further exploration in this research area is warranted to elucidate the complex role of the circadian clock in adrenal aldosterone production.

Supplementary Materials: Supplementary materials can be found at http://www.mdpi.com/1422-0067/19/6/1675/s1.

Author Contributions: M.T., C.M. and S.M. performed the in-vitro experiments (cell culture, gene expression studies and sequencing, silencing studies) and wrote the manuscript; I.C. performed the histopathological analyses and the immunohistochemistry; F.V. (Francesca Veneziano) performed the immunohistochemistry; F.V. (Franco Veglio) critically revised the manuscript; S.M. and P.M. conceived the design of the study, wrote part of the manuscript and critically revised it.

Funding: This research was funded by MIUR Ricerca locale ex-60% 2015.

Conflicts of Interest: The authors declare no conflict of interest.

References

1. Monticone, S.; Burrello, J.; Tizzani, D.; Bertello, C.; Viola, A.; Buffolo, F.; Gabetti, L.; Mengozzi, G.; Williams, T.A.; Rabbia, F.; et al. Prevalence and Clinical Manifestations of Primary Aldosteronism Encountered in Primary Care Practice. *J. Am. Coll. Cardiol.* **2017**, *69*, 1811–1820. [CrossRef] [PubMed]
2. Käyser, S.C.; Dekkers, T.; Groenewoud, H.J.; van der Wilt, G.J.; Carel Bakx, J.; van der Wel, M.C.; Hermus, A.R.; Lenders, J.W.; Deinum, J. Study Heterogeneity and Estimation of Prevalence of Primary Aldosteronism: A Systematic Review and Meta-Regression Analysis. *J. Clin. Endocrinol. Metab.* **2016**, *101*, 2826–2835. [CrossRef] [PubMed]
3. Calhoun, D.A.; Nishizaka, M.K.; Zaman, M.A.; Harding, S.M. Aldosterone excretion among subjects with resistant hypertension and symptoms of sleep apnea. *Chest* **2004**, *125*, 112–117. [CrossRef] [PubMed]
4. Choi, M.; Scholl, U.I.; Yue, P.; Björklund, P.; Zhao, B.; Nelson-Williams, C.; Ji, W.; Cho, Y.; Patel, A.; Men, C.J.; et al. K$^+$ channel mutations in adrenal aldosterone-producing adenomas and hereditary hypertension. *Science* **2011**, *331*, 768–772. [CrossRef] [PubMed]
5. Azizan, E.A.; Poulsen, H.; Tuluc, P.; Zhou, J.; Clausen, M.V.; Lieb, A.; Maniero, C.; Garg, S.; Bochukova, E.G.; Zhao, W.; et al. Somatic mutations in *ATP1A1* and *CACNA1D* underlie a common subtype of adrenal hypertension. *Nat. Genet.* **2013**, *45*, 1055–1060. [CrossRef] [PubMed]
6. Beuschlein, F.; Boulkroun, S.; Osswald, A.; Wieland, T.; Nielsen, H.N.; Lichtenauer, U.D.; Penton, D.; Schack, V.R.; Amar, L.; Fischer, E.; et al. Somatic mutations in *ATP1A1* and *ATP2B3* lead to aldosterone-producing adenomas and secondary hypertension. *Nat. Genet.* **2013**, *45*, 440–444. [CrossRef] [PubMed]
7. Scholl, U.I.; Goh, G.; Stölting, G.; de Oliveira, R.C.; Choi, M.; Overton, J.D.; Fonseca, A.L.; Korah, R.; Starker, L.F.; Kunstman, J.W.; et al. Somatic and germline *CACNA1D* calcium channel mutations in aldosterone-producing adenomas and primary aldosteronism. *Nat. Genet.* **2013**, *45*, 1050–1054. [CrossRef] [PubMed]
8. Doi, M.; Takahashi, Y.; Komatsu, R.; Yamazaki, F.; Yamada, H.; Haraguchi, S.; Emoto, N.; Okuno, Y.; Tsujimoto, G.; Kanematsu, A.; et al. Salt-sensitive hypertension in circadian clock-deficient *Cry*-null mice involves dysregulated adrenal Hsd3b6. *Nat. Med.* **2010**, *16*, 67–74. [CrossRef] [PubMed]
9. Van Laake, L.W.; Lüscher, T.F.; Young, M.E. The circadian clock in cardiovascular regulation and disease: Lessons from the Nobel Prize in Physiology or Medicine 2017. *Eur. Heart J.* **2017**. [CrossRef] [PubMed]
10. Dunlap, J.C. Molecular bases for circadian clocks. *Cell* **1999**, *22*, 271–290. [CrossRef]
11. Hattangady, N.G.; Olala, L.O.; Bollag, W.B.; Rainey, W.E. Acute and chronic regulation of aldosterone production. *Mol. Cell. Endocrinol.* **2012**, *350*, 151–162. [CrossRef] [PubMed]
12. Simard, J.; Ricketts, M.L.; Gingras, S.; Soucy, P.; Feltus, F.A.; Melner, M.H. Molecular biology of the 3β-hydroxysteroid dehydrogenase/Δ^5-Δ^4 isomerase gene family. *Endocr. Rev.* **2005**, *26*, 525–582. [CrossRef] [PubMed]
13. Doi, M.; Satoh, F.; Maekawa, T.; Nakamura, Y.; Fustin, J.M.; Tainaka, M.; Hotta, Y.; Takahashi, Y.; Morimoto, R.; Takase, K.; et al. Isoform-specific monoclonal antibodies against 3β-hydroxysteroid dehydrogenase/isomerase family provide markers for subclassification of human primary aldosteronism. *J. Clin. Endocrinol. Metab.* **2014**, *99*, E257–E262. [CrossRef] [PubMed]
14. Gomez-Sanchez, C.E.; Lewis, M.; Nanba, K.; Rainey, W.E.; Kuppusamy, M.; Gomez-Sanchez, E.P. Development of monoclonal antibodies against the human 3β-hydroxysteroid dehydrogenase/isomerase isozymes. *Steroids* **2017**, *127*, 56–61. [CrossRef] [PubMed]
15. Konosu-Fukaya, S.; Nakamura, Y.; Satoh, F.; Felizola, S.J.; Maekawa, T.; Ono, Y.; Morimoto, R.; Ise, K.; Takeda, K.; Katsu, K.; et al. 3β-Hydroxysteroid dehydrogenase isoforms in human aldosterone-producing adenoma. *Mol. Cell. Endocrinol.* **2015**, *408*, 205–212. [CrossRef] [PubMed]
16. Pamar, J.; Key, R.E.; Rainey, W.E. Development of an adrenocorticotropin-responsive human adrenocortical carcinoma cell line. *J. Clin. Endocrinol. Metab.* **2008**, *93*, 4542–4546. [CrossRef] [PubMed]
17. Monticone, S.; Else, T.; Mulatero, P.; Williams, T.A.; Rainey, W.E. Understanding primary aldosteronism: Impact of next generation sequencing and expression profiling. *Mol. Cell. Endocrinol.* **2015**, *399*, 311–320. [CrossRef] [PubMed]
18. Aragao-Santiago, L.; Gomez-Sanchez, C.E.; Mulatero, P.; Spyroglou, A.; Reincke, M.; Williams, T.A. Mouse Models of Primary Aldosteronism: From Physiology to Pathophysiology. *Endocrinology* **2017**, *158*, 4129–4138. [CrossRef] [PubMed]

19. Okamura, H.; Doi, M.; Goto, K.; Kojima, R. Clock genes and salt-sensitive hypertension: A new type of aldosterone-synthesizing enzyme controlled by the circadian clock and angiotensin II. *Hypertens. Res.* **2016**, *39*, 681–687. [CrossRef] [PubMed]

20. Nakamura, Y.; Felizola, S.; Satoh, F.; Konosu-Fukaya, S.; Hironobu, S. Dissecting the molecular pathways of primary aldosteronism. *Pathol. Int.* **2014**, *64*, 482–489. [CrossRef] [PubMed]

21. Romero, D.G.; Rilli, S.; Plonczynski, M.W.; Yanes, L.L.; Zhou, M.Y.; Gomez-Sanchez, E.P.; Gomez-Sanchez, C.E. Adrenal transcription regulatory genes modulated by angiotensin II and their role in steroidogenesis. *Physiol. Genom.* **2007**, *30*, 26–34. [CrossRef] [PubMed]

22. Richards, J.; Cheng, K.Y.; All, S.; Skopis, G.; Jeffers, L.; Lynch, I.J.; Wingo, C.S.; Gumz, M.L. A role for the circadian clock protein Per1 in the regulation of aldosterone levels and renal Na+ retention. *Am. J. Physiol. Renal Physiol.* **2013**, *305*, F1697–F1704. [CrossRef] [PubMed]

23. Ota, T.; Doi, M.; Yamazaki, F.; Yarimizu, D.; Okada, K.; Murai, I.; Hayashi, H.; Kunisue, S.; Nakagawa, Y.; Okamura, H. Angiotensin II triggers expression of the adrenal gland zona glomerulosa-specific 3β-hydroxysteroid dehydrogenase isoenzyme through de novo protein synthesis of the orphan nuclear receptors NGFIB and NURR1. *Mol. Cell. Biol.* **2014**, *34*, 3880–3894. [CrossRef] [PubMed]

24. Zhang, E.E.; Liu, A.C.; Hirota, T.; Miraglia, L.J.; Welch, G.; Pongsawakul, P.Y.; Liu, X.; Atwood, A.; Huss, J.W., 3rd; Janes, J.; et al. A genome-wide RNAi screen for modifiers of the circadian clock in human cells. *Cell* **2009**, *139*, 199–210. [CrossRef] [PubMed]

25. Funder, J.W.; Carey, R.M.; Fardella, C.; Gomez-Sanchez, C.E.; Mantero, F.; Stowasser, M.; Young, W.F., Jr.; Montori, V.M. Endocrine Society. Case detection, diagnosis, and treatment of patients with primary aldosteronism: An endocrine society clinical practice guideline. *J. Clin. Endocrinol. Metab.* **2008**, *93*, 3266–3281. [CrossRef] [PubMed]

26. Monticone, S.; Satoh, F.; Giacchetti, G.; Viola, A.; Morimoto, R.; Kudo, M.; Iwakura, Y.; Ono, Y.; Turchi, F.; Paci, E.; et al. Effect of adrenocorticotropic hormone stimulation during adrenal vein sampling in primary aldosteronism. *Hypertension* **2012**, *59*, 840–846. [CrossRef] [PubMed]

27. Williams, T.A.; Lenders, J.W.M.; Mulatero, P.; Burrello, J.; Rottenkolber, M.; Adolf, C.; Satoh, F.; Amar, L.; Quinkler, M.; Deinum, J.; et al. Primary Aldosteronism Surgery Outcome (PASO) investigators. Outcomes after adrenalectomy for unilateral primary aldosteronism: An international consensus on outcome measures and analysis of remission rates in an international cohort. *Lancet Diabetes Endocrinol.* **2017**, *5*, 689–699. [CrossRef]

28. Williams, T.A.; Monticone, S.; Schack, V.R.; Stindl, J.; Burrello, J.; Buffolo, F.; Annaratone, L.; Castellano, I.; Beuschlein, F.; Reincke, M.; et al. Somatic *ATP1A1*, *ATP2B3*, and *KCNJ5* mutations in aldosterone-producing adenomas. *Hypertension* **2014**, *63*, 188–195. [CrossRef] [PubMed]

29. Fernandes-Rosa, F.L.; Williams, T.A.; Riester, A.; Steichen, O.; Beuschlein, F.; Boulkroun, S.; Strom, T.M.; Monticone, S.; Amar, L.; Meatchi, T.; et al. Genetic spectrum and clinical correlates of somatic mutations in aldosterone-producing adenoma. *Hypertension* **2014**, *64*, 354–361. [CrossRef] [PubMed]

30. Neville, A.M.; O'Hare, M.J. Histopathology of the human adrenal cortex. *Clin. Endocrinol. Metab.* **1985**, *14*, 791–820. [CrossRef]

31. Monticone, S.; Castellano, I.; Versace, K.; Lucatello, B.; Veglio, F.; Gomez-Sanchez, C.E.; Williams, T.A.; Mulatero, P. Immunohistochemical, genetic and clinical characterization of sporadic aldosterone-producing adenomas. *Mol. Cell. Endocrinol.* **2015**, *411*, 146–154. [CrossRef] [PubMed]

32. Gomez-Sanchez, C.E.; Qi, X.; Velarde-Miranda, C.; Plonczynski, M.W.; Parker, C.R.; Rainey, W.; Satoh, F.; Maekawa, T.; Nakamura, Y.; Sasano, H.; et al. Development of monoclonal antibodies against human CYP11B1 and CYP11B2. *Mol. Cell. Endocrinol.* **2014**, *383*, 111–117. [CrossRef] [PubMed]

33. Monticone, S.; Bandulik, S.; Stindl, J.; Zilbermint, M.; Dedov, I.; Mulatero, P.; Allgaeuer, M.; Lee, C.C.; Stratakis, C.A.; Williams, T.A.; et al. A case of severe hyperaldosteronism caused by a de novo mutation affecting a critical salt bridge Kir3.4 residue. *J. Clin. Endocrinol. Metab.* **2015**, *100*, E114–E118. [CrossRef] [PubMed]

34. Catt, K.J.; Cran, E.; Zimmet, P.Z.; Best, J.B.; Cain, M.D.; Coghlan, J.P. Angiotensin II blood-levels in human hypertension. *Lancet* **1971**, *1*, 459–464. [CrossRef]

International Journal of *Molecular Sciences*

MDPI

Review

Genetics of Hypertension in African Americans and Others of African Descent

Mihail Zilbermint [1,2,3,4,*,†], **Fady Hannah-Shmouni** [1,†] **and Constantine A. Stratakis** [1]

[1] Section on Endocrinology and Genetics, The Eunice Kennedy Shriver National Institute of Child Health and Human Development, National Institutes of Health, BG 31 RM 2A46, 31 Center Dr, Bethesda, MD 20892, USA; Fady.Hannah-Shmouni@nih.gov (F.H.-S.); stratakc@cc1.nichd.nih.gov (C.A.S.)
[2] Division of Endocrinology, Diabetes, and Metabolism, Johns Hopkins University School of Medicine, Baltimore, MD 21287, USA
[3] Johns Hopkins Community Physicians at Suburban Hospital, Bethesda, MD 20814, USA
[4] Johns Hopkins University Carey Business School, Baltimore, MD 21202, USA
* Correspondence: mihail.zilbermint@nih.gov; Tel.: +1-301-594-5984; Fax: +1-301-402-0574
† These authors contributed equally to this work.

Received: 5 January 2019; Accepted: 21 February 2019; Published: 2 March 2019

Abstract: Hypertension is the leading cause of cardiovascular disease in the United States, affecting up to one-third of adults. When compared to other ethnic or racial groups in the United States, African Americans and other people of African descent show a higher incidence of hypertension and its related comorbidities; however, the genetics of hypertension in these populations has not been studied adequately. Several genes have been identified to play a role in the genetics of hypertension. They include genes regulating the renin-aldosterone-angiotensin system (RAAS), such as Sodium Channel Epithelial 1 Beta Subunit (*SCNN1B*), Armadillo Repeat Containing 5 (*ARMC5*), G Protein-Coupled Receptor Kinase 4 (*GRK4*), and Calcium Voltage-Gated Channel Subunit Alpha1 D (*CACNA1D*). In this review, we focus on recent genetic findings available in the public domain for potential differences between African Americans and other populations. We also cover some recent and relevant discoveries in the field of low-renin hypertension from our laboratory at the National Institutes of Health. Understanding the different genetics of hypertension among various groups is essential for effective precision-guided medical therapy of high blood pressure.

Keywords: genetics; hypertension; African American; low-renin; *ARMC5*; *SCNN1B*; *GRK4*; *CACNA1D*; endocrine hypertension

---

1. Introduction

Hypertension is the leading cause of cardiovascular disease in the United States, affecting 29% of adults, or approximately 75 million people [1]. The economic burden of hypertension in America is enormous [2], often raising concerns about its major impact on health disparity [3–5]. When compared to other ethnic or racial groups in the United States, African American and others of African descent show a higher incidence of hypertension and its related comorbidities, including cardiovascular and end-stage renal diseases [2]. Moreover, African Americans and others of African descent may have higher blood pressure beginning in childhood, as well as a higher incidence and prevalence of hypertension across the lifespan [6–11]. The age-adjusted prevalence of hypertension in African Americans is ~45%, significantly higher than in other ethnicities, including ~32% among non-Hispanic whites and ~30% among Hispanics [2].

Major predictors of hypertension in African Americans exist. For example, impaired arterial elasticity [12–14] has been demonstrated to be more prevalent than in whites [5,15–17]. Brown et al. found in a large cohort study of untreated normotensive participants (Multi-Ethnic Study of

Atherosclerosis) that subjects who were found to have a suppressed renin phenotype were more likely to be African Americans, and had higher systolic blood pressure [18]. In a recent observational study on normotensive African Americans, aldosterone sensitivity (magnitude of the association between plasma aldosterone concentration and blood pressure) increased with age and was associated with plasma aldosterone concentration and the aldosterone-to-renin ratio, suggesting that mineralocorticoid receptor activity may increase with age, especially in African Americans [19]. Other factors that distinguish hypertension in African Americans from other ethnicities include increased awareness of diagnosis, increased intensity of treatment, poor BP control, and more resistant hypertension [20]. On the other hand, hypertensive diagnostic inertia, defined as a failure to investigate the underlying cause of hypertension, is an ongoing issue in African Americans with cardiovascular disease [21,22]. Moreover, adrenocortical hyperplasia was found to be more common in African Americans and other patients of African descent [23], suggesting the possibility of aldosterone and/or cortisol excess as an important contributor to the pathogenesis of hypertension in this ethnic group.

The regulation of blood pressure is complex. Research that examines the association of the various pathophysiological factors with incident hypertension among African Americans and others of African descent is limited, as detailed in Table 1. Despite heredity contributing 40–50% to the pathogenesis of essential hypertension and genome-wide association studies identifying ~6% of the genetic contribution [24,25], little is known about the genetic diversity of hypertension in African American populations, particularly in relation to the factors listed in Table 1. Some researchers have focused on genes implicated in altered renal sodium handling in the kidneys and volume loading, which are key players in the development of low-renin hypertension in this at-risk population [26,27]. Studies that failed to discover any relationship between African American and hypertension were limited by several factors, including variation in allele frequency, small statistical power, and the possibility of weak ancestral effects [28–30]. Large sample size studies could exclude > 95% of the genome as harboring risk loci of > 1.3 due to African or European ancestry, further suggesting the complexity of understanding the underpinning of hypertension across various ethnicities. Ongoing studies are examining genetic susceptibility and environmental factors as determinants of hypertension in African Americans [31].

Table 1. Examples of pathophysiological mechanisms of hypertension in individuals of African descent.

Mechanism
Psychosocial factors [32]
Endothelial dysfunction [33,34]
Kidney injury and function [35]
Renin-angiotensin system activation [36]
Insulin resistance [37]
Impaired baroreflex [38]
Oxidative stress [39]
Genetic factors [20]
Adrenomedullary/cortical hormones [23]
Blood volume [40]
Salt retention [20,41]
Socioeconomic determinants [42]

Diversity may exist within African descent population. Clinical and genetic data of African-Americans and others of African descent should be interpreted and compared with caution. Chor et al. studied blood pressure profiles of 15,103 civil servants in Brazil and found a high prevalence of high blood pressure among browns (38.2%) and blacks (49.3%) [11]. Importantly, they found that hypertensive characteristics of Brazilian brown populations were like that seen in the individuals of African descent.

Genetic sequencing has gained enormous popularity in the scientific field. Affordability and fast throughput are promising to deliver "Genetic Health Risk and Carrier Status" for as little as $99, through direct-to-consumer salivary collection kits [43]. However, the clinical interpretation of an individual's genome, its utility in clinical practice, and the overall cost has yet to be implemented as a standard of care approach in universal clinical management guidelines. One of the major goals of understanding the genetics of hypertension includes the transfer of genomic knowledge into daily clinical practice [10], for potential gene-targeted medical therapies [5], among other useful utilities.

In this review, we briefly highlight the mechanistic and genetic underpinnings of hypertension in African Americans and other populations of African ancestry. We have focused our discussion on the biologically plausible hypertension candidate gene loci in African Americans [44]

2. Hypertension in African Americans: Clinical Differences and Genetics

2.1. Renin and Aldosterone

African Americans excrete a sodium load more slowly and less completely than whites [45]. This results in suppression of the renin-aldosterone-angiotensin system (RAAS) due to volume-loading that typically begins in childhood [46–49]. Ultimately, a low-renin state, which compensates for the relative tendency to retain sodium, ensues [50]. Low-renin hypertension is a frequent cause of hypertension, with a prevalence of 20%–30%, and higher in African Americans [51–53]. One study demonstrated lower levels of plasma renin activity and aldosterone in normotensive African Americans across all ages, with BP positively correlating with plasma aldosterone, an effect that increased as plasma renin activity decreased [48]. Thus, a typical biochemical profile in an African American person with hypertension is a low or high plasma aldosterone concentration, a low or suppressed plasma renin activity or direct renin concentration, and suppressed angiotensin I and II [36,50,54,55]. This results in a normal or elevated aldosterone-to-renin ratio, which can be categorized into two broad hypertension phenotypes: low or suppressed renin and elevated aldosterone (primary aldosteronism type, or hyporeninemic hyperaldosteronism) and low renin and low aldosterone (Liddle syndrome type, or hyporeninemic hypoaldosteronism) [56]. The threshold set to diagnose a low renin state is assay specific but generally defined as a plasma renin activity < 0.65 ng/mL/h or a direct renin concentration < 15μU/mL [52].

2.2. Regulation of Sodium Reabsorbtion

It has been speculated that several genes implicated in the regulation of sodium reabsorption in the kidneys were likely selected as an adaptation to high temperature environments, particularly in people from Sub-Saharan Africa [20,57]. These include genes regulating RAAS, such as (Sodium Channel Epithelial 1 Beta Subunit (*SCNN1B*) and Neural Precursor Cell Expressed, Developmentally Down-Regulated 4 (*NEDD4*) that alter sodium retention from the kidneys, and possibly armadillo repeat containing 5 (*ARMC5*) that might be responsible for increased aldosterone production from the adrenal cortex (see below).

These genetic factors may have played an important physiological adaptation ("natural selection") to the low sodium environments and survival of African Americans during their passage from Africa to America on ships [20,57], where they witnessed extreme conditions including severe heat, hyperhidrosis, and fluid loss through sickness. Indeed, this selection process may have contributed to the increased prevalence of hypertension in this population [58,59]. In this section, we cover the most important known genetic contributions to hypertension in African Americans.

2.3. ENaC Function

The amiloride-sensitive epithelial sodium channel (ENaC) is in the distal nephron and responsible for regulating the amount of sodium reabsorbed by the kidneys, primarily through the action of aldosterone. ENaC is composed of 3 homologous subunits of similar structure and encoded by separate genes: Sodium Channel Epithelial 1 Alpha Subunit (*SCNN1A*) on 12p13.31, Sodium Channel Epithelial 1 Beta Subunit (*SCNN1B*) on 16p12.2, and Sodium Channel Epithelial 1 Gamma Subunit (*SCNN1G*) on

16p12.2. There are two transmembrane domains (TM1 and TM2) and two short intracellular domains (*C*- and *N*-terminus); the C-termini contain a binding site for Nedd4 (neural precursor cell expressed, developmentally down-regulated 4), a ubiquitin E3 ligase protein encoded by *NEED4* on 15q21.3 and responsible for the internalization and the proteasomal degradation of ENaC [60–62].

2.3.1. Liddle's Syndrome

Constitutive activation variants of *SCNN1B* or *SCNN1G* result in salt-sensitive hypertension known as Liddle's syndrome, an autosomal dominant form of monogenic hypertension that is characterized by early-onset of low-renin hypertension [63]. Patients with Liddle's syndrome are resistant to mineralocorticoid antagonist therapy but respond to an ENaC inhibitor, such as amiloride therapy [63,64]. Gain-of-function variants in the genes encoding for ENaC are in the carboxyterminal cytoplasmic tail of the protein, which is involved in down-regulation of channel number or activity [50,65]. This area of the nephron is the final regulator of sodium balance and activating variants in ENaC leads to sodium retention, potassium excretion, low renin/aldosterone (hyporeninemic hypoaldosteronism), and volume overload [20,66].

2.3.2. Hyporeninemic Hypoaldosteronism (Liddle Phenotype)

Hyporeninemic hypoaldosteronism not due to Liddle's syndrome, also referred to as the Liddle phenotype, is more common in African Americans for multiple reasons, including the interplay of certain genes that lead to ethnic differences in proximal and distal tubular sodium reabsorption [67]. Tu et al. confirmed this association between ENaC overreactivity and hypertension in African Americans by demonstrating increased retention of sodium and water after stimulation with 2 weeks of 9-α fludrocortisone [48]. Moreover, ENaC over-activation could also be due to altered internalization and degradation by *NEDD4* and acquired or inherited causes of aldosterone excess. On the other hand, loss-of-function variants in other segments of ENaC cause pseudohypoaldosteronism, an autosomal recessive condition that is characterized by salt-loss and mineralocorticoid resistance [68].

2.3.3. *SCNN1B* and *NEED4*

Activation of ENaC, either due to structural variants of the channel subunits (e.g.,: *SCNN1B*) or altered activity of regulatory processes (including *NEED4*), could underlie low-renin hypertension in African Americans. One study showed that ENaC channel activity, as assessed by nasal transmucosal electrical potential difference, was greater in African Americans than in whites [69]. In a mixed ancestry population of South Africans, a variant in *SCNN1B* (p.R563Q) was present in 18 people, of whom 17 were hypertensive [50]. In another study, this variant was present in 6% of Africans from urban South Africa that responded to treatment with amiloride therapy [70]. This variant was associated with a resistant form of hyporeninemic hypoaldosteronism hypertension, analogous to Liddle syndrome. Moreover, the p.T594M but not p.G442V (which causes lower aldosterone secretion, suggesting increased ENaC activity) variants in *SCNN1B* may also contribute to hypertension in African Americans [71]. In another study, individuals with variants in *NEDD4* were linked to increased BP and adverse cardiovascular outcomes [20,72,73].

2.3.4. ENaC Function, CYP4A11 and Responsiveness to Amiloride Therapy

Individuals with variants affecting ENaC function and hypertension may respond preferentially to amiloride therapy. Studies from salt-sensitive hypertensive rodent models showed decreased expression of *Cyp4a* and increased ENaC activity responsive to amiloride [74,75]. Human studies on African American patients with resistant hypertension demonstrated homozygosity for the C allele at rs3890011 of Cytochrome P450 Family 4 Subfamily A Member 11 (*CYP4A11*) (1p33), which has been previously associated with blood pressure in various populations [30,76], and a positive response to amiloride therapy [77]. A large high-density admixture scan in 1670 African Americans with hypertension identified this locus as a candidate gene for hypertension [30]. *CYP4A11* encodes a member of the cytochrome P450 superfamily of enzymes, a monooxygenase which catalyze reactions involved in the synthesis of cholesterol, steroids, and other lipids and localized to the endoplasmic reticulum. Several lines of evidence suggest that this gene serves as a modulator of ENaC function [75,77], likely through decreased epoxygenase activity and renal synthesis of epoxyeicosatrienoic acids [20,77]. Collectively, although these variants are more frequent in African Americans, their association with hypertension has been weak and/or inconsistent [50,78,79]. Further studies with a larger population size are required to study their effects on hypertension in the African American populations.

2.4. Adrenocortical Hyperplasia, Tumors, and Primary Aldosteronism

2.4.1. *KCNJ5*, *CACNA1D*, *ATP1A1*, *ATP2B3*, and *CTNNB1*

Primary aldosteronism is the most common cause of endocrine hypertension [80]. Autonomous secretion of aldosterone from the adrenal glands suppresses endogenous renin production, which results in an increased volume status. The most frequent cause of primary aldosteronism is bilateral adrenal hyperplasia [81], often referred to as idiopathic adrenal hyperplasia. Recent evidence suggests that these lesions harbor areas of hyperplasia due to cytochrome P450 family 11 subfamily B member 2 (CYP11B2)-expressing cells from an unknown germline variants, and at least 1 CYP11B2-positive aldosterone-producing cell cluster (APCC, that typically develops with aging) or micro-aldosterone-producing adenomas, in part due to calcium or potassium channel variants (Calcium Voltage-Gated Channel Subunit Alpha1 D (*CACNA1D*), 58%; Potassium Voltage-Gated Channel Subfamily J Member 5 (*KCNJ5*), 1%) [82].

Recently, Nanba et al. studied the genetic characteristics of 73 aldosterone-producing adrenocortical adenomas in 79 subjects of African American decent who had primary aldosteronism; 65 subjects had somatic alterations in driver genes. The genetic landscape of these tumors was different than in non-African Americans: alteartions in *CACNA1D* (*n* = 42%), *KCNJ5* (34%), ATPase Na+/K+ Transporting Subunit Alpha 1 (*ATP1A1*) (8%), and ATPase Plasma Membrane Ca2+ Transporting 3 (*ATP2B3*) (4%) represented the spectrum [83]. No variants in *ARMC5* were found in this study. These results suggest that *CACNA1D* could be one of the most frequently mutated aldosterone-driver gene in African Americans, suggesting a possible primary role for calcium channel blockers in the management of these individuals.

Familial or inherited causes of primary aldosteronism are rare and caused by disease-causing germline activating variants in several genes as detailed elsewhere [81].

2.4.2. Bilateral Adrenocortical Hyperplasia

Bilateral adrenocortical hyperplasias are grossly divided into the micronodular and macronodular disease. The micronodular subtypes are usually diagnosed in children and young adults and are either pigmented (primary pigmented nodular adrenocortical disease [PPNAD] as seen in Carney complex) or not pigmented. The macronodular subtypes, which are usually diagnosed in adults over the age of 40, may be sporadic or familial and caused by disease-causing variants in *ARMC5*, Adenomatous Polyposis Coli (*APC*), Multiple Endocrine Neoplasia type 1 (*MEN1*), and Fumarate Hydratase (*FH*) [84–86]. African Americans with hypertension and a biochemical phenotype of hyporeninemic hyperaldosteronism are more likely to have bilateral adrenal hyperplasia, with or without nodules [84,85,87–91].

2.4.3. *ARMC5*

The *ARMC5* gene is a putative tumor-suppressor that is located on chromosome 16p11.2 and belongs to the family of armadillo (ARM)-repeat-containing proteins. In humans, *ARMC5* consists of 8 exons and has an unknown function. The *ARMC5* gene has been recently implicated in endogenous hypercortisolemia due to a rare form of adrenocortical hyperplasia, termed primary bilateral macronodular adrenal hyperplasia (PBMAH) [84,92,93]. This condition is characterized by multiple macronodules (>1 cm) in the adrenal cortex and hypercortisolemia; it is also rarely associated with primary aldosteronism [85]. Biallelic inactivating variants in *ARMC5* (germline and somatic) are required for the development of adrenocortical hyperplasia, which is consistent with the two-hit hypothesis of tumorigenesis [84,85,92]. Most variants in *ARMC5* are frameshift and/or nonsense, and lead to loss of function of the gene. Our group has recently shown that *Armc5* knockout mice died during early embryonic development, while a third of heterozygotes developed hypercorticosteronemia at 18 months of age [94]. Several pathways may be involved in *Armc5* haploinsufficiency, including cyclic AMP (protein kinase A, its catalytic subunit Cα) and the Wnt/β-catenin pathways [94].

Our laboratory has recently identified an association between biallelic variants of *ARMC5* in African Americans and primary aldosteronism [85]. We hypothesized that these variants likely act as a selective advantage for people of African descent to excrete water more slowly as a survival mechanism in hot climates through enhanced excretion of aldosterone from the adrenal cortex [95]. Our initial studies showed that 20 unrelated and two related study subjects (39.3%) harbored 12 germline *ARMC5* variants that were predicted to be damaging by in silico analysis. Interestingly, all patients carrying a variant predicted to be damaging were African Americans (Table 2).

Table 2. Genes associated with pulse pressure, systolic, diastolic blood pressure and hypertension. Adapted by permission from Springer Nature: [Springer Nature] [Journal of Cardiovascular Translational Research] [Hall, J.L.; Duprez, D.A.; Barac, A.; Rich, S.S. A review of genetics, arterial stiffness, and blood pressure in African Americans, 5, 302-308.e), [Springer Science+Business media, LLC] (2012).

Study	Race-ethnicity	Traits	Discovery
Primary aldosteronism NIH study [85]	*ARMC5* variants in AA with PA ($n = 22$)/*ARMC5* variants ESPS6500 Sample ($n = 2203$)	Primary aldosteronism and hypertension	*ARMC5* maps to 16p11.2. rs35461188 (Benign), rs200655247 (Damaging), rs142376949 (Benign),
University of Michigan study and NIH [83]	Black patients from USA, $n = 79$ (73 patients had aldosterone-producing adenoma)	Primary aldosteronism and hypertension	73 adrenocortical tumors from 79 PA patients expressed *CACNA1D* (42%), *KCNJ5* (34%), *ATP1A1* ($n = 8$%), *ATP2B3* (4%).
University of California San Diego/Kaiser/VA/Loyola [96]	AA (n = 383), Combined cohort Kaiser/VA (n = 527), Nigerian cohort Loyola	Systolic Blood Pressure, hypertension	SCG2 (two transactivating factors were identified, ARIX (PHOX2A) and PHOX2B
The State University of NY Health Science Center [97]	AA (n = 342), Caucasians (n = 263)	Systolic Blood pressure	AGT (-217A variant)
Family Blood Pressure Program [98]	AA ($n = 3962$)/Whites ($n = 3667$)/Hispanics ($n = 1612$)/Asian ($n = 1557$) U.S.	Pulse pressure	Chromosome 7 at 75 cM, LOD 3.1 in AA, chromosome 19 at 0 cM LOD 3.1 in combined sample whites and AA, and a region on chromosome 18 at 71 cM LOD score of 3.2 in whites, AA, and Hispanics.
HyperGEN [13]	AA ($n = 1251$) U.S.	Pulse Pressure	Chromosome 1, 215 cM, LOD 3.08, Chromosome 14, 85 cM, LOD 2.42
Howard University Family Study [99]	AA ($n = 1017$) U.S. A 2nd cohort of 980 West Africans	Systolic BP	rs5743185, rs16877320, rs11160059, rs17365948, rs12279202, rs3751664. *SLC24A4* (a sodium/potassium/calcium exchanger) and *CACNA1H* (a voltage-dependent calcium channel),
Family Blood Pressure Program and Nigerian cohort [27]	AA ($n = 737$) (cases) European Americans ($n = 573$) (controls)	Hypertension	5 markers on chromosome 6q (near region 6q24); 2 markers on chromosome 21 (near region 21q21) may contain genes influencing risk of hypertension in AA
Dallas Heart Study [100]	AA ($n = 1743$), White ($n = 1000$), Mexican American ($n = 581$)	Hypertension	rs2272996
Candidate Gene Association Resource Consortum [101]	AA ($n = 6303$), replication Cohorts: Women's Health Initiative, Maywood, GENOA, Howard University Family Study, native Nigerian African Sample ($n = 11,882$)	Systolic and diastolic blood pressure	rs7726475 between genes *SUB1* and *NPR3*

Table 2. *Cont.*

Study	Race-ethnicity	Traits	Discovery
International Consortium for Blood Pressure Association Studies [102]	AA (*n* = 19,775), Europeans (*n* = 200,000), East Asians (*n* = 29,719), South Asians	Systolic and Diastolic blood pressure, hypertension	rs13082711 (*SLC4A7*) (SBP/DBP); rs419076 (*MECOM*), (SBP); rs13107325 (*SLC39A8*) (SBP/DBP); rs13139571 (*GUCY1A3-GUCY1B3*) (SBP/DBP); rs1173771 (*NPR3*-C5orf23) (SBP); rs11953630 (*EBF1*) (SBP/DBP); rs805303 (*BAT2-BAT5*) (DBP); rs7129220 (*ADM*) (SBP/DBP); rs633185 (*FLJ32810/TMEM133*) (SBP); rs2521501 (*FURIN-FES*) (SBP/DBP); rs17608766 (*GOSR2*) (SBP/DBP); rs1327235 (*JAG1*) (SBP/DBP); rs6015450 (*GNAS-EDN3*) (SBP/DBP); rs17367504 (*MTHFR-NPPB*) (SBP/DBP); rs3774372 (*ULK4*) (SBP/DBP); rs1458038 (*FGF5*) (SBP/DBP); rs1813353 (*CACNB2*(3′) (SBP/DBP); rs11191548 (*CYP17A1-NT5C2*) (SBP/DBP); rs381815 (*PLEKHA7*) (SBP/DBP); rs3184504 (SH2B3) (SBP/DBP); rs1378942 (*CYP1A1-ULK3*) (SBP): rs12940887 (*ZNF652*) (SBP/DBP)

In a different study, we investigated a large cohort of African Americans in the Minority Health Genomics and Translational Research Bio-Repository Database (MH-GRID) study. The MH-GRID genomic database comprises a large group of subjects with hypertension, both resistant and severe subtypes. We hypothesized that a direct association between *ARMC5* variants and increased risk of hypertension in African American exists. 1377 subjects (mean age: 48.25 (SD $\pm$ 6.06), controls 43.35 (SD $\pm$ 7.23), P = 1.17×10-40) and 44 variants within *ARMC5* (3 common, 4 low frequency and 37 rare variants) were considered for analysis. *ARMC5* variant rs116201073 reached nominal significance (P = 0.044) and odds ratio (OR) = 0.7, suggesting a protective effect for this variant. A set of 16 rare variants significantly associated with hypertension was identified and combined with the common variant, associated with hypertension in the single-variant analysis, representing a variant set associated with hypertension (*P* = 0.0121). These results confirmed our previous report of increased germline *ARMC5* variants that may be associated with hypertension. Further genetic and molecular studies are needed to confirm these findings [103].

2.5. Other genes

This section highlights several variants in genes that have been associated with hypertension in African Americans, which are also listed in Table 2.

2.5.1. GRK4

G protein-coupled receptor kinases (GRKs) participate in the desensitization of G protein-coupled receptors, including D1 receptors, in the proximal renal tubules; variants in *GRK4* (4p16.3) were shown to improperly excrete sodium in rodents and humans with hypertension [104]. A recent report using genetic sequencing of genes implicated in sodium and water retention in African Americans revealed variants associated with amino acid changes were implicated with low renin resistant hypertension [105]. Three variants in *GRK4* (pR65L, p.A142V, and p. A486V) were 94.4% predictive of salt-sensitive hypertension. Additionally, the number of *GRK4* variants was inversely related to salt excretion [105], suggesting that *GRK4* is an important gene in the regulation of the hyporeninemic hypoaldosteronism phenotype in African Americans with hypertension. In this study, other variants that were reported in association with hypertension in the same population included *CYP11B2*, which encodes for aldosterone synthase.

2.5.2. SCG2, PHOX2A and PHOX2B

Wen et al. reported on the association between a regulatory variant in secretogranin II (*SCG2*) in African Americans [96]. They sequenced the entire gene from 180 diverse ethnic group and showed that variant 736 was common among subjects of African descent. Two transactivating factors were identified, including ARIX (*PHOX2A*) and *PHOX2B*, while a positive selection of the protective allele within the human lineage was observed [96]. In another study, a single variant that converts methionine to threonine at amino acid 235 (M235T) of the angiotensinogen gene (*AGT*) was found to be associated with hypertension in Caucasians [106]. Kumar et al. reported an A/G polymorphism at position -217 of the *AGT* gene promoter and found that the frequency of allele A is increased in African Americans [97].

2.5.3. GPR25 and SMOC1

The Family Blood Pressure Program, which included 10,798 participants in 3320 families (3962 African Americans) observed a significant linkage on chromosome 7 (logarithm of odds [LOD] = 3.1) in African Americans people from the GenNet Network [98]. Sherva et al. identified loci contributing to arterial stiffness in a cohort 1251 Black people (HyperGEN study) [13]. Scientists identified two regions which were highly suggestive of linkage on chromosome 1 (LOD = 3.08), and another one on chromosome 14 (LOD = 2.42). Two candidate genes ((G Protein-Coupled Receptor 25 (*GPR25*), SPARC Related Modular Calcium Binding 1 (*SMOC1*)) were in the linked regions [13].

2.5.4. *SLC24A4* and *CACNA1H*

The Howard University Family Study studied a panel of over 800,000 variants from 1,017 African Americans from the Washington, D.C., metropolitan region and found two genes (Solute Carrier Family 24 Member 4 (*SLC24A4*) and Calcium Voltage-Gated Channel Subunit Alpha1 H (*CACNA1H*)) as potential candidates for blood pressure regulation [99].

2.5.5. *CYP11B2*

The *CYP11B2* gene encodes for a cytochrome P450 protein, a monooxygenase which catalyzes the synthesis of cholesterol, steroids, and other lipids in the inner mitochondrial membrane; also known as aldosterone synthase, this enzyme has an 18-hydroxylase activity to synthesize aldosterone and 18-oxocortisol as well as steroid 11 beta-hydroxylase activity and is the rate limiting step in aldosterone production [107].

The largest study to date from 3 African countries examined several variants of genes implicated in low renin-resistant hypertension in Africans with suppressed renin and increased aldosterone. Six candidate genes were sequenced, including *CYP11B2*, *SCNN1B*, *NEDD4L*, *GRK4*, Uromodulin (*UMOD*), and Natriuretic Peptide A (*NPPA*) based on the renin-aldosterone status [105]. Fourteen nonsynonymous variants of *CYP11B2* were found, with 3 previously described and associated with alterations in aldosterone synthase production (R87G, V386A, and G435S) [105]. Further studies are required to ascertain these findings.

Apparent mineralocorticoid excess (AME) refers to a rare autosomal recessive disorder leading to low renin hypertension due to alterations in *HSD11B2* (16q22.1), which encodes for the corticosteroid 11-beta-dehydrogenase. This is a microsomal enzyme complex responsible for the interconversion of cortisol and cortisone in the kidneys, thus preventing the activation of the mineralocorticoid receptor by glucocorticoids. Variants in this gene have been associated with essential hypertension [108–110], likely through decreased cortisol inactivation to cortisone which is seen with aging, and biochemically confirmed as elevated urinary tetrahydrocortisol (THF, A-ring-reduced cortisol metabolite) + alloTHF to tetrahydrocortisone (THE, cortisol metabolite) [(THF+alloTHF)/THE] [111]. One study demonstrated an association between microsatellite markers close to the *HSD11B2* gene and hypertension in African Americans that also suffered from end-stage renal disease likely due to hypertension [53,112]. Further studies are required to confirm this association with the increased predisposition of African Americans to low-renin, salt-sensitive hypertension.

2.5.6. Other Genetic Variants

Zhu et al. performed admixture mapping for hypertension loci with genome-scan markers from individuals of African descent and European Americans (Family Blood Pressure Program). They found that chromosome 6q24 and 21q21 may contain genes associated with risk of hypertension in African Americans [27]. Another admixture mapping in African Americans was done in the Dallas Heart Study [100]. Researchers genotyped a panel of 2270 variants in a random sample of 1743 African Americans and found a missense variant in Vanin 1 (*VNN1*) (rs2272996) that was significantly associated with hypertension in African Americans. In the Candidate Gene Association Resource (CARe) consortium, a novel variant (rs7726475) on chromosome 5 (between the SUB1 Homolog, Transcriptional Regulator (*SUB1*) and Natriuretic Peptide Receptor 3 (*NPR3*) genes) and rs7726475 was found to be associated with both systolic and diastolic hypertension [101].

Finally, in the International Consortium for Blood Pressure Genome-Wide Association (19,775 subjects of African ancestry) [102], scientists identified a large number of variants associated with either systolic or diastolic hypertension in Africans, including: rs13082711 (*SLC4A7*) (SBP/DBP); rs419076 (*MECOM*), (SBP); rs13107325 (*SLC39A8*) (SBP/DBP); rs13139571 (*GUCY1A3-GUCY1B3*) (SBP/DBP); rs1173771 (*NPR3-C5orf23*) (SBP); rs11953630 (*EBF1*) (SBP/DBP); rs805303 (*BAT2-BAT5*) (DBP); rs7129220 (*ADM*) (SBP/DBP); rs633185 (*FLJ32810/TMEM133*) (SBP); rs2521501 (*FURIN-FES*)

(SBP/DBP); rs17608766 (*GOSR2*) (SBP/DBP); rs1327235 (*JAG1*) (SBP/DBP); rs6015450 (*GNAS-EDN3*) (SBP/DBP); rs17367504 (*MTHFR-NPPB*) (SBP/DBP); rs3774372 (*ULK4*) (SBP/DBP); rs1458038 (*FGF5*) (SBP/DBP); rs1813353 (*CACNB2*) (SBP/DBP); rs11191548 (*CYP17A1-NT5C2*) (SBP/DBP); rs381815 (*PLEKHA7*) (SBP/DBP); rs3184504 (*SH2B3*) (SBP/DBP); rs1378942 (*CYP1A1-ULK3*) (SBP): rs12940887 (*ZNF652*) (SBP/DBP).

3. Genetic Counselling

To date, the identification of genetic variants that predispose to hypertension in African Americans has not enabled genetic diagnosis and early identification of patients and their at-risk family members. Thus, genetic testing is not currently routine in clinical practice. Indeed, with the exception of *ARMC5* and *CACNA1D* (as outlined above), the other genes discussed in this review have no current clinical implications for the management of hypertension in African Americans. When a clinician encounters a patient with a pathogenic and damaging *ARMC5* variant, screening for Cushing syndrome and primary aldosteronism is encouraged. From our experience, *ARMC5*-related adrenal pathology does not clinically present in early childhood. *ARMC5*-related endocrine hypertension diseases typically develop in adulthood as either subclinical Cushing syndrome, with or without primary aldosteronism, or overt Cushing syndrome. Carriers may not show signs of these conditions until later in adulthood, typically over 40 years of age. Genetic testing and counselling of family members should be considered as the conditions associated with *ARMC5* follow an autosomal dominant inheritance pattern with decreased penetrance.

4. Conclusions

Hypertension in African Americans is the leading cause of cardiovascular disease in this population. The complex interactions between genetic and environmental determinants are yet to be identified. Several genes implicated in RAAS activation have been studied in African American populations and have revealed a surprising number of novel variants and pathways possibly implicated in the pathogenesis of hypertension. Among them, variants in the *ARMC5* gene appear to be a rare but inherited cause of primary aldosteronism and consequently low-renin hypertension in African Americans. Further studies are needed to determine the significance of the genes discussed in this review and respective pathways, which will guide personalized precision therapy for hypertension.

Author Contributions: M.Z. and F.H.-S. co-authored the manuscript. C.A.S. provided oversight and edited the manuscript.

Acknowledgments: This research was supported in part by the Intramural Research Program of The *Eunice Kennedy Shriver* National Institute of Child Health and Human Development, National Institutes of Health (NIH), clinical trials NCT00005927 (Clinical and Molecular Analysis of ACTH-Independent Steroid Hormone Production in Adrenocortical Tissue) and NCT00001595 (A Clinical and Genetic Investigation of Pituitary Tumors and Related Hypothalamic Disorders). We thank Diane Cooper, MSLS, NIH Library, for assistance in writing this manuscript.

Conflicts of Interest: The authors declare no conflict of interest. The founding sponsors had no role in the design of the study; in the collection, analyses, or interpretation of data; in the writing of the manuscript, and in the decision to publish the results.

Abbreviations

AA   African Americans
PP   pulse pressure
PA   Primary aldosteronism
SBP   Systolic blood pressure
DBP   Diastolic blood pressure

References

1. Merai, R.; Siegel, C.; Rakotz, M.; Basch, P.; Wright, J.; Wong, B.; Thorpe, P. CDC Grand Rounds: A Public Health Approach to Detect and Control Hypertension. *MMWR Morb. Mortal. Wkly. Rep.* **2016**, *65*, 1261–1264. [CrossRef] [PubMed]

2. Mozaffarian, D.; Benjamin, E.J.; Go, A.S.; Arnett, D.K.; Blaha, M.J.; Cushman, M.; de Ferranti, S.; Després, J.-P.; Fullerton, H.J.; Howard, V.J.; et al. Heart disease and stroke statistics—2015 update: A report from the American Heart Association. *Circulation* **2015**, *131*, e29–e322. [CrossRef] [PubMed]

3. Quinlan, J.; Pearson, L.N.; Clukay, C.J.; Mitchell, M.M.; Boston, Q.; Gravlee, C.C.; Mulligan, C.J. Genetic Loci and Novel Discrimination Measures Associated with Blood Pressure Variation in African Americans Living in Tallahassee. *PLoS ONE* **2016**, *11*, e0167700. [CrossRef] [PubMed]

4. Hall, J.L.; Duprez, D.A.; Barac, A.; Rich, S.S. A review of genetics, arterial stiffness, and blood pressure in African Americans. *J. Cardiovasc. Transl. Res.* **2012**, *5*, 302–308. [CrossRef] [PubMed]

5. Redmond, N.; Baer, H.J.; Hicks, L.S. Health behaviors and racial disparity in blood pressure control in the national health and nutrition examination survey. *Hypertension* **2011**, *57*, 383–389. [CrossRef] [PubMed]

6. Mozaffarian, D.; Benjamin, E.J.; Go, A.S.; Arnett, D.K.; Blaha, M.J.; Cushman, M.; Das, S.R.; de Ferranti, S.; Despres, J.P.; Fullerton, H.J.; et al. Heart Disease and Stroke Statistics-2016 Update: A Report From the American Heart Association. *Circulation* **2016**, *133*, e38–e360. [CrossRef] [PubMed]

7. Berenson, G.S.; Wattigney, W.A.; Webber, L.S. Epidemiology of hypertension from childhood to young adulthood in black, white, and Hispanic population samples. *Public Health Rep.* **1996**, *111* (Suppl. 2), 3–6.

8. Bao, W.; Threefoot, S.A.; Srinivasan, S.R.; Berenson, G.S. Essential hypertension predicted by tracking of elevated blood pressure from childhood to adulthood: The Bogalusa Heart Study. *Am. J. Hypertens.* **1995**, *8*, 657–665. [CrossRef]

9. Carson, A.P.; Howard, G.; Burke, G.L.; Shea, S.; Levitan, E.B.; Muntner, P. Ethnic differences in hypertension incidence among middle-aged and older adults: The multi-ethnic study of atherosclerosis. *Hypertension* **2011**, *57*, 1101–1107. [CrossRef] [PubMed]

10. Calhoun, D.A.; Nishizaka, M.K.; Zaman, M.A.; Thakkar, R.B.; Weissmann, P. Hyperaldosteronism among black and white subjects with resistant hypertension. *Hypertension* **2002**, *40*, 892–896. [CrossRef] [PubMed]

11. Chor, D.; Pinho Ribeiro, A.L.; Sa Carvalho, M.; Duncan, B.B.; Andrade Lotufo, P.; Araujo Nobre, A.; Aquino, E.M.; Schmidt, M.I.; Griep, R.H.; Molina Mdel, C.; et al. Prevalence, Awareness, Treatment and Influence of Socioeconomic Variables on Control of High Blood Pressure: Results of the ELSA-Brasil Study. *PLoS ONE* **2015**, *10*, e0127382. [CrossRef] [PubMed]

12. Ge, D.; Young, T.W.; Wang, X.; Kapuku, G.K.; Treiber, F.A.; Snieder, H. Heritability of arterial stiffness in black and white American youth and young adults. *Am. J. Hypertens.* **2007**, *20*, 1065–1072. [CrossRef] [PubMed]

13. Sherva, R.; Miller, M.B.; Lynch, A.I.; Devereux, R.B.; Rao, D.C.; Oberman, A.; Hopkins, P.N.; Kitzman, D.W.; Atwood, L.D.; Arnett, D.K. A whole genome scan for pulse pressure/stroke volume ratio in African Americans: The HyperGEN study. *Am. J. Hypertens.* **2007**, *20*, 398–402. [CrossRef] [PubMed]

14. Chen, W.; Srinivasan, S.R.; Boerwinkle, E.; Berenson, G.S. Beta-adrenergic receptor genes are associated with arterial stiffness in black and white adults: The Bogalusa Heart Study. *Am. J. Hypertens.* **2007**, *20*, 1251–1257. [CrossRef] [PubMed]

15. Heffernan, K.S.; Jae, S.Y.; Wilund, K.R.; Woods, J.A.; Fernhall, B. Racial differences in central blood pressure and vascular function in young men. *Am. J. Physiol. Heart Circ. Physiol.* **2008**, *295*, H2380–H2387. [CrossRef] [PubMed]

16. Din-Dzietham, R.; Couper, D.; Evans, G.; Arnett, D.K.; Jones, D.W. Arterial stiffness is greater in African Americans than in whites: Evidence from the Forsyth County, North Carolina, ARIC cohort. *Am. J. Hypertens.* **2004**, *17*, 304–313. [CrossRef] [PubMed]

17. Ferreira, A.V.; Viana, M.C.; Mill, J.G.; Asmar, R.G.; Cunha, R.S. Racial differences in aortic stiffness in normotensive and hypertensive adults. *J. Hypertens.* **1999**, *17*, 631–637. [CrossRef] [PubMed]

18. Brown, J.M.; Robinson-Cohen, C.; Luque-Fernandez, M.A.; Allison, M.A.; Baudrand, R.; Ix, J.H.; Kestenbaum, B.; de Boer, I.H.; Vaidya, A. The Spectrum of Subclinical Primary Aldosteronism and Incident Hypertension: A Cohort Study. *Ann. Intern. Med.* **2017**, *167*, 630–641. [CrossRef] [PubMed]

19. Tu, W.; Li, R.; Bhalla, V.; Eckert, G.J.; Pratt, J.H. Age-Related Blood Pressure Sensitivity to Aldosterone in Blacks and Whites. *Hypertension* **2018**, *72*, 247–252. [CrossRef] [PubMed]

20. Spence, J.D.; Rayner, B.L. Hypertension in Blacks: Individualized Therapy Based on Renin/Aldosterone Phenotyping. *Hypertension* **2018**, *72*, 263–269. [CrossRef] [PubMed]
21. Spence, J.D.; Rayner, B.L. J Curve and Cuff Artefact, and Diagnostic Inertia in Resistant Hypertension. *Hypertension* **2016**, *67*, 32–33. [CrossRef] [PubMed]
22. Spence, J.D. Blood pressure control in Canada: The view from a stroke prevention clinic. *Can. J. Cardiol.* **2015**, *31*, 593–595. [CrossRef] [PubMed]
23. Spence, J.D. Physiologic tailoring of therapy for resistant hypertension: 20 Years' experience with stimulated renin profiling. *Am. J. Hypertens.* **1999**, *12 Pt 1*, 1077–1083.
24. Kupper, N.; Ge, D.; Treiber, F.A.; Snieder, H. Emergence of novel genetic effects on blood pressure and hemodynamics in adolescence: The Georgia Cardiovascular Twin Study. *Hypertension* **2006**, *47*, 948–954. [CrossRef] [PubMed]
25. Zhang, K.; Weder, A.B.; Eskin, E.; O'Connor, D.T. Genome-wide case/control studies in hypertension: Only the 'tip of the iceberg'. *J. Hypertens.* **2010**, *28*, 1115–1123. [CrossRef] [PubMed]
26. Darlu, P.; Sagnier, P.P.; Bois, E. Genealogical and genetical African admixture estimations, blood pressure and hypertension in a Caribbean community. *Ann. Hum. Biol.* **1990**, *17*, 387–397. [CrossRef] [PubMed]
27. Zhu, X.; Luke, A.; Cooper, R.S.; Quertermous, T.; Hanis, C.; Mosley, T.; Gu, C.C.; Tang, H.; Rao, D.C.; Risch, N.; et al. Admixture mapping for hypertension loci with genome-scan markers. *Nat. Genet.* **2005**, *37*, 177–181. [CrossRef] [PubMed]
28. Reiner, A.P.; Ziv, E.; Lind, D.L.; Nievergelt, C.M.; Schork, N.J.; Cummings, S.R.; Phong, A.; Burchard, E.G.; Harris, T.B.; Psaty, B.M.; et al. Population structure, admixture, and aging-related phenotypes in African American adults: The Cardiovascular Health Study. *Am. J. Hum. Genet.* **2005**, *76*, 463–477. [CrossRef] [PubMed]
29. Tang, H.; Jorgenson, E.; Gadde, M.; Kardia, S.L.; Rao, D.C.; Zhu, X.; Schork, N.J.; Hanis, C.L.; Risch, N. Racial admixture and its impact on BMI and blood pressure in African and Mexican Americans. *Hum. Genet.* **2006**, *119*, 624–633. [CrossRef] [PubMed]
30. Deo, R.C.; Patterson, N.; Tandon, A.; McDonald, G.J.; Haiman, C.A.; Ardlie, K.; Henderson, B.E.; Henderson, S.O.; Reich, D. A high-density admixture scan in 1670 African Americans with hypertension. *PLoS Genet.* **2007**, *3*, e196. [CrossRef] [PubMed]
31. Ward, R.H. Genetic and sociocultural components of high blood pressure. *Am. J. Phys. Anthropol.* **1983**, *62*, 91–105. [CrossRef] [PubMed]
32. Ford, C.D.; Sims, M.; Higginbotham, J.C.; Crowther, M.R.; Wyatt, S.B.; Musani, S.K.; Payne, T.J.; Fox, E.R.; Parton, J.M. Psychosocial Factors Are Associated With Blood Pressure Progression Among African Americans in the Jackson Heart Study. *Am. J. Hypertens.* **2016**, *29*, 913–924. [CrossRef] [PubMed]
33. Eirin, A.; Zhu, X.Y.; Woollard, J.R.; Herrmann, S.M.; Gloviczki, M.L.; Saad, A.; Juncos, L.A.; Calhoun, D.A.; Rule, A.D.; Lerman, A.; et al. Increased circulating inflammatory endothelial cells in blacks with essential hypertension. *Hypertension* **2013**, *62*, 585–591. [CrossRef] [PubMed]
34. Mulukutla, S.R.; Venkitachalam, L.; Bambs, C.; Kip, K.E.; Aiyer, A.; Marroquin, O.C.; Reis, S.E. Black race is associated with digital artery endothelial dysfunction: Results from the Heart SCORE study. *Eur. Heart J.* **2010**, *31*, 2808–2815. [CrossRef] [PubMed]
35. Norris, K.C.; Agodoa, L.Y. Unraveling the racial disparities associated with kidney disease. *Kidney Int.* **2005**, *68*, 914–924. [CrossRef] [PubMed]
36. van Rooyen, J.M.; Poglitsch, M.; Huisman, H.W.; Mels, C.; Kruger, R.; Malan, L.; Botha, S.; Lammertyn, L.; Gafane, L.; Schutte, A.E. Quantification of systemic renin-angiotensin system peptides of hypertensive black and white African men established from the RAS-Fingerprint(R). *J. Renin-Angiotensin-Aldosterone Syst.* **2016**, *17*, 1470320316669880. [CrossRef] [PubMed]
37. Osei, K.; Schuster, D.P. Effects of race and ethnicity on insulin sensitivity, blood pressure, and heart rate in three ethnic populations: Comparative studies in African-Americans, African immigrants (Ghanaians), and white Americans using ambulatory blood pressure monitoring. *Am. J. Hypertens.* **1996**, *9 Pt 1*, 1157–1164. [CrossRef]
38. Holwerda, S.W.; Fulton, D.; Eubank, W.L.; Keller, D.M. Carotid baroreflex responsiveness is impaired in normotensive African American men. *Am. J. Physiol. Heart Circ. Physiol.* **2011**, *301*, H1639–H1645. [CrossRef] [PubMed]

39. Feairheller, D.L.; Park, J.Y.; Sturgeon, K.M.; Williamson, S.T.; Diaz, K.M.; Veerabhadrappa, P.; Brown, M.D. Racial differences in oxidative stress and inflammation: In vitro and in vivo. *Clin. Transl. Sci.* **2011**, *4*, 32–37. [CrossRef] [PubMed]

40. Luft, F.C.; Grim, C.E.; Fineberg, N.; Weinberger, M.C. Effects of volume expansion and contraction in normotensive whites, blacks, and subjects of different ages. *Circulation* **1979**, *59*, 643–650. [CrossRef] [PubMed]

41. Akintunde, A.; Nondi, J.; Gogo, K.; Jones, E.S.W.; Rayner, B.L.; Hackam, D.G.; Spence, J.D. Physiological Phenotyping for Personalized Therapy of Uncontrolled Hypertension in Africa. *Am. J. Hypertens.* **2017**, *30*, 923–930. [CrossRef] [PubMed]

42. Norris, K.; Nissenson, A.R. Race, gender, and socioeconomic disparities in CKD in the United States. *J. Am. Soc. Nephrol.* **2008**, *19*, 1261–1270. [CrossRef] [PubMed]

43. Annas, G.J.; Elias, S. 23andMe and the FDA. *N. Engl. J. Med.* **2014**, *370*, 985–988. [CrossRef] [PubMed]

44. Mulatero, P.; Verhovez, A.; Morello, F.; Veglio, F. Diagnosis and treatment of low-renin hypertension. *Clin. Endocrinol. (Oxf.)* **2007**, *67*, 324–334. [CrossRef] [PubMed]

45. Brier, M.E.; Luft, F.C. Sodium kinetics in white and black normotensive subjects: Possible relevance to salt-sensitive hypertension. *Am. J. Med. Sci.* **1994**, *307* (Suppl. 1), S38–S42. [PubMed]

46. Helmer, O.M.; Judson, W.E. Metabolic studies on hypertensive patients with suppressed plasma renin activity not due to hyperaldosternosm. *Circulation* **1968**, *38*, 965–976. [CrossRef] [PubMed]

47. Wilson, D.K.; Bayer, L.; Krishnamoorthy, J.S.; Ampey-Thornhill, G.; Nicholson, S.C.; Sica, D.A. The prevalence of salt sensitivity in an African-American adolescent population. *Ethn. Dis.* **1999**, *9*, 350–358. [PubMed]

48. Tu, W.; Eckert, G.J.; Hannon, T.S.; Liu, H.; Pratt, L.M.; Wagner, M.A.; Dimeglio, L.A.; Jung, J.; Pratt, J.H. Racial differences in sensitivity of blood pressure to aldosterone. *Hypertension* **2014**, *63*, 1212–1218. [CrossRef] [PubMed]

49. Li, R.; Richey, P.A.; DiSessa, T.G.; Alpert, B.S.; Jones, D.P. Blood aldosterone-to-renin ratio, ambulatory blood pressure, and left ventricular mass in children. *J. Pediatr.* **2009**, *155*, 170–175. [CrossRef] [PubMed]

50. Rayner, B.L.; Owen, E.P.; King, J.A.; Soule, S.G.; Vreede, H.; Opie, L.H.; Marais, D.; Davidson, J.S. A new mutation, R563Q, of the beta subunit of the epithelial sodium channel associated with low-renin, low-aldosterone hypertension. *J. Hypertens.* **2003**, *21*, 921–926. [CrossRef] [PubMed]

51. Gillum, R.F. Pathophysiology of hypertension in blacks and whites. A review of the basis of racial blood pressure differences. *Hypertension* **1979**, *1*, 468–475. [CrossRef] [PubMed]

52. Monticone, S.; Losano, I.; Tetti, M.; Buffolo, F.; Veglio, F.; Mulatero, P. Diagnostic approach to low-renin hypertension. *Clin. Endocrinol. (Oxf.)* **2018**, *89*, 385–396. [CrossRef] [PubMed]

53. Hannah-Shmouni, F.; Melcescu, E.; Koch, C.A. Testing for Endocrine Hypertension. In *Endotext*; De Groot, L.J., Chrousos, G., Dungan, K., Feingold, K.R., Grossman, A., Hershman, J.M., Koch, C., Korbonits, M., McLachlan, R., et al., Eds.; MDText.com, Inc.: South Dartmouth, MA, USA, 2000.

54. Rayner, B.L.; Myers, J.E.; Opie, L.H.; Trinder, Y.A.; Davidson, J.S. Screening for primary aldosteronism—normal ranges for aldosterone and renin in three South African population groups. *S. Afr. Med. J.* **2001**, *91*, 594–599. [PubMed]

55. Gafane, L.F.; Schutte, R.; Van Rooyen, J.M.; Schutte, A.E. Plasma renin and cardiovascular responses to the cold pressor test differ in black and white populations: The SABPA study. *J. Hum. Hypertens.* **2016**, *30*, 346–351. [CrossRef] [PubMed]

56. Tetti, M.; Monticone, S.; Burrello, J.; Matarazzo, P.; Veglio, F.; Pasini, B.; Jeunemaitre, X.; Mulatero, P. Liddle Syndrome: Review of the Literature and Description of a New Case. *Int. J. Mol. Sci.* **2018**, *19*, 812. [CrossRef] [PubMed]

57. Grim, C.E.; Robinson, M. Salt, slavery and survival- hypertension in the African diaspora. *Epidemiology* **2003**, *14*, 120–122; discussion 124–126. [CrossRef] [PubMed]

58. Wilson, T.W.; Grim, C.E. Biohistory of slavery and blood pressure differences in blacks today. A hypothesis. *Hypertension* **1991**, *17* (Suppl. 1), I122–I128. [CrossRef] [PubMed]

59. Tishkoff, S.A.; Reed, F.A.; Friedlaender, F.R.; Ehret, C.; Ranciaro, A.; Froment, A.; Hirbo, J.B.; Awomoyi, A.A.; Bodo, J.M.; Doumbo, O.; et al. The genetic structure and history of Africans and African Americans. *Science* **2009**, *324*, 1035–1044. [CrossRef] [PubMed]

60. Rotin, D.; Staub, O. Nedd4-2 and the regulation of epithelial sodium transport. *Front. Physiol.* **2012**, *3*, 212. [CrossRef] [PubMed]

61. Hanukoglu, I.; Hanukoglu, A. Epithelial sodium channel (ENaC) family: Phylogeny, structure-function, tissue distribution, and associated inherited diseases. *Gene* **2016**, *579*, 95–132. [CrossRef] [PubMed]

62. Schild, L.; Lu, Y.; Gautschi, I.; Schneeberger, E.; Lifton, R.P.; Rossier, B.C. Identification of a PY motif in the epithelial Na channel subunits as a target sequence for mutations causing channel activation found in Liddle syndrome. *EMBO J.* **1996**, *15*, 2381–2387. [CrossRef] [PubMed]

63. Hansson, J.H.; Schild, L.; Lu, Y.; Wilson, T.A.; Gautschi, I.; Shimkets, R.; Nelson-Williams, C.; Rossier, B.C.; Lifton, R.P. A de novo missense mutation of the beta subunit of the epithelial sodium channel causes hypertension and Liddle syndrome, identifying a proline-rich segment critical for regulation of channel activity. *Proc. Natl. Acad. Sci. USA* **1995**, *92*, 11495–11499. [CrossRef] [PubMed]

64. Hansson, J.H.; Nelson-Williams, C.; Suzuki, H.; Schild, L.; Shimkets, R.; Lu, Y.; Canessa, C.; Iwasaki, T.; Rossier, B.; Lifton, R.P. Hypertension caused by a truncated epithelial sodium channel gamma subunit: Genetic heterogeneity of Liddle syndrome. *Nat. Genet.* **1995**, *11*, 76–82. [CrossRef] [PubMed]

65. Garty, H.; Palmer, L.G. Epithelial sodium channels: Function, structure, and regulation. *Physiol. Rev.* **1997**, *77*, 359–396. [CrossRef] [PubMed]

66. Baudrand, R.; Vaidya, A. The Low-Renin Hypertension Phenotype: Genetics and the Role of the Mineralocorticoid Receptor. *Int. J. Mol. Sci.* **2018**, *19*, 546. [CrossRef] [PubMed]

67. Bochud, M.; Staessen, J.A.; Maillard, M.; Mazeko, M.J.; Kuznetsova, T.; Woodiwiss, A.; Richart, T.; Norton, G.; Thijs, L.; Elston, R.; et al. Ethnic differences in proximal and distal tubular sodium reabsorption are heritable in black and white populations. *J. Hypertens.* **2009**, *27*, 606–612. [CrossRef] [PubMed]

68. Nobel, Y.R.; Lodish, M.B.; Raygada, M.; Rivero, J.D.; Faucz, F.R.; Abraham, S.B.; Lyssikatos, C.; Belyavskaya, E.; Stratakis, C.A.; Zilbermint, M. Pseudohypoaldosteronism type 1 due to novel variants of SCNN1B gene. *Endocrinol. Diabetes Metab. Case Rep.* **2016**, *2016*, 150104. [CrossRef] [PubMed]

69. Baker, E.H.; Ireson, N.J.; Carney, C.; Markandu, N.D.; MacGregor, G.A. Transepithelial sodium absorption is increased in people of African origin. *Hypertension* **2001**, *38*, 76–80. [CrossRef] [PubMed]

70. Jones, E.S.; Owen, E.P.; Rayner, B.L. The association of the R563Q genotype of the ENaC with phenotypic variation in Southern Africa. *Am. J. Hypertens.* **2012**, *25*, 1286–1291. [CrossRef] [PubMed]

71. Dong, Y.B.; Zhu, H.D.; Baker, E.H.; Sagnella, G.A.; MacGregor, G.A.; Carter, N.D.; Wicks, P.D.; Cook, D.G.; Cappuccio, F.P. T594M and G442V polymorphisms of the sodium channel beta subunit and hypertension in a black population. *J. Hum. Hypertens.* **2001**, *15*, 425–430. [CrossRef] [PubMed]

72. Dahlberg, J.; Sjogren, M.; Hedblad, B.; Engstrom, G.; Melander, O. Genetic variation in NEDD4L, an epithelial sodium channel regulator, is associated with cardiovascular disease and cardiovascular death. *J. Hypertens.* **2014**, *32*, 294–299. [CrossRef] [PubMed]

73. Luo, F.; Wang, Y.; Wang, X.; Sun, K.; Zhou, X.; Hui, R. A functional variant of NEDD4L is associated with hypertension, antihypertensive response, and orthostatic hypotension. *Hypertension* **2009**, *54*, 796–801. [CrossRef] [PubMed]

74. Kakizoe, Y.; Kitamura, K.; Ko, T.; Wakida, N.; Maekawa, A.; Miyoshi, T.; Shiraishi, N.; Adachi, M.; Zhang, Z.; Masilamani, S.; et al. Aberrant ENaC activation in Dahl salt-sensitive rats. *J. Hypertens.* **2009**, *27*, 1679–1689. [CrossRef] [PubMed]

75. Nakagawa, K.; Holla, V.R.; Wei, Y.; Wang, W.H.; Gatica, A.; Wei, S.; Mei, S.; Miller, C.M.; Cha, D.R.; Price, E., Jr.; et al. Salt-sensitive hypertension is associated with dysfunctional Cyp4a10 gene and kidney epithelial sodium channel. *J. Clin. Investig.* **2006**, *116*, 1696–1702. [CrossRef] [PubMed]

76. Zhang, R.; Lu, J.; Hu, C.; Wang, C.; Yu, W.; Ma, X.; Bao, Y.; Xiang, K.; Guan, Y.; Jia, W. A common polymorphism of CYP4A11 is associated with blood pressure in a Chinese population. *Hypertens. Res.* **2011**, *34*, 645–648. [CrossRef] [PubMed]

77. Laffer, C.L.; Elijovich, F.; Eckert, G.J.; Tu, W.; Pratt, J.H.; Brown, N.J. Genetic variation in CYP4A11 and blood pressure response to mineralocorticoid receptor antagonism or ENaC inhibition: An exploratory pilot study in African Americans. *J. Am. Soc. Hypertens.* **2014**, *8*, 475–480. [CrossRef] [PubMed]

78. Persu, A.; Coscoy, S.; Houot, A.M.; Corvol, P.; Barbry, P.; Jeunemaitre, X. Polymorphisms of the gamma subunit of the epithelial Na+ channel in essential hypertension. *J. Hypertens.* **1999**, *17*, 639–645. [CrossRef] [PubMed]

79. Ambrosius, W.T.; Bloem, L.J.; Zhou, L.; Rebhun, J.F.; Snyder, P.M.; Wagner, M.A.; Guo, C.; Pratt, J.H. Genetic variants in the epithelial sodium channel in relation to aldosterone and potassium excretion and risk for hypertension. *Hypertension* **1999**, *34 Pt 1*, 631–637. [CrossRef]

80. Zilbermint, M.; Stratakis, C.A. Primary Aldosteronism. In *Endocrine Surgery*, 2nd ed.; Pertsemlidis, D., Inabnet, W.B., Gagner, M., Eds.; CRC Press: Boca Raton, FL USA, 2017.

81. Funder, J.W.; Carey, R.M.; Mantero, F.; Murad, M.H.; Reincke, M.; Shibata, H.; Stowasser, M.; Young, W.F., Jr. The Management of Primary Aldosteronism: Case Detection, Diagnosis, and Treatment: An Endocrine Society Clinical Practice Guideline. *J. Clin. Endocrinol. Metab.* **2016**, *101*, 1889–1916. [CrossRef] [PubMed]

82. Omata, K.; Satoh, F.; Morimoto, R.; Ito, S.; Yamazaki, Y.; Nakamura, Y.; Anand, S.K.; Guo, Z.; Stowasser, M.; Sasano, H.; et al. Cellular and Genetic Causes of Idiopathic Hyperaldosteronism. *Hypertension* **2018**, *72*, 874–880. [CrossRef] [PubMed]

83. Nanba, K.; Omata, K.; Gomez-Sanchez, C.E.; Stratakis, C.A.; Demidowich, A.P.; Suzuki, M.; Thompson, L.D.R.; Cohen, D.L.; Luther, J.M.; et al. Genetic Characteristics of Aldosterone-Producing Adenomas in Blacks. *Hypertension* **2019**. [CrossRef] [PubMed]

84. Faucz, F.R.; Zilbermint, M.; Lodish, M.B.; Szarek, E.; Trivellin, G.; Sinaii, N.; Berthon, A.; Libe, R.; Assie, G.; Espiard, S.; et al. Macronodular adrenal hyperplasia due to mutations in an armadillo repeat containing 5 (ARMC5) gene: A clinical and genetic investigation. *J. Clin. Endocrinol. Metab.* **2014**, *99*, E1113–E1119. [CrossRef] [PubMed]

85. Zilbermint, M.; Xekouki, P.; Faucz, F.R.; Berthon, A.; Gkourogianni, A.; Schernthaner-Reiter, M.H.; Batsis, M.; Sinaii, N.; Quezado, M.M.; Merino, M.; et al. Primary Aldosteronism and ARMC5 Variants. *J. Clin. Endocrinol. Metab.* **2015**, *100*, E900–E909. [CrossRef] [PubMed]

86. Stratakis, C.A.; Boikos, S.A. Genetics of adrenal tumors associated with Cushing's syndrome: A new classification for bilateral adrenocortical hyperplasias. *Nat. Clin. Pract. Endocrinol. Metab.* **2007**, *3*, 748–757. [CrossRef] [PubMed]

87. Spence, J.D. The current epidemic of primary aldosteronism: Causes and consequences. *J. Hypertens.* **2004**, *22*, 2038–2039; author reply 2039. [CrossRef] [PubMed]

88. Russell, R.P.; Masi, A.T. The prevalence of adrenal cortical hyperplasia at autopsy and its association with hypertension. *Ann. Intern. Med.* **1970**, *73*, 195–205. [CrossRef] [PubMed]

89. Kidambi, S.; Kotchen, J.M.; Grim, C.E.; Raff, H.; Mao, J.; Singh, R.J.; Kotchen, T.A. Association of adrenal steroids with hypertension and the metabolic syndrome in blacks. *Hypertension* **2007**, *49*, 704–711. [CrossRef] [PubMed]

90. Russell, R.P.; Masi, A.T. Significant associations of adrenal cortical abnormalities with "essential" hypertension. *Am. J. Med.* **1973**, *54*, 44–51. [CrossRef]

91. Russell, R.P.; Masi, A.T.; Richter, E.D. Adrenal cortical adenomas and hypertension. A clinical pathologic analysis of 690 cases with matched controls and a review of the literature. *Medicine (Baltimore)* **1972**, *51*, 211–225. [CrossRef] [PubMed]

92. Correa, R.; Zilbermint, M.; Berthon, A.; Espiard, S.; Batsis, M.; Papadakis, G.Z.; Xekouki, P.; Lodish, M.B.; Bertherat, J.; Faucz, F.R.; et al. The ARMC5 gene shows extensive genetic variance in primary macronodular adrenocortical hyperplasia. *Eur. J. Endocrinol.* **2015**, *173*, 435–440. [CrossRef] [PubMed]

93. Assie, G.; Libe, R.; Espiard, S.; Rizk-Rabin, M.; Guimier, A.; Luscap, W.; Barreau, O.; Lefevre, L.; Sibony, M.; Guignat, L.; et al. ARMC5 mutations in macronodular adrenal hyperplasia with Cushing's syndrome. *N. Engl. J. Med.* **2013**, *369*, 2105–2114. [CrossRef] [PubMed]

94. Berthon, A.; Faucz, F.R.; Espiard, S.; Drougat, L.; Bertherat, J.; Stratakis, C.A. Age-dependent effects of Armc5 haploinsufficiency on adrenocortical function. *Hum. Mol. Genet.* **2017**, *26*, 3495–3507. [CrossRef] [PubMed]

95. Weder, A.B.; Gleiberman, L.; Sachdeva, A. Whites excrete a water load more rapidly than blacks. *Hypertension* **2009**, *53*, 715–718. [CrossRef] [PubMed]

96. Wen, G.; Wessel, J.; Zhou, W.; Ehret, G.B.; Rao, F.; Stridsberg, M.; Mahata, S.K.; Gent, P.M.; Das, M.; Cooper, R.S.; et al. An ancestral variant of Secretogranin II confers regulation by PHOX2 transcription factors and association with hypertension. *Hum. Mol. Genet.* **2007**, *16*, 1752–1764. [CrossRef] [PubMed]

97. Kumar, A.; Li, Y.; Patil, S.; Jain, S. A haplotype of the angiotensinogen gene is associated with hypertension in african americans. *Clin. Exp. Pharmacol. Physiol.* **2005**, *32*, 495–502. [CrossRef] [PubMed]

98. Bielinski, S.J.; Lynch, A.I.; Miller, M.B.; Weder, A.; Cooper, R.; Oberman, A.; Chen, Y.D.; Turner, S.T.; Fornage, M.; Province, M.; et al. Genome-wide linkage analysis for loci affecting pulse pressure: The Family Blood Pressure Program. *Hypertension* **2005**, *46*, 1286–1293. [CrossRef] [PubMed]

99. Adeyemo, A.; Gerry, N.; Chen, G.; Herbert, A.; Doumatey, A.; Huang, H.; Zhou, J.; Lashley, K.; Chen, Y.; Christman, M.; et al. A genome-wide association study of hypertension and blood pressure in African Americans. *PLoS Genet.* **2009**, *5*, e1000564. [CrossRef] [PubMed]

100. Zhu, X.; Cooper, R.S. Admixture mapping provides evidence of association of the VNN1 gene with hypertension. *PLoS ONE* **2007**, *2*, e1244. [CrossRef] [PubMed]

101. Zhu, X.; Young, J.H.; Fox, E.; Keating, B.J.; Franceschini, N.; Kang, S.; Tayo, B.; Adeyemo, A.; Sun, Y.V.; Li, Y.; et al. Combined admixture mapping and association analysis identifies a novel blood pressure genetic locus on 5p13: Contributions from the CARe consortium. *Hum. Mol. Genet.* **2011**, *20*, 2285–2295. [CrossRef] [PubMed]

102. International Consortium for Blood Pressure Genome-Wide Association Studies; Ehret, G.B.; Munroe, P.B.; Rice, K.M.; Bochud, M.; Johnson, A.D.; Chasman, D.I.; Smith, A.V.; Tobin, M.D.; Verwoert, G.C.; et al. Genetic variants in novel pathways influence blood pressure and cardiovascular disease risk. *Nature* **2011**, *478*, 103–109. [CrossRef] [PubMed]

103. Zilbermint, M.; Gaye, A.; Berthon, A.; Hannah-Shmouni, F.; Faucz, F.; Davis, A.; Gibbons, G.; Lodish, M.; Stratakis, C. ARMC5 variants and risk of hypertension in African Americans: Minority Health-GRID study. In Proceedings of the 20th European Congress of Endocrinology, Barcelona, Spain, 19–22 May 2018.

104. Felder, R.A.; Sanada, H.; Xu, J.; Yu, P.Y.; Wang, Z.; Watanabe, H.; Asico, L.D.; Wang, W.; Zheng, S.; Yamaguchi, I.; et al. G protein-coupled receptor kinase 4 gene variants in human essential hypertension. *Proc. Natl. Acad. Sci. USA* **2002**, *99*, 3872–3877. [CrossRef] [PubMed]

105. Jones, E.S.; Spence, J.D.; McIntyre, A.D.; Nondi, J.; Gogo, K.; Akintunde, A.; Hackam, D.G.; Rayner, B.L. High Frequency of Variants of Candidate Genes in Black Africans with Low Renin-Resistant Hypertension. *Am. J. Hypertens.* **2017**, *30*, 478–483. [CrossRef] [PubMed]

106. Jeunemaitre, X.; Soubrier, F.; Kotelevtsev, Y.V.; Lifton, R.P.; Williams, C.S.; Charru, A.; Hunt, S.C.; Hopkins, P.N.; Williams, R.R.; Lalouel, J.M.; et al. Molecular basis of human hypertension: Role of angiotensinogen. *Cell* **1992**, *71*, 169–180. [CrossRef]

107. Gomez-Sanchez, C.E.; Kuppusamy, M.; Reincke, M.; Williams, T.A. Disordered CYP11B2 Expression in Primary Aldosteronism. *Horm. Metab. Res.* **2017**, *49*, 957–962. [CrossRef] [PubMed]

108. Walker, B.R.; Stewart, P.M.; Shackleton, C.H.; Padfield, P.L.; Edwards, C.R. Deficient inactivation of cortisol by 11 beta-hydroxysteroid dehydrogenase in essential hypertension. *Clin. Endocrinol. (Oxf.)* **1993**, *39*, 221–227. [CrossRef] [PubMed]

109. Soro, A.; Ingram, M.C.; Tonolo, G.; Glorioso, N.; Fraser, R. Evidence of coexisting changes in 11 beta-hydroxysteroid dehydrogenase and 5 beta-reductase activity in subjects with untreated essential hypertension. *Hypertension* **1995**, *25*, 67–70. [CrossRef] [PubMed]

110. Alikhani-Koopaei, R.; Fouladkou, F.; Frey, F.J.; Frey, B.M. Epigenetic regulation of 11 beta-hydroxysteroid dehydrogenase type 2 expression. *J. Clin. Investig.* **2004**, *114*, 1146–1157. [CrossRef] [PubMed]

111. Shackleton, C.H.; Rodriguez, J.; Arteaga, E.; Lopez, J.M.; Winter, J.S. Congenital 11 beta-hydroxysteroid dehydrogenase deficiency associated with juvenile hypertension: Corticosteroid metabolite profiles of four patients and their families. *Clin. Endocrinol. (Oxf.)* **1985**, *22*, 701–712. [CrossRef] [PubMed]

112. Watson, B., Jr.; Bergman, S.M.; Myracle, A.; Callen, D.F.; Acton, R.T.; Warnock, D.G. Genetic association of 11 beta-hydroxysteroid dehydrogenase type 2 (HSD11B2) flanking microsatellites with essential hypertension in blacks. *Hypertension* **1996**, *28*, 478–482. [CrossRef] [PubMed]

International Journal of
Molecular Sciences

Review

Endogenous Ouabain and Related Genes in the Translation from Hypertension to Renal Diseases

Marco Simonini, Paola Casanova, Lorena Citterio, Elisabetta Messaggio, Chiara Lanzani and Paolo Manunta *

Genomics of Renal Disease and Hypertension Unit, IRCCS San Raffaele Scientific Institute, Università Vita Salute San Raffaele, 20132 Milan, Italy; simonini.marco@hsr.it (M.S.); casanova.paola@hsr.it (P.C.); citterio.lorena@hsr.it (L.C.); messaggio.elisabetta@hsr.it (E.M.); lanzani.chiara@hsr.it (C.L.)
* Correspondence: manunta.paolo@hsr.it; Tel.: +39-02-2643-3890 (ext. 5330)

Received: 22 May 2018; Accepted: 29 June 2018; Published: 3 July 2018

Abstract: The endogenous ouabain (EO) is a steroid hormone secreted by the adrenal gland with cardio-tonic effects. In this article, we have reviewed and summarized the most recent reports about EO, particularly with regard to how it may interact with specific genetic backgrounds. We have focused our attention on the EO's potential pathogenic role in several diseases, including renal failure, essential hypertension and heart failure. Notably, these reports have demonstrated that EO acts as a pro-hypertrophic and growth-promoting hormone, which might lead to a cardiac remodeling affecting cardiovascular functions and structures. In addition, a possible role of EO in the development of acute kidney injury has been hypothesized. During the last decays, many important improvements permitted a deeper understanding of EO's metabolisms and functions, including the characteristics of its receptor and the effects of its activation. Such progresses indicated that EO has significant implications in the pathogenesis of many common diseases. The patho-physiological role of EO in the development of hypertension and other cardiac and renal complications have laid the basis for the development of a new selective compound that could selectively modulate the genetic and molecular mechanisms involved in EO's action. It is evident that the knowledge of EO has incredibly increased; however, many important areas remain to be further investigated.

Keywords: cardio-tonic steroids; endogenous ouabain; adducin; hypertension; renal damage

---

1. Introduction

The cardiac glycosides are a class of drugs derived from the leaves of the *Digitalis purpurea* with a positive inotropic effect on the heart. For a very long time, they were successfully used as a primary treatment for congestive heart failure and arrhythmias [1]. The mechanism of action of the cardiotonic steroids in the human heart has been widely studied, and it is now accepted that it consists in the inhibition of the Na^+/K^+-ATPase—a transmembrane enzyme that regulates the gradient of sodium and potassium across the plasma membrane. It was assessed that the inhibition is made through the binding to a highly conserved extracellular recognition sequence of the Na^+/K^+ pump [2]. The favorable result of the use of the plant-derived cardiotonic steroids led to hypothesizing the existence of an endogenous cardiac glycoside counterpart in mammals and to assuming that its functional receptor might be the Na^+/K^+-ATPase [3]. This enzyme consists of an α- and a β-subunits. The α-subunit is made of 10 transmembrane segments that include, on the extracellular loops, the binding region for the cardiac glycoside, known as the ouabain-binding site. This region is highly conserved in the evolution among species from drosophila to rodents, sheep and humans [2,4].

2. Endogenous Ouabain (EO) and Na$^+$/K$^+$-ATPase Interaction

The hypothesis of the existence of an endogenous cardiac glycoside became solid, when several animals [5–10] and human [11,12] studies showed that, in the context of a volume-expanded condition, it was possible to find an endogenous humoral substance that could counterbalance the increased renal reabsorption of sodium and water by the inhibition of Na$^+$ transport through vascular and tubular cell membranes. The volume expansion might be a stimulus for the release of this substance, called natriuretic hormone, which could control sodium homeostasis through the inhibition of the key enzyme in the process of its tubular reabsorption, the Na$^+$/K$^+$-ATPase [13,14]. The effect of the endogenous cardiac glycoside is not limited to the kidney, but also involves the Na$^+$/K$^+$ pumps in other regions, such as the neuro-vascular system. The consequence of this enzyme's inhibition is the increase of the intracellular sodium concentration, which is exchanged for calcium through the Na$^+$/Ca$^+$ exchanger, particularly active in cardiac mussels and smooth vascular muscles [15]. The increase of the intracellular concentration of these two ions, promoted by elevated levels of this endogenous inhibitor, might augment vascular tone determining peripheral vasoconstriction and might lead to a rise in blood pressure [16,17]. Notably, increased levels of cardiotonic glycosides were found in low renin (i.e., volume expanded) hypertension [18]. The vasopressor effect of the cardiotonic steroids has acute and chronic aspects. The acute pressor effect is mediated by the increase in the calcium concentration that causes vasoconstriction, when the short-term cardio-vascular reflexes are blocked. In the case of sustained and chronic elevation of circulating cardiotonic steroids, the pressor effect is maintained by the activation of a signaling pathway that up-regulates the expression of several ion transports in arterial myocytes [19]. It was initially hard to demonstrate that an endogenous digitalis actually exists, and during past years, many research groups have tried to identify it, particularly under physiological and pathological conditions, such as hypertension, pregnancy and neonatal period. Consistent findings obtained over the years suggested that it was possible to isolate different candidate inhibitors of the Na$^+$-pump in mammalian tissues, urine and plasma, and several natriuretic substances able to inhibit sodium pumps were identified [4]. Among this compound, it was isolated and characterized the one presenting the majority of the functional properties of the plant-derived cardiac glycoside. Hamlyn's and Haupert's groups were the first to describe the presence of a cardiotonic steroid indistinguishable from ouabain in human plasma in 1991 [20]. The endogenous ouabain (EO) was then isolated from bovine adrenal glands [21], human adrenal glands [22], bovine hypothalamus [23], rat adrenomedullary cells [24] and biological fluids [25,26] by using the high performance liquid chromatography (HPLC) and immunoassay methods. The presence of EO was finally demonstrated with mass spectrometry, nuclear magnetic resonance (NMR) and chromatography, confirming its existence and its identicalness to the plant-derived ouabain [27,28]. Furthermore, it was possible to identify the most important production site of EO in the adrenal cortex [29,30]. All this evidence led to the identification in mammals of numerous endogenous cardiotonic steroids as cardenolides and bufodienolides (as marinobufagenine) [31]. However, these endogenous compounds are different from each other and it is fair to assume that they could play a distinct patho-physiological role, acting as tissue-specific regulators of different isoforms of the Na$^+$/K$^+$-pump [19].

3. Endogenous Ouabain Pressor Mechanism and Genes Involved in the Pathogenesis of Hypertension

During the past years, many research groups investigated the molecular basis of essential hypertension, focusing their attention on renal, endocrine, nervous and humoral dysfunction. In particular, they hypothesized that alterations in renal sodium management could have a key role in its pathogenesis. The Na$^+$/K$^+$-ATPase activity in the kidneys is regulated by hormonal and genetic factors including EO and the gene coding for α-adducin (*ADD1*). Adducin is a cytoskeletal protein consisting of two heterodimers (α/β or α/γ). *ADD1*, *ADD2* and *ADD3* are the three coding genes for these subunits. It was shown that a polymorphism in the gene *ADD1* (determining the presence of a tryptophan instead of a glycine in the amino-acid position 460, Gly460Trp) was associated with a higher expression of Na^+/K^+-ATPase in the surface of the cell and an enhancement of its activity. To better comprehend the mechanisms undergoing primary hypertension, researchers developed several rat models of genetic hypertension, including the Milan hypertensive strain (MHS) of rats that represents a suitable model for a subgroup of human patients with hypertension. It was shown, in both MHS rats and humans, that increased concentrations of EO corresponded to an increased tubular sodium reabsorption and, consequently, hypertension [32]. Furthermore, a prolonged infusion of low doses of plant-derived ouabain in normotensive rats and rat renal tubular cultured cells was associated to an enhanced expression and activity of the Na^+/K^+ pump, leading to a reversible form of hypertension. Starting from these evidences, it was hypothesized that ouabain itself may be considered as a pressor agent in vivo. The same phenomenon was documented in cells transfected with genetic variants of the MHS adducin [33]. In the following years, researchers also tried to understand the mechanism, by which both ouabain and mutated adducin could modify the expression of the Na^+/K^+-ATPase in the kidneys. Under these two conditions, it was possible to evidence a slower recycling of the sodium pump from the surface of the cell and, consequently, an excessive expression of the Na^+/K^+-ATPase in light of a tighter anchoring to the cytoskeletal proteins. This is the biochemical alteration present in both the ouabain- and adducin-dependent forms of hypertension [34]. These findings apparently contradict the traditional natriuretic hypothesis that considers the cardiotonic steroids as inhibitors of the Na^+/K^+-ATPase. According to this hypothesis, volume expansion conditions might induce the release of an endogenous hormone (EH) able to promote natriuresis. High levels of EO should lead to a decrease, rather than an increase, of Na^+/K^+-ATPase activity. To clarify this issue, the relation between EO and changes in sodium balance was studied in both rats and patients with essential hypertension. The results showed that an acute and chronic restriction of salt intake (but not the acute salt loading) was associated with a significant rise in EO plasmatic levels [35]. Consequently, conditions of salt and water reductions might provoke the elevation of the EO humoral concentration, meaning that EO does not act as a natriuretic hormone in vivo [4,36,37]. An important augmented EO level during physical exercise was demonstrated, which is a state characterized by an increased sympathetic activity and a decline in renal blood flow [4,38], and is found in patients undergoing cardiac surgery [39]. These results evidenced that EO induces a variety of important mechanisms, which augment vascular tone promoting renal sodium retention [4]. It might be a fair assumption that EO, through the enhancement of the renal Na^+/K^+-ATPase activity, plays a role in body sodium homeostasis and in the re-establishment and maintenance of the hydro-saline equilibrium [3,19]. Previous studies demonstrated that adducin has a direct role in the modulation of the renal Na^+/K^+-ATPase (Figure 1). Following this evidence, it was important to understand whether adducin polymorphism (Gly460Trp) might influence EO's response of the adaptation to a low-salt diet.

In hypertensive patients with mutated *ADD1* gene (*Trp-460*), a chronic low-salt diet is associated to an important augment of EO plasmatic concentration. Contrarily, it is not possible to recognize the same condition in wild-type patients (Gly-460), thus counteracting the hypotensive effect of the low-salt diet. Similarly, there is an increase of EO levels in MHS rats and the congenic rat strain NA (obtained by the introgression of the MHS *ADD1* locus into the normotensive genetic background), but not in normotensive rats. A high-salt diet does not modify plasma EO in rats, nor in humans. These data suggest that adducin genotype might predict the changes of EO plasmatic levels under salt restriction conditions [40,41]. Another study examined the linkage between EO and blood pressure in the general population obtaining several new evidences. Notably: (1) people with the Gly460Trp polymorphism of α-adducin have higher EO plasmatic levels compared to the carriers of the wild-type genotype; (2) the EO plasmatic concentration is directly proportional to urinary potassium excretion; and (3) there is an important interaction between blood pressure, EO levels and urinary sodium excretion [42]. We can finally affirm that EO acts as a positive regulator of blood pressure during chronic low-salt diet, whilst it prevents high salt-induced blood pressure when the salt intake is elevated. There is a correlation between EO and genetics in the homeostatic regulation of blood pressure in response to changes in salt intake (Figure 2). However, this complex relationship requires further investigation in order to be fully clarified.

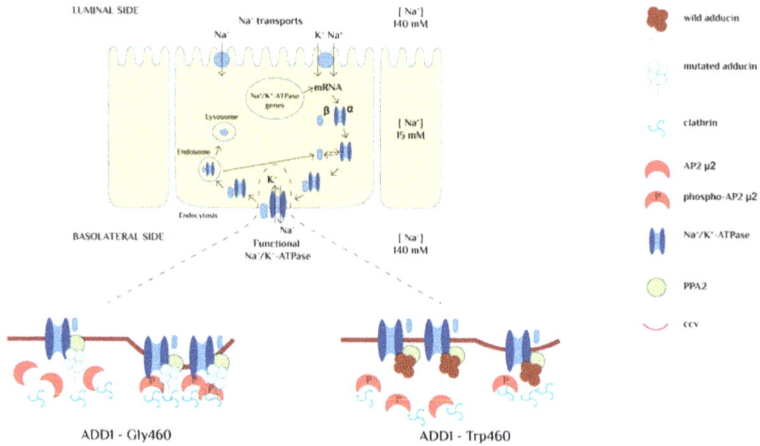

Adapted from Giuseppe Bianchi et al., Hypertension 2005

Figure 1. Effect of α-adducin of Na⁺/K⁺-ATPase pump. The mutated form of α-adducin reduces endocytosis, leading to an over-expression of the Na⁺/K⁺ pump molecules on the basolateral membrane and to an increased sodium reabsorption. In the basal condition, the association between phosphatase A2 (PPA2) and adducin is reduced in tubular cells transfected with mutated adducin. The impairment of this cycle may represent the molecular mechanism underlying the reduced endocytosis observed in the presence of mutated adducin (figure adapted from Bianchi et al., Hypertnsion 2005) [3].

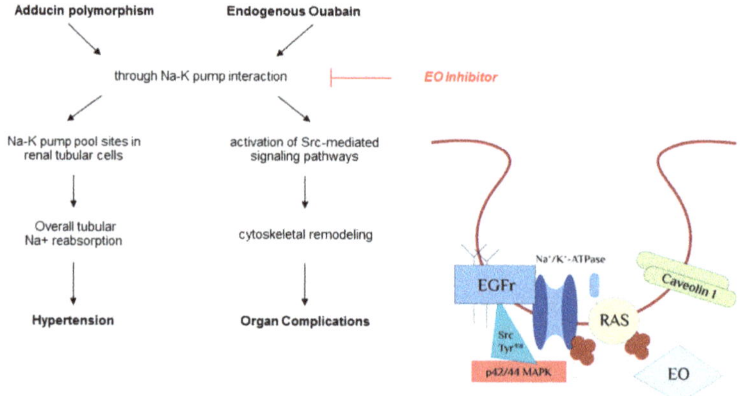

Adapted from Mara Ferrandi et al., JBC 2004

Figure 2. Physio-pathological interaction between Endogenous Ouabain (EO) and of α-adducin. This interaction can lead to the development of hypertension and organ maladaptive remodeling and potential target of an anti-ouabain compound (as rostafuroxin) (figure adapted from Ferrandi et al., JBC 2004) [43].

4. The Correlation between Endogenous Ouabain and Organ Damage

4.1. Endogenous Ouabain and Cardiovascular Disease

Studies on rat (normotensive, MNS and MHS rats) and human models have demonstrated that EO contributes not only to the pathogenesis of hypertension, but also to the development of cardiac complications, such as ventricular hypertrophy, heart failure and acute myocardial infarction [12–14]. Approximately half of patients affected by essential hypertension have high circulating levels of EO, and previous studies demonstrated a direct relation between increased EO levels, left ventricular mass indices and stroke volumes, whilst a negative correlation of increased EO levels with heart rates was seen [15–17]. EO might be actually considered as a growth-promoting hormone. Indeed, many authors showed that EO is involved in pro-hypertrophic and pro-fibrotic pathways that lead to cardiac remodeling with a negative impact on both cardiac structures and cardiovascular functions [30,44]. The authors also tried to understand how EO might perform the role of a signal transducer. Studies on cultured vascular cells showed that EO, after the binding to its receptor (Na^+/K^+-ATPase), could trigger a tyrosine-kinase protein starting an intracellular-signaling cascade that constitutes a stimulus for the epidermal growth factor receptor (EGFR). This pathway finally permits the transcription of gene encoding for pro-fibrotic factors [45], promoting cardiac hypertrophy. Relevant evidences showed that a chronic activation of this complex protein-kinase cascade could finally lead to heart failure [46]. It was hypothesized that the molecular pathway that leads to organ hypertrophy in vivo was similar to the one described in cultured vascular cells [16]. In a study conducted on more than 800 patients undergoing elective cardiac surgery, it was supposed that EO might be also used as a valuable biomarker of heart failure. This study confirmed the already hypothesized negative relation that intercourses between increased EO levels and left ventricular ejection fraction and the positive correlation between EO levels and cardiac end-diastolic diameters. EO were dosed in all of these patients after and before the surgery, and it was seen that higher EO circulating levels both in the pre-operation and the immediate post-operation were associated with worst cardiovascular presentation, higher morbidity and increased risks of perioperative mortality after cardiac surgery [47].

4.2. Endogenous Ouabain and Renal Disease

The main actor in cardiac glucosides' metabolism is the liver; however, it is now clear that kidneys also have an important role in their clearance. To validate this evidence, experiments conducted on the rat demonstrated that partial nephrectomy was associated with higher EO circulating levels [12]. Similarly, it was seen that the progression of kidney disease, in particular the end-stage renal failure, was associated with an increase of EO levels [19]. High concentrations of EO are comparable to an excessive and not controllable digitalization that might lead to a vasopressor effect and other important cardiac side effects. Indeed, studies on patients with Chronic Kidney Disease (CKD) or undergoing dialysis demonstrated that elevated levels of EO were strongly associated with alterations in ventricular mass and geometry [48,49], independently from blood pressure and other determinants of left ventricular hypertrophy. The comorbidities that characterize patients with severe renal diseases and dialyzed patients, such as hypertension and cardiac hypertrophy, might thus be potentially related to EO. These data stimulated the interest of the impact of EO on the renal function. To understand this aspect, a selective podocyte marker protein called nephrin was used. Studies on rats showed that a prolonged exposure to high levels of EO was associated with a less expression of nephrin in the podocyte, a reduction in creatinine clearance and increased proteinuria. This last finding was also replicated ex vivo with the incubation of low dose of ouabain in podocyte primary cell cultures [50]. EO modulates the Na^+/K^+-ATPase that is involved in tubular ischemic damage, and it is responsible for the initial nephrinuria and the glomerular damage. The EO's promotion of kidneys' damage shown in rat models suggested a possible role of EO in acute kidney injury (AKI). In a recent observational study, a blood sample for the dosage of EO was taken during the induction of anesthesia in patients destined to elective cardiac surgery. Interestingly, patients with higher preoperative levels of EO were characterized by worst renal outcomes and higher mortality rates [47,50,51]. Starting from these evidences, it is a fair assumption that EO might be considered to be a valuable biomarker of individual susceptibility to the development of AKI after cardiac surgery [52,53].

Recent studies individuated an additional effect of EO, showing that it might also act as a pro-cystogenic agent in the development of autosomal dominant polycystic kidney disease (ADPKD). It was shown that the Na^+/K^+-ATPase of ADPKD cells has an increased affinity with EO, and even if the mechanism undergoing this abnormal affinity remains uncertain, it might enhance ADPKD cell's susceptibility to circulating EO. The exposure of primary cultures of cells isolated from renal cysts of ADPKD patients to nanomolar concentrations of EO provokes the proliferation of cyst epithelial cells. In contrast, similar concentrations of EO had a pour influence on the proliferation of normal human kidney cells [52,53].

4.3. Endogenous Ouabain and Brain Disease

The presence of EO inside the central nervous system (CNS) is well-known since the early 1990s [54,55]. It was demonstrated that it is an integrated component of a hypothalamic renin-angiotensin-aldosterone system (RAAS) and also plays an important role in regulation of systemic Blood Pressure (BP) [37,40,56]. The relation between the central and peripheral RAASes with central and peripheral EO is still not completely understood [55], but it was discovered that EO-induced signaling in neurons had positive and direct consequences on rat brain in terms of brain cells survival [57]. Similarly, in mouse models, the reduction of endogenous steroids seems to have a protective effect on oxidative stress for CNS [58]. Endogenous steroids, in particular EO-like compounds, were also described as potential risk factors involved in the etiology of bipolar disorder [59], mania [60] and depression [61]. Moreover, the use of anti-ouabain antibodies showed a reduction of maniac [60], as well as depressive [62] status. The development of EO associated behavior disorders seems directly associated with a dys-regulation of Na^+/K^+-ATPase inside CNSes [59,60].

4.4. Endogenous Ouabain as Therapeutic Target

As already said in this review, several studies during the last decays individuated a central core in the increased EO levels that could connect the pathogenesis of several cardiac and renal diseases, including hypertension, cardiac hypertrophy, heart failure, renal failure and ADPKD. This fundamental evidence stimulated the birth of a new research branch aiming to individuate a selective competitor of EO, paving the way for the formulation of new antihypertensive agents that could selectively correct the molecular mechanisms behind. On this basis, authors' research groups developed a new digitoxigening-derived compound, called Rostafuroxin. Its mechanism of action expects to display EO from its specific binding sites on the Na^+/K^+-ATPase, modulating its abnormal expression in the cell surface without inhibiting other renal sodium transporters and without influencing other hormonal pathways [19,43,63]. As already mentioned, the interaction between EO and its receptor triggers a complex cascade of intracellular second-messengers that ends with the generation of hypertrophic stimuli. It was seen that nanomolar concentrations of Rostafuroxin could antagonize the interaction between EO and the Na^+/K^+ pump, normalizing this signal response and blocking the excessive activation of EGFR [19]. In both normotensive and MHS rat models, low oral doses of Rostafuroxin could normalize the up-regulation of renal Na^+-pump, leading to a reduction of blood pressure levels [19,63]. A similar effect was obtained with the use of Rostafuroxin in other rat models, such as the deoxycorticosterone acetate–salt rat and the reduced-renal-mass hypertensive rat. Both models were characterized by a condition of low plasmatic renin, volume expansion and increased levels of EO [64]. In contrast, it was interesting to evaluate that Rostafuroxin does not act as an antihypertensive agent in these models, in which EO and α-adducin polymorphism are not implicated in the pathogenesis of hypertension, such as the normotensive control rats and spontaneous hypertensive rats. Moreover, preliminary results suggest that this molecule, used at oral doses of 7–10 µg/kg/day, might revert the ouabain-induced hypertrophic activity [35,37]. In a trial of unselected patients, it was shown that

Rostafuroxin does not influence blood pressure in these patients with no elevated EO levels, assuming that its effect on blood pressure is strictly related to the genetic background that regulates the synthesis and the clearance of EO. It was assessed that in Rostafuroxin-sensitive patients, there is a decline of systolic blood pressure of 14 mmHg after 4 weeks of treatment [65]. Nowadays, we do not have data about the effect of Rostafuroxin on the treatment of renal failure and the effects of hypertension on patients with end-stage renal disease, but this compound might have the effect of amelioration of the grade of hypertrophy and heart failure [46]. The pharmacological profile and the selective mechanism of action of Rostafuroxin make it the prototype of a compound devoid of the cardiovascular and hormonal side effects associated with digitalis and diuretic. Indeed, there was no evidence of intrinsic cardiac inotropic effects or arrhythmogenic activity.

5. Conclusions

During the last years, numerous important advances have led to a better understanding of EO's metabolisms and functions, including the characteristics of its receptor and the molecular effects of its activation. Although many important areas need to be further investigated, compelling data reinforce the concept that high EO plasmatic levels and adducin polymorphism (Gly460Trp) are associated with an increased risk of developing diseases including hypertension, ADPKD and organ complications, such as podocyte injury, cardiac and kidney hypertrophy. It also contributes to the development and the maintenance of AKI in critically ill patients.

The importance of these evidences is that the understanding of the patho-physiological mechanism undergoing complex diseases represents a valid substrate to individuate specific pharmacological targets. This is mainly important in complex multifactorial disease, when a tailored approach based on individuals' phenotypes and genotypes should be chosen to treat individual patients carrying a specific genetic background. Rostafuroxin might be considered to be a revolutionary therapy for hypertension when increased EO circulating levels and adducin polymorphism exert a pathogenetic role. Rostafuroxin perfectly fits the new concept of personalized medicine, and it can be considered to be a safe drug without the most common side effects of previous used compounds.

Author Contributions: Conceptualization: P.M., M.S., L.C., C.L.; methodology: E.M., P.C.; software: L.C.; validation: E.M.; formal analysis: C.L.; investigation: M.S.; resources and data curation: all the authors; writing of the original draft preparation: P.M., M.S., P.C.; writing of review and editing: P.M., P.C.; visualization: all the authors; supervision: P.M.; project administration: P.M.

Funding: This research received no external funding.

Conflicts of Interest: The authors declare no conflicts of interest.

References

1. Silva, E.; Soares-da-Silva, P. *New Insights into the Regulation of Na+,K+-ATPase by Ouabain; International Review of Cell and Molecular Biology;* Elsevier: New York, NY, USA, 2012; Volume 294, pp. 99–132.

2. Lingrel, J.B. The Physiological Significance of the Cardiotonic Steroid/Ouabain-Binding Site of the Na,K-ATPase. *Annu. Rev. Physiol.* **2010**, *72*, 395–412. [CrossRef] [PubMed]

3. Ferrandi, M.; Manunta, P.; Ferrari, P.; Bianchi, G. The endogenous ouabain: Molecular basis of its role in hypertension and cardiovascular complications. *Front. Biosci.* **2005**, *10*, 2472–2477. [CrossRef] [PubMed]

4. Hamlyn, J.M. Natriuretic hormones, endogenous Ouabain, and related sodium transport inhibitors. *Front. Endocrinol.* **2014**, *5*, 199. [CrossRef] [PubMed]

5. Fedorova, O.V.; Shapiro, J.I.; Bagrov, A.Y. Endogenous cardiotonic steroids and salt-sensitive hypertension. *Biochim. Biophys. Acta* **2010**, *1802*, 1230–1236. [CrossRef] [PubMed]

6. Fedorova, O.V.; Talan, M.I.; Agalakova, N.I.; Lakatta, E.G.; Bagrov, A.Y. Endogenous ligand of alpha(1) sodium pump, marinobufagenin, is a novel mediator of sodium chloride—Dependent hypertension. *Circulation* **2002**, *105*, 1122–1127. [CrossRef] [PubMed]

7. Fedorova, O.V.; Kolodkin, N.I.; Agalakova, N.I.; Namikas, A.R.; Bzhelyansky, A.; St.-Louis, J.; Lakatta, E.G.; Bagrov, A.Y. Antibody to marinobufagenin lowers blood pressure in pregnant rats on a high NaCl intake. *J. Hypertens.* **2005**, *23*, 835–842. [CrossRef] [PubMed]

8. Vu, H.; Ianosi-Irimie, M.; Danchuk, S.; Rabon, E.; Nogawa, T.; Kamano, Y.; Pettit, G.R.; Wiese, T.; Puschett, J.B. Resibufogenin corrects hypertension in a rat model of human preeclampsia. *Exp. Biol. Med.* **2006**, *231*, 215–220. [CrossRef]

9. Puschett, J.B.; Agunanne, E.; Uddin, M.N. Marinobufagenin, resibufogenin and preeclampsia. *Biochim. Biophys. Acta* **2010**, *1802*, 1246–1253. [CrossRef] [PubMed]

10. Bagrov, A.Y.; Agalakova, N.I.; Kashkin, V.A.; Fedorova, O.V. Endogenous cardiotonic steroids and differential patterns of sodium pump inhibition in NaCl-loaded salt-sensitive and normotensive rats. *Am. J. Hypertens.* **2009**, *22*, 559–563. [CrossRef] [PubMed]

11. Fridman, A.I.; Matveev, S.A.; Agalakova, N.I.; Fedorova, O.V.; Lakatta, E.G.; Bagrov, A.Y. Marinobufagenin, an endogenous ligand of alpha-1 sodium pump, is a marker of congestive heart failure severity. *J. Hypertens.* **2002**, *20*, 1189–1194. [CrossRef] [PubMed]

12. Gonick, H.C.; Ding, Y.; Vaziri, N.D.; Bagrov, A.Y.; Fedorova, O.V. Simultaneous measurement of marinobufagenin, ouabain, and hypertension-associated protein in various disease states. *Clin. Exp. Hypertens.* **1998**, *20*, 617–627. [CrossRef] [PubMed]

13. Manunta, P.; Messaggio, E.; Casamassima, N.; Gatti, G.; Carpini, S.D.; Zagato, L.; Hamlyn, J.M. Endogenous ouabain in renal Na(+) handling and related diseases. *Biochim. Biophys. Acta* **2010**, *1802*, 1214–1218. [CrossRef] [PubMed]

14. Blaustein, M.P. Sodium ions, calcium ions, blood pressure regulation, and hypertension: A reassessment and a hypothesis. *Am. J. Physiol.* **1977**, *232*, C165–C173. [CrossRef] [PubMed]

15. Schoner, W. Endogenous cardiac glycosides, a new class of steroid hormones. *Eur. J. Biochem.* **2002**, *269*, 2440–2448. [CrossRef] [PubMed]

16. Hamlyn, J.M.; Blaustein, M.P. Endogenous Ouabain: Recent Advances and Controversies. *Hypertension* **2016**, *68*, 526–532. [CrossRef] [PubMed]

17. Blaustein, M.P.; Chen, L.; Hamlyn, J.M.; Leenen, F.H.H.; Lingrel, J.B.; Wier, W.G.; Zhang, J. Pivotal role of $\alpha2$ Na$^+$ pumps and their high affinity ouabain binding site in cardiovascular health and disease. *J. Physiol.* **2016**, *594*, 6079–6103. [CrossRef] [PubMed]

18. Haddy, F.J.; Pamnani, M.; Clough, D.; Huot, S. Role of a humoral sodium-potassium pump inhibitor in experimental low renin hypertension. *Life Sci.* **1982**, *30*, 571–575. [CrossRef]

19. Hamlyn, J.M.; Manunta, P. Endogenous Cardiotonic Steroids in Kidney Failure: A Review and an Hypothesis. *Adv. Chronic Kidney Dis.* **2015**, *22*, 232–244. [CrossRef] [PubMed]

20. Hamlyn, J.M.; Blaustein, M.P.; Bova, S.; DuCharme, D.W.; Harris, D.W.; Mandel, F.; Mathews, W.R.; Ludens, J.H. Identification and characterization of a ouabain-like compound from human plasma. *Proc. Natl. Acad. Sci. USA* **1991**, *88*, 6259–6263. [CrossRef] [PubMed]

21. Schneider, R.; Wray, V.; Nimtz, M.; Lehmann, W.D.; Kirch, U.; Antolovic, R.; Schoner, W. Bovine adrenals contain, in addition to ouabain, a second inhibitor of the sodium pump. *J. Biol. Chem.* **1998**, *273*, 784–792. [CrossRef] [PubMed]

22. El-Masri, M.A.; Clark, B.J.; Qazzaz, H.M.; Valdes, R. Human adrenal cells in culture produce both ouabain-like and dihydroouabain-like factors. *Clin. Chem.* **2002**, *48*, 1720–1730. [PubMed]

23. Tymiak, A.A.; Norman, J.A.; Bolgar, M.; DiDonato, G.C.; Lee, H.; Parker, W.L.; Lo, L.C.; Berova, N.; Nakanishi, K.; Haber, E. Physicochemical characterization of a ouabain isomer isolated from bovine hypothalamus. *Proc. Natl. Acad. Sci. USA* **1993**, *90*, 8189–8193. [CrossRef] [PubMed]

24. Komiyama, Y.; Nishimura, N.; Munakata, M.; Mori, T.; Okuda, K.; Nishino, N.; Hirose, S.; Kosaka, C.; Masuda, M.; Takahashi, H. Identification of endogenous ouabain in culture supernatant of PC12 cells. *J. Hypertens.* **2001**, *19*, 229–236. [CrossRef] [PubMed]

25. Ferrandi, M.; Minotti, E.; Salardi, S.; Florio, M.; Bianchi, G.; Ferrari, P. Characteristics of a ouabain-like factor from Milan hypertensive rats. *J. Cardiovasc. Pharmacol.* **1993**, *22* (Suppl. 2), S75–S78. [CrossRef] [PubMed]

26. Di Bartolo, V.; Balzan, S.; Pieraccini, L.; Ghione, S.; Pegoraro, S.; Biver, P.; Revoltella, R.; Montali, U. Evidence for an endogenous ouabain-like immunoreactive factor in human newborn plasma coeluted with ouabain on HPLC. *Life Sci.* **1995**, *57*, 1417–1425. [CrossRef]

27. Bagrov, A.Y.; Shapiro, J.I.; Fedorova, O.V. Endogenous cardiotonic steroids: Physiology, pharmacology, and novel therapeutic targets. *Pharmacol. Rev.* **2009**, *61*, 9–38. [CrossRef] [PubMed]

28. Zhao, N.; Lo, L.C.; Berova, N.; Nakanishi, K.; Tymiak, A.A.; Ludens, J.H.; Haupert, G.T. Na,K-ATPase inhibitors from bovine hypothalamus and human plasma are different from ouabain: Nanogram scale CD structural analysis. *Biochemistry* **1995**, *34*, 9893–9896. [CrossRef] [PubMed]

29. Kawamura, A.; Guo, J.; Itagaki, Y.; Bell, C.; Wang, Y.; Haupert, G.T.; Magil, S.; Gallagher, R.T.; Berova, N.; Nakanishi, K. On the structure of endogenous ouabain. *Proc. Natl. Acad. Sci. USA* **1999**, *96*, 6654–6659. [CrossRef] [PubMed]

30. Blaustein, M.P. The pump, the exchanger, and the Holy Spirit: Origins and 40-year evolution of ideas about the ouabain-Na$^+$ pump endocrine system. *Am. J. Physiol. Cell Physiol.* **2018**, *314*, C3–C26. [CrossRef] [PubMed]

31. Dmitrieva, R.I.; Bagrov, A.Y.; Lalli, E.; Sassone-Corsi, P.; Stocco, D.M.; Doris, P.A. Mammalian bufadienolide is synthesized from cholesterol in the adrenal cortex by a pathway that Is independent of cholesterol side-chain cleavage. *Hypertension* **2000**, *36*, 442–448. [CrossRef] [PubMed]

32. Ferrari, P.; Ferrandi, M.; Valentini, G.; Bianchi, G. Rostafuroxin: An ouabain antagonist that corrects renal and vascular Na$^+$-K$^+$-ATPase alterations in ouabain and adducin-dependent hypertension. *AJP Regul. Integr. Comp. Physiol.* **2006**, *290*, R529–R535. [CrossRef] [PubMed]

33. Manunta, P.; Rogowski, A.C.; Hamilton, B.P.; Hamlyn, J.M. Ouabain-induced hypertension in the rat: Relationships among plasma and tissue ouabain and blood pressure. *J. Hypertens.* **1994**, *12*, 549–560. [CrossRef] [PubMed]

34. Citterio, L.; Lanzani, C.; Manunta, P.; Bianchi, G. Genetics of primary hypertension: The clinical impact of adducin polymorphisms. *BBA Mol. Basis Dis.* **2010**, *1802*, 1285–1298. [CrossRef] [PubMed]

35. Tripodi, G.; Citterio, L.; Kouznetsova, T.; Lanzani, C.; Florio, M.; Modica, R.; Messaggio, E.; Hamlyn, J.M.; Zagato, L.; Bianchi, G.; et al. Steroid biosynthesis and renal excretion in human essential hypertension: Association with blood pressure and endogenous ouabain. *Am. J. Hypertens.* **2009**, *22*, 357–363. [CrossRef] [PubMed]

36. Manunta, P.; Manunta, P.; Hamilton, B.P.; Hamlyn, J.M. Salt intake and depletion increase circulating levels of endogenous ouabain in normal men. *AJP Regul. Integr. Comp. Physiol.* **2006**, *290*, R553–R559. [CrossRef] [PubMed]

37. Hamlyn, J.M.; Blaustein, M.P. Salt sensitivity, endogenous ouabain and hypertension. *Curr. Opin. Nephrol. Hypertens.* **2013**, *22*, 51–58. [CrossRef] [PubMed]

38. Bauer, N.; Müller-Ehmsen, J.; Krämer, U.; Hambarchian, N.; Zobel, C.; Schwinger, R.H.G.; Neu, H.; Kirch, U.; Grünbaum, E.-G.; Schoner, W. Ouabain-like compound changes rapidly on physical exercise in humans and dogs: Effects of beta-blockade and angiotensin-converting enzyme inhibition. *Hypertension* **2005**, *45*, 1024–1028. [CrossRef] [PubMed]

39. Bignami, E.; Casamassima, N.; Frati, E. Endogenous Ouabain Changes Rapidly during Cardiac Pulmonary by Pass. *J. Steroids Hormon. Sci.* **2011**. [CrossRef]

40. Blaustein, M.P.; Leenen, F.H.H.; Chen, L.; Golovina, V.A.; Hamlyn, J.M.; Pallone, T.L.; Van Huysse, J.W.; Zhang, J.; Wier, W.G. How NaCl raises blood pressure: A new paradigm for the pathogenesis of salt-dependent hypertension. *Am. J. Physiol. Heart Circ. Physiol.* **2012**, *302*, H1031–H1049. [CrossRef] [PubMed]

41. Linde, C.I.; Karashima, E.; Raina, H.; Zulian, A.; Wier, W.G.; Hamlyn, J.M.; Ferrari, P.; Blaustein, M.P.; Golovina, V.A. Increased arterial smooth muscle Ca^{2+} signaling, vasoconstriction, and myogenic reactivity in Milan hypertensive rats. *Am. J. Physiol. Heart Circ. Physiol.* **2012**, *302*, H611–H620. [CrossRef] [PubMed]

42. Manunta, P.; Maillard, M.; Tantardini, C.; Simonini, M.; Lanzani, C.; Citterio, L.; Stella, P.; Casamassima, N.; Burnier, M.; Hamlyn, J.M.; et al. Relationships among endogenous ouabain, alpha-adducin polymorphisms and renal sodium handling in primary hypertension. *J. Hypertens.* **2008**, *26*, 914–920. [CrossRef] [PubMed]

43. Ferrandi, M.; Molinari, I.; Barassi, P.; Minotti, E.; Bianchi, G.; Ferrari, P. Organ hypertrophic signaling within caveolae membrane subdomains triggered by ouabain and antagonized by PST 2238. *J. Biol. Chem.* **2004**, *279*, 33306–33314. [CrossRef] [PubMed]

44. Bagrov, A.Y.; Shapiro, J.I. Endogenous digitalis: Pathophysiologic roles and therapeutic applications. *Nat. Clin. Pract. Nephrol.* **2008**, *4*, 378–392. [CrossRef] [PubMed]

45. Huang, L.; Huang, L.; Li, H.; Li, H.; Xie, Z. Ouabain-induced Hypertrophy in Cultured Cardiac Myocytes is Accompanied by Changes in Expression of Several Late Response Genes. *J. Mol. Cell. Cardiol.* **1997**, *29*, 429–437. [CrossRef] [PubMed]

46. Blaustein, M.P. How does pressure overload cause cardiac hypertrophy and dysfunction? High-ouabain affinity cardiac Na$^+$ pumps are crucial. *Am. J. Physiol. Heart Circ. Physiol.* **2017**, *313*, H919–H930. [CrossRef] [PubMed]

47. Simonini, M.; Pozzoli, S.; Bignami, E.; Casamassima, N.; Messaggio, E.; Lanzani, C.; Frati, E.; Botticelli, I.M.; Rotatori, F.; Alfieri, O.; et al. Endogenous Ouabain: An Old Cardiotonic Steroid as a New Biomarker of Heart Failure and a Predictor of Mortality after Cardiac Surgery. *Biomed. Res. Int.* **2015**, *2015*, 714793. [CrossRef] [PubMed]

48. Pavlovic, D. The role of cardiotonic steroids in the pathogenesis of cardiomyopathy in chronic kidney disease. *Nephron Clin. Pract.* **2014**, *128*, 11–21. [CrossRef] [PubMed]

49. Stella, P.; Manunta, P.; Mallamaci, F.; Melandri, M.; Spotti, D.; Tripepi, G.; Hamlyn, J.M.; Malatino, L.S.; Bianchi, G.; Zoccali, C. Endogenous ouabain and cardiomyopathy in dialysis patients. *J. Intern. Med.* **2008**, *263*, 274–280. [CrossRef] [PubMed]

50. Bignami, E.; Casamassima, N.; Frati, E.; Lanzani, C.; Corno, L.; Alfieri, O.; Gottlieb, S.; Simonini, M.; Shah, K.B.; Mizzi, A.; et al. Preoperative endogenous ouabain predicts acute kidney injury in cardiac surgery patients. *Crit. Care Med.* **2013**, *41*, 744–755. [CrossRef] [PubMed]

51. Simonini, M.; Lanzani, C.; Bignami, E.; Casamassima, N.; Frati, E.; Meroni, R.; Messaggio, E.; Alfieri, O.; Hamlyn, J.; Body, S.C.; et al. A new clinical multivariable model that predicts postoperative acute kidney injury: Impact of endogenous ouabain. *Nephrol. Dial. Transplant.* **2014**, *29*, 1696–1701. [CrossRef] [PubMed]

52. Blanco, G.; Wallace, D.P. Novel role of ouabain as a cystogenic factor in autosomal dominant polycystic kidney disease. *Am. J. Physiol. Renal Physiol.* **2013**, *305*, F797–F812. [CrossRef] [PubMed]

53. Venugopal, J.; Blanco, G. On the Many Actions of Ouabain: Pro-Cystogenic Effects in Autosomal Dominant Polycystic Kidney Disease. *Molecules* **2017**, *22*, 729. [CrossRef]

54. Dostanic-Larson, I.; Van Huysse, J.W.; Lorenz, J.N.; Lingrel, J.B. The highly conserved cardiac glycoside binding site of Na,K-ATPase plays a role in blood pressure regulation. *Proc. Natl. Acad. Sci. USA* **2005**, *102*, 15845–15850. [CrossRef] [PubMed]

55. Blaustein, M.P. Why isn't endogenous ouabain more widely accepted? *AJP Heart Circ. Physiol.* **2014**, *307*, H635–H639. [CrossRef] [PubMed]

56. Leenen, F.H.H. The central role of the brain aldosterone-"ouabain" pathway in salt-sensitive hypertension. *Biochim. Biophys. Acta* **2010**, *1802*, 1132–1139. [CrossRef] [PubMed]

57. Lopachev, A.V.; Lopacheva, O.M.; Osipova, E.A.; Vladychenskaya, E.A.; Smolyaninova, L.V.; Fedorova, T.N.; Koroleva, O.V.; Akkuratov, E.E. Ouabain-induced changes in MAP kinase phosphorylation in primary culture of rat cerebellar cells. *Cell Biochem. Funct.* **2016**, *34*, 367–377. [CrossRef] [PubMed]

58. Hodes, A.; Lifschytz, T.; Rosen, H.; Cohen Ben-Ami, H.; Lichtstein, D. Reduction in endogenous cardiac steroids protects the brain from oxidative stress in a mouse model of mania induced by amphetamine. *Brain Res. Bull.* **2018**, *137*, 356–362. [CrossRef] [PubMed]

59. Christo, P.J.; el-Mallakh, R.S. Possible role of endogenous ouabain-like compounds in the pathophysiology of bipolar illness. *Med. Hypotheses* **1993**, *41*, 378–383. [CrossRef]

60. Hodes, A.; Rosen, H.; Deutsch, J.; Lifschytz, T.; Einat, H.; Lichtstein, D. Endogenous cardiac steroids in animal models of mania. *Bipolar Disord.* **2016**, *18*, 451–459. [CrossRef] [PubMed]

61. Goldstein, I.; Levy, T.; Galili, D.; Ovadia, H.; Yirmiya, R.; Rosen, H.; Lichtstein, D. Involvement of Na(+), K(+)-ATPase and endogenous digitalis-like compounds in depressive disorders. *Biol. Psychiatry* **2006**, *60*, 491–499. [CrossRef] [PubMed]

62. Goldstein, I.; Lax, E.; Gispan-Herman, I.; Ovadia, H.; Rosen, H.; Yadid, G.; Lichtstein, D. Neutralization of endogenous digitalis-like compounds alters catecholamines metabolism in the brain and elicits anti-depressive behavior. *Eur. Neuropsychopharmacol.* **2012**, *22*, 72–79. [CrossRef] [PubMed]

63. Ferrandi, M.; Manunta, P.; Rivera, R.; Bianchi, G.; Ferrari, P. Role of the ouabain-like factor and Na-K pump in rat and human genetic hypertension. *Clin. Exp. Hypertens.* **1998**, *20*, 629–639. [CrossRef] [PubMed]

64. Wenceslau, C.F.; Rossoni, L.V. Rostafuroxin ameliorates endothelial dysfunction and oxidative stress in resistance arteries from deoxycorticosterone acetate-salt hypertensive rats. *J. Hypertens.* **2014**, *32*, 542–554. [CrossRef] [PubMed]
65. Lanzani, C.; Citterio, L.; Glorioso, N.; Manunta, P.; Tripodi, G.; Salvi, E.; Carpini, S.D.; Ferrandi, M.; Messaggio, E.; Staessen, J.A.; et al. Adducin- and ouabain-related gene variants predict the antihypertensive activity of rostafuroxin, part 2: Clinical studies. *Sci. Transl. Med.* **2010**, *2*, 59ra87. [CrossRef] [PubMed]

International Journal of
Molecular Sciences

Review

Genomics of Fibromuscular Dysplasia

Silvia Di Monaco [1,2], **Adrien Georges** [3], **Jean-Philippe Lengelé** [1,4], **Miikka Vikkula** [5] and
Alexandre Persu [1,6,*]

[1]   Division of Cardiology, Cliniques Universitaires Saint-Luc, Université Catholique de Louvain,
      1200 Brussels, Belgium; silvia.dimonaco@gmail.com (S.D.M.); Jean_Philippe.LENGELE@ghdc.be (J.-P.L.)
[2]   Department of Medical Sciences, Internal Medicine and Hypertension Division, AOU Città della Salute e
      della Scienza, University of Turin, 10124 Turin, Italy
[3]   INSERM, UMR970 Paris Cardiovascular Research Center (PARCC), F-75015 Paris, France;
      adrien.georges@inserm.fr
[4]   Department of Nephrology, Grand Hôpital De Charleroi, 6060 Gilly, Belgium
[5]   Human Molecular Genetics, de Duve Institute, Université catholique de Louvain, 1200 Brussels, Belgium;
      miikka.vikkula@uclouvain.be
[6]   Pole of Cardiovascular Research, Institut de Recherche Expérimentale et Clinique, Université catholique de
      Louvain, 1200 Brussels, Belgium
*   Correspondence: alexandre.persu@uclouvain.be; Tel.: +32-2-764-6306 or 32-2-764-2533; Fax: +32-2-764-8980

Received: 2 May 2018; Accepted: 16 May 2018; Published: 21 May 2018

Abstract: Fibromuscular Dysplasia (FMD) is "an idiopathic, segmental, non-atherosclerotic and non-inflammatory disease of the musculature of arterial walls, leading to stenosis of small and medium-sized arteries" (Persu, et al; 2014). FMD can lead to hypertension, arterial dissections, subarachnoid haemorrhage, stroke or mesenteric ischemia. The pathophysiology of the disease remains elusive. While familial cases are rare (<5%) in contemporary FMD registries, there is evidence in favour of the existence of multiple genetic factors involved in this vascular disease. Recent collaborative efforts allowed the identification of a first genetic locus associated with FMD. This intronic variant located in the phosphatase and actin regulator 1 gene (*PHACTR1*) may influence the transcription activity of the endothelin-1 gene (*EDN1*) located nearby on chromosome 6. Interestingly, the *PHACTR1* locus has also been involved in vascular hypertrophy in normal subjects, carotid dissection, migraine and coronary artery disease. National and international registries of FMD patients, with deep and harmonised phenotypic and genetic characterisation, are expected to be instrumental to improve our understanding of the genetic basis and pathophysiology of this intriguing vascular disease.

Keywords: fibromuscular dysplasia; non atherosclerotic vascular stenosis; *PHACTR1*; genetic association; cervical artery dissection; spontaneous coronary arteries dissection

---

1. Definition and Main Features of Fibromuscular Dysplasia (FMD)

Fibromuscular Dysplasia (FMD) has been defined as "an idiopathic, segmental, non-atherosclerotic and non-inflammatory disease of the musculature of arterial walls, leading to stenosis of small and medium-sized arteries" [1]. In the last years, the perception of FMD has evolved from a rare cause of secondary hypertension involving renal arteries to a more diffuse vascular disease, affecting also cervico-cephalic, visceral, coronary and iliac arteries [1,2].

Renal and cervico-cephalic arteries are the most frequently affected vascular beds, and both lesions often coexist (65% in the US registry) [3]. In the Assessment of Renal and Cervical Artery Dysplasia (ARCADIA) Belgian-French registry [4], systematic, state of the art exploration of the main arterial beds in 469 patients with FMD revealed that 48% of patients have focal or multifocal FMD lesions of two vascular beds or more and 15% of three vascular beds or more (multivessel FMD).

According to the affected arterial bed, FMD can lead to hypertension (renal artery FMD), as well as to severe complications including arterial aneurysms and dissections, subarachnoid haemorrhage, stroke or mesenteric ischemia [1,5]. While FMD was initially considered as a rare disease, silent FMD lesions have been detected in 3–6% of potential kidney donors [6,7].

Recently, multifocal FMD lesions of extracoronary vascular beds have been also identified in up to 75–80% of patients with spontaneous coronary artery dissection (SCAD). Patients who have had a SCAD share many similarities with FMD patients, including female predominance (>90%), age at diagnosis (≈50 years old) and few cardiovascular risk factors [8]. FMD affects predominantly women: 91% in the US registry [3], 84% in the ARCADIA study [4], and 83% in the European registry (A. Persu, personal communication). In contemporary cohorts, the age at diagnosis is in the range of 50–55 years, with a wide distribution from small infants [9] to octogenarians [4,5].

Based on angiographic patterns two subtypes have been defined for renal FMD [1,5,10], and subsequently for cervico-cephalic FMD [1,11], i.e., (i) multifocal FMD: "string-of-beads" appearance, alternation of stenosis and aneurysmal dilations (>80% of cases), corresponding to the medial form according to the former histological classification [12]; and (ii) focal FMD: isolated stenosis, whatever its length, in young patients (usually <40 year old) with few or no cardiovascular risk factors, in the absence of atherosclerotic or inflammatory lesions. While the pattern of multifocal FMD is almost pathognomonic, in order to establish the diagnosis of focal FMD, it is necessary to rule out early localized atherosclerosis, as well as a series of inflammatory and genetic arteriopathies (see Table 1).

Table 1. Differential diagnosis of focal Fibromuscular Dysplasia (FMD) in a nutshell (modified from [1]).

Focal atherosclerotic lesions
Inflammatory arterial diseases:
• Takayasu arteritis;   • Giant cell arteritis.
Arterial diseases of monogenic origin:
• Type 1 neurofibromatosis;   • Alagille syndrome;   • Williams syndrome.

Until recently, the aetiology of FMD remains elusive. The classical aetiological hypothesis are summarized in the next section.

2. Classical Aetiological Hypothesis

2.1. Female Hormones

In view of the predominance of women among FMD patients, a role for female hormones has been assumed. In particular, it has been hypothesized that the estrogen-induced production of extracellular matrix proteins from vascular fibroblasts and vascular smooth muscle cells [13], may be responsible for some typical histopathological alterations of FMD [12,14]. Unfortunately, up to now, we lack consistent and powerful data to assess for associations between FMD and e.g., the number of pregnancies, age at menarche, history of hysterectomy, gynaecological problems, spontaneous abortion and/or oral contraceptive therapy [12,15]. In addition, all available studies are retrospective, small and underpowered. Therefore, large prospective registries are needed to test the female hormonal environment hypothesis.

2.2. Traumatic Theory

Other triggers of FMD may include arterial wall stress [12,14], as it was observed that cyclic vascular wall stretch increased synthesis of matrix components by smooth muscle cells [16,17]. It has also been hypothesized that FMD might result from renal artery traction due to nephroptosis [15] or repetitive micro-traumas of the vascular wall induced by excessive arterial pulsatility [18]. However, longitudinal studies that would allow proving or disproving these hypotheses are lacking.

2.3. Intramural Ischemia

In 1970s, it was proposed that the first determinant of FMD lesions is occlusion of the vasa vasorum, which may induce hypoxia in the arterial wall and transformation of smooth muscle cells into myofibroblasts, typically found in histopathological samples of arteries affected by FMD [19,20]. In particular, Sottiurai and co-workers demonstrated an increase in myofibroblasts and extra-cellular connective tissue of the media, as a result of vasa vasorum obliteration by injection of a thrombine–gelatine mixture into dog femoral arteries [19]. Furthermore, the extracranial internal carotid and external iliac arteries have fewer vasa vasorum than other muscular arteries of similar size, which may make them more susceptible to intramural ischemia [12,15]. However, occlusion of vasa vasorum in patients with FMD has not been demonstrated [20].

2.4. Smoking

Since 1979, smoking has been associated with genesis of FMD and/or its progression in at least 6 studies [15,21–25]. In particular, in a French cohort including 337 patients with FMD, the proportion of current and ever smokers was 30% and 50%, respectively, compared to 18% and 37% in 337 essential hypertensive controls ($p \leq 0.001$) [24]. Furthermore, current smoking compared to non-current smoking was associated with an earlier FMD onset (43 vs. 51 years, $p < 0.001$) [24], and a higher proportion of renal asymmetry (21% vs. 4%, $p = 0.001$) and number of renal artery revascularization interventions (57% vs. 31%, $p \leq 0.001$) [24]. Finally, in the US Registry, ever smokers were characterized by an increased proportion of aortic (4.8% vs. 1.5%, $p < 0.01$) [26] but also intracerebral (4.8% vs. 1.7%, $p < 0.01$) aneurysms [25] compared to never smokers.

3. FMD as a Genetic Disease

3.1. Familial Forms of FMD

Since 1960s to 1970s, a number of cases of familial FMD, i.e., occurrence of FMD in two or more siblings or relatives, have been reported [26–31]. We report a case of 3 siblings affected by hypertension and renal artery FMD followed at the Grand Hôpital de Charleroi, Gilly, Belgium (Figure 1), as an illustration. Interestingly, while two of them had typical string-of-beads lesions, the third had an isolated stenosis compatible with focal FMD.

In the absence of systematic exploration of first-degree relatives of patients with FMD, the prevalence of familial FMD remains unclear. In a French cohort of 100 unrelated patients with renal artery FMD, FMD lesions were documented by angiography in at least one first-degree relative in 11% of patients [32]. In the first report of the US registry, 7.3% of 447 patients reported a history of FMD in at least another 1st or 2nd degree relative [3]. However, in recent reports this percentage tends to be substantially lower: 2.9% in the ARCADIA study [4] and 2.8% in European registry (A. Persu, personal communication). The prevalence of familial FMD may be higher in paediatric than in adult patients (17.2% of 29 children vs. 4.7% of 864 adult patients with FMD in a recent report of the US registry) [33].

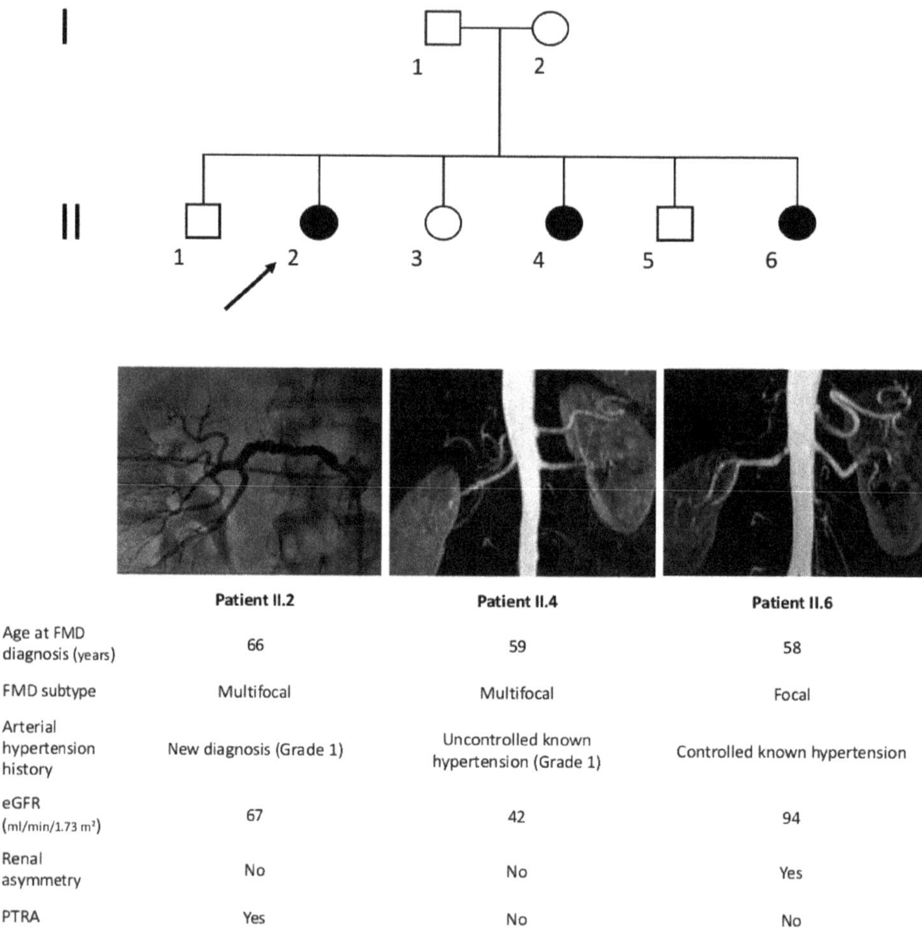

	Patient II.2	Patient II.4	Patient II.6
Age at FMD diagnosis (years)	66	59	58
FMD subtype	Multifocal	Multifocal	Focal
Arterial hypertension history	New diagnosis (Grade 1)	Uncontrolled known hypertension (Grade 1)	Controlled known hypertension
eGFR (ml/min/1.73 m²)	67	42	94
Renal asymmetry	No	No	Yes
PTRA	Yes	No	No

Figure 1. Example of familial FMD from Belgian Multicentric FMD Cohort (BEL-FMD). Patient II.2 came to clinical attention at the age of 64 years for de novo arterial hypertension. The abdominal Computed Tomography Angiogram showed an aspect compatible with multifocal FMD of both renal arteries, significant on the right side. She underwent right percutaneous transluminal renal angioplasty (PTRA), and the hypertensive crises regressed. Patient II.4 was hypertensive from the age of 53 years and was referred at the age of 59 years for worsening of blood pressure control, associated to decreased renal function. Abdominal Magnetic Resonance Angiography (MRA) showed mild irregularities suggestive of FMD in the right renal artery, in the absence of significant stenosis. Patient II.6 came to clinical attention at the age of 58 years for renal asymmetry (kidney length: 9.5 cm on the right side vs. 12 cm on the left). The abdominal MRA identified a focal stenosis of 70% of the right renal artery. Patient II.6 underwent renal arteriography, which disclosed an irregular aspect of the arterial wall on the left side and only a 30% stenosis on the right side. Notably, none of these three patients had lesions of cervico-cephalic or others vascular beds. Estimated Glomerular Filtration Rate (eGFR) was calculated using the Modification of Diet in Renal Disease (MDRD) equation.

3.2. Genetic Susceptibility to FMD

While clear Mendelian familial presentation of FMD appears to be rare, genetic factors may be involved in the pathophysiology of the more common, apparently sporadically occurring forms of

FMD. Already in 1980, Rushton suggested a familial pattern of FMD in 12 out of 20 unrelated FMD patients, with indication of autosomal dominant inheritance [34]. Nevertheless, the presence of FMD in relatives was assumed on the basis a history of vascular diseases occurring before 50 years, many of which may have been due to atherosclerosis [34].

More than 20 years later, using high resolution echotracking, Boutouyrie and co-workers described typical abnormalities of the arterial wall of the carotid bifurcation and radial artery of patients with renal artery FMD, including a characteristic "triple signal" pattern [23]. Interestingly, these abnormalities were significantly more frequent in patients with FMD (echotracking arterial score 4.02) but also in apparently healthy first-degree relatives of patients with FMD (echotracking arterial score 4.17) compared to unrelated normotensive controls (echotracking arterial score 2.52), with a pattern of inheritance suggesting once again an autosomal-dominant transmission [35]. Beyond the existence of a small proportion of clearly familial forms, these data suggest the existence of an inherited component in FMD at large, and provide a rationale for recent studies aiming at identifying susceptibility genes.

3.3. Overlap with Genetic Syndromes

Vascular abnormalities associated with FMD include arterial stenosis, tortuosity, aneurysms and dissections in one or more arterial beds [1]. Patients with rare inherited arteriopathies and connective tissue diseases (CTD) may harbour similar lesions. Furthermore, a proportion of patients with FMD harbour non-vascular features associated with CTD, suggesting an overlap between these entities and a possible role of the corresponding genes in non-syndromic FMD. In a cohort of 47 multifocal FMD patients, without family history of CTD, 95.7% of subjects presented radiological features, such as cerebral aneurysm (12.8%), early onset degenerative spine disease (95.7%), increased incidence of Arnold-Chiari I malformation (6.4%) and dural ectasia (42.6%) [36]. However, genetic screening failed to identify mutations in genes involved in hereditary CTD (*COL3A1, FBN1, PLOD1, TGFβR1, TGFβR2, TGFβ2, SMAD3, ACTA2,* and *COL5A1*) [36]. The most frequent clinical CTD features in a cohort of 139 female patients with FMD from the US registry were early onset (before age 50) osteoarthritis (15.6%), palatal abnormalities (56.1%), moderately severe myopia (29.1%), pectus excavatum or carinatum (7.2%), and dental crowding (29.7%) [37]. However, the prevalence of classical CTD features was estimated at 18.7%, similar to the general female population [37]. Finally, no mutation in *TGFβR2, COL3A1, FBN1, ACTA2,* or *SMAD3* gene was detected in a series of 35 FMD patients, again from the US registry. In this cohort, two distinct variants were identified in transforming growth factor beta receptor 1 (*TGFβR1*) gene in two unrelated patients. Both variants induced an amino acid substitution in a highly conserved region of *TGFβR1*, however their role is unknown [38]. Still, these findings may be relevant in view of the identification of increased plasma transforming growth factor beta-1 (TGF-β1) and transforming growth factor beta-2 (TGF-β2) secretion in dermal fibroblast lines from patients with FMD compared to age and gender-matched controls [36]. Overall, while FMD and CTD share some similarities, arguments in favour of a common genetic basis are scarce.

4. Challenges of the Genetic Investigation of FMD

Several reasons make genetic investigation of FMD challenging and limit access to multi-generational families or large cohorts of patients: (i) as most patients are no longer operated nowadays, histopathologic confirmation of the diagnosis is lacking; (ii) the surrogate "gold standard", catheter-based angiography, which was at the basis of the current angiographic diagnostic criteria and classification cannot be proposed as a first-line screening test, in view of its invasiveness and potential risks [1]; (iii) while the string of beads is almost pathognomonic for multifocal FMD, the diagnosis of focal FMD requires exclusion of a number of FMD mimics, such as localized atherosclerosis, and inflammatory or inherited arteriopathies (Table 1), which implies a risk of misclassification and subsequent decrease in statistical power in case-control studies; (iv) in the absence of systematic exploration of controls and relatives of FMD patients, study results may be unreliable, as it has been estimated that 3–6% of subjects from the general population may harbour silent FMD lesion [6,7];

and (v) few large, multigenerational families are currently available for analysis. These hurdles may be partly overcome by the development of large national and international registries [3,4,39] including well characterized patients, with image archiving and biobanking, and the use of standardized, state-of-art imaging, as well as identification and validation of specific biomarkers, such as subtle changes in the arterial wall detected by high resolution echography [23,35,40,41].

5. Current Knowledge from Genetic Studies

5.1. Candidate Gene Studies

Literature on the genetics of FMD is scarce and, when available, studies are generally underpowered to assess the role of the tested genes. The main reason is the lack of large pedigrees to perform linkage analyses or large cohorts of well-characterized FMD patients for association studies. Variants in genes belonging to pathways related to the extracellular matrix of the arterial wall failed to show significant or even promising associations with FMD. Single Nucleotide Polymorphisms (SNPs) in the elastin gene (*ELN*), which harbours causative mutations for supravalvular aortic stenosis in Williams–Beuren syndrome (OMIM reference: #194050) (Perdu & Jeunemaitre, Unpublished data), as well as in the alpha1-antitrypsin gene (*AAT*) [42], were distributed evenly in patients with FMD and healthy controls or patients with essential hypertension. The modest sample size (N cases = 161) and lack of comprehensive coverage of the genes of interest, including regulatory regions, may partially explain these negative findings. Screening for causative mutations in the actin α-2 gene (*ACTA2*), involved in rare cases of aortic aneurysms was also negative [43]. Finally, as indicated above, screening for mutations involved in rare vascular syndromes such as Marfan, Loeys-Dietz and vascular Ehlers-Danlos Syndrome, in the TGFβ signalling pathway and/or extracellular matrix components (e.g., fibrillin and collagen genes) was disappointing [26,37]. Fortunately, the development of Next Generation Sequencing (NGS) opened the possibility to conduct hypothesis free experiments, such as exome sequencing, to explore families.

5.2. Exome Studies

5.2.1. Genetic Investigation of FMD Using Exome Sequencing in Families

The first exome sequencing study published on FMD was conducted in 16 related patients from 7 families including at least two first-degree relatives with confirmed FMD recruited in the French database of the Rare Vascular Diseases Referral Centre of the European Hospital Georges Pompidou (HEGP), Paris, France [31]. Under the hypothesis of implication of rare variants, the authors applied a filter on minor allele frequency <0.01, and only analysed coding variants that changed amino acids. Affected siblings share 50% of their genetic variants by chance. However, if several genetic variants located in the same gene are shared by siblings from several families, one can argue for genetic causality for this gene. A threshold of three families sharing variants in the same gene was arbitrary chosen in order to consider a gene as a putative candidate for FMD. Unfortunately, this condition has not be satisfied by any of the 3971 genes analysed [31], suggesting the lack of major common genetic determinant for FMD, at least under the assumption of classical Mendelian inheritance.

In the same study, the authors selected some genes harbouring rare coding variants without intra-familial segregation, and looked for the frequency of variants in these genes in 249 unrelated FMD patients, also ascertained from the Rare Vascular Diseases Referral Centre, HEGP and 689 controls from the SUpplémentation en VItamines et Minéraux AntioXydants (SU.VI.MAX) study, a national sample of healthy volunteers. They applied a gene-based collapsing analysis named Optimal Sequence Kernel Association Test (SKAT-O), which is recommended for measuring association of multiple rare variants in a given gene instead of testing each individual variant separately, thereby providing more power [43]. Despite overall negative findings, this study unravelled nominally significant association between multifocal FMD and myosin light chain kinase (*MYLK*; previously involved in thoracic aortic aneurysms), dynein cytoplasmic heavy chain 1 gene (*DYNC2H1*), obscurin (*OBSCN*; a sarcomeric protein) and *RNF213*, previously associated with Moyamoya disease ($p = 0.01$) [43]. While these findings need replication, they highlight the importance of accurate phenotyping and classification, to distinguish multifocal FMD from focal FMD, which may have different pathophysiologies, and/or to avoid including phenocopies/FMD mimics.

5.2.2. Potential Genetic Link between the Grange Syndrome and FMD

Recently, Guo and co-workers [44] conducted a whole-exome-sequencing analysis on DNA from two affected siblings of the first reported family affected by Grange syndrome. This is a unique autosomal recessive syndrome, characterized by arterial stenosis that is similar to focal FMD in angiographic appearance and distribution, congenital cardiac defects, brachydactyly, syndactyly, bone fragility and learning disabilities [45]. Using very stringent filtering criteria, and by retaining only loss-of-function mutations, whole exome sequencing led to the identification of two compound heterozygous loss-of-function mutations in the YY1 associated protein 1 gene (*YY1AP1*) in the two siblings [44]. Sanger sequencing confirmed the presence of both *YY1AP1* mutations in the third affected sibling. Interestingly, their mother, who harboured one of the variants, had a stenosis of the proximal renal artery without other syndromic manifestations. Finally, three other unrelated patients with the Grange syndrome were found to harbour distinct *YY1AP1* mutations at the homozygous state.

The *YY1AP1* gene encodes a 88 kDa protein that is part of the INO80 chromatin-remodelling complex, and may play a role in transcriptional regulation and cell cycle control. The authors showed that *YY1AP1* expression is induced during smooth muscle cell differentiation in vitro and that its knockdown prevents the expression of smooth muscle markers and the differentiation of smooth muscle cells. *YY1AP1* loss-of-function mutations may thus lead to an increased presence of proliferating smooth muscle cell precursors and reduced amounts of differentiated quiescent, contractile smooth muscle cells. In view of these elements, the authors hypothesized that heterozygous *YY1AP1* variants may be a rare predisposing allele of FMD [44]. While analysis of whole-exome-sequencing data of 282 patients with FMD and 286 matched controls failed to show an increased burden of *YYAP1* variants in FMD cases vs. controls, a single heterozygous *YY1AP1* frameshift mutation was identified in one out of the 282 cases—a female patient with involvement of both renal and carotid arteries—and none in the controls subjects [44]. Further genetic studies are required to determine the possible contribution of *YYAP1* variants in non-syndromic FMD.

5.3. Non-Hypothesis Driven Genetic Association Study

After the limited success of whole-exome-sequencing in related FMD patients and poor yield of screening for rare variants [31,44], alternative hypotheses needed to be explored. Using a genetic association study in 1154 patients with (mostly renal artery) FMD and 3895 controls of European ancestry, Kiando and co-workers reported the first consistent and replicated association. They found that allele A of a genetic variant (rs9349379) of the phosphatase and actin regulator 1 gene (*PHACTR1*), highly prevalent in the general population (~60%), was associated with a 40% increase in relative risk of FMD (OR = 1.39, $p < 7.36 \times 10^{-10}$) [46]. Moreover, in 2458 healthy volunteers, the authors documented an association between the *PHACTR1* at-risk allele and increased carotid intima-media thickness (IMT) ($p < 1.65 \times 10^{-4}$) and wall-to-lumen ratio ($p = 0.002$) [46]. Furthermore, while expression of *PHACTR1* did not differ in cultured human fibroblasts from 51 patients with FMD compared to 39 controls, stratification by rs9349379 genotype indicated increased expression in subjects with the at-risk allele, rs9349379[A] ($p < 0.003$) [46]. Finally, *zPhactr1* knockdown was associated with a mild (~8%) but significant ($p < 0.05$) vessel dilatation in zebrafish [46].

Already in 2015, the Cervical Artery Dissections and Ischemic Stroke Patients (CADISP) study, a consortium dedicated to the genetics of cervical artery dissection (CeAD), had identified *PHACTR1* as a risk locus for stroke in young adults. In particular, the rs9349379[A] allele was associated with a higher CeAD risk (OR = 1.33, $p = 4.46 \times 10^{-10}$) [47]. Furthermore, in a meta-analysis of

29 genome-wide association studies (GWAS), including a total of 23285 individuals with migraine and 95,425 population-matched controls, rs9349379[A] was associated with an increased risk of migraine without aura ($p < 2.81 \times 10^{-10}$) [48], clinical presentation shared by FMD and CeAD. These findings are of interest in view of the overlap between FMD, carotid artery dissection and migraine. Indeed, in a sample of 732 patients with CeAD, the prevalence of documented FMD was 5.6%, with a significant difference between patients with multiple and single CeAD (15% vs. 3.8%, $p < 0.0001$) [49]. Furthermore, headaches are a common symptom in patients with FMD (60% of patients in the US registry, with classical migraine-type headaches reported in 32%) [3]. More intriguingly, in a meta-analysis of four large genome-wide association studies including 15420 individuals with coronary artery disease (CAD) and 15,062 controls, the same allele, rs9349379[A], was associated with a decreased risk of CAD ($p = 5.8 \times 10^{-19}$) [50]. The associations of the rs9349379 locus with cardio- and cerebrovascular diseases are summarized in Figure 2.

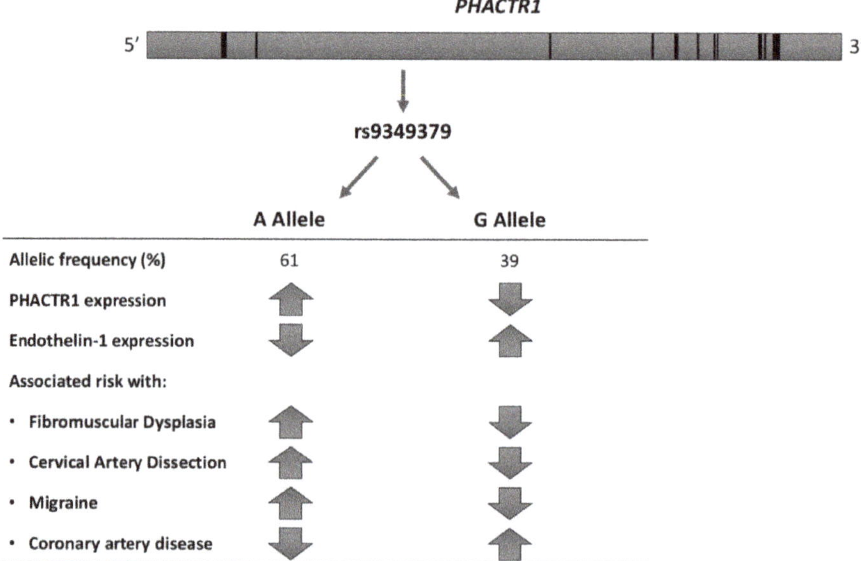

Figure 2. Association of rs9349379 locus with PHACTR1 and Endothelin-1 expression, cardio- and cerebrovascular diseases [46,51].

5.4. Insights and Limitations from Functional Genomics at PHACTR1 Locus

Although rs9349379 is seemingly positioned in a *PHACTR1* intron, it is actually located about 54 kbp upstream of *PHACTR1* transcription start site used in endothelial and smooth muscle cells of the arteries, and about 265 kbp upstream of another *PHACTR1* transcription start site used in macrophages [52]. The homozygous A allele at rs9349379 was associated with higher *PHACTR1* expression in skin fibroblasts and macrophages in healthy donors [46,53] (Figure 2). Publicly available datasets show an enrichment of histone acetylation marks (H3K27ac) close to rs9349379 in arteries, but not in other analysed tissues, suggesting that the region may serve as a tissue-specific enhancer [51]. The rs9349379[G] allele is predicted to disrupt a bona fide binding site for myocyte enhancer factor 2 (*MEF2*) transcription factors, and binding of MEF2s to this site was indeed shown in vitro [54]. However, knockdown of either *MEF2A* or *MEF2C* in human umbilical vein endothelial cells did not result in a decrease of *PHACTR1* expression.

A recent study used induced pluripotent stem cells and CRISPR-Cas9 modification to study the effect of homozygous rs9349379[A] or [G] on gene expression [51]. Modified cells were differentiated into vascular endothelial or vascular smooth muscle cell lineages, and gene expression was analysed using microarrays and qPCR. The results suggested that the rs9349379[G] genotype increases the expression of endothelin-1 (ET-1), whose precursor is encoded by the *EDN1* gene, located about 600 kbp upstream of rs9349379. This observation was corroborated by the finding of higher levels of Big ET-1, a precursor of endothelin-1, in the plasma of healthy subjects harbouring G allele, at least at the homozygous state. ET-1 is a potent vasoconstrictor, which acts on smooth muscle cells to favour their contraction, proliferation and migration. It also plays a role in vascular relaxation through endothelial cells [36]. Accordingly, endothelin-1 imbalance may play a role in the association of rs9349379 to various cardiovascular diseases, although further exploration of the mechanisms specifically involved in FMD is still required.

6. Ongoing Studies

Current efforts to improve diagnostics of FMD aim to identify screening methods that would have a high sensitivity and specificity for FMD, and be easy to use, cost-effective, and available in the clinical practice. Preliminary analysis of the very high-Frequency Ultrasonography for arterial phenotyping in patients with Cervico-cerebral artery dissection, Hypertension, Spontaneous coronary artery dissection and fibromuscular dysplasIA (FUCHSIA) study showed an eutrophic remodelling of the radial artery wall in 11 hypertensive female FMD patients with a peculiar "blurred" pattern [41]. Similar images have also been identified in 5 female SCAD patients [55]. New information is also expected from the Pathophysiological Mechanisms of Fibromuscular Dysplasia (MeDyA) study (ClinicalTrials.gov Identifier: NCT01935752), a pathophysiology study designed to assess endothelial function in patients with FMD compared to hypertensive and healthy controls, and to identify possible plasmatic biomarkers (circulating microparticles, circulating microRNAs, endothelial proteins), and vascular echo-tracking abnormalities related to the FMD.

From a genetic and pathophysiologic point of view, the Defining the Basis of Fibromuscular Dysplasia (DEFINE) study (ClinicalTrials.gov Identifier: NCT01967511) is in the process of recruitment. The researcher's aim is to establish functional, molecular and genetic profiles of fibroblasts from FMD, SCAD and CeAD patients as compared to matched control subjects, in order to dissect the core biologic mechanisms underlying these disorders. Such data should allow to identify similarities and differences in pathophysiology, which could help stratify patients by using biological markers.

7. Future Research Directions and Clinical Perspectives

In view of the relatively modest association of rs9349379[A] with FMD [46], compatible with polygenic inheritance, and the poor yield of screening for mutations in genes involved in CTD [36–38], genetic testing has currently no place in the diagnosis and management of the disease outside a research context [1]. Still, progressive dissection of the genetic basis of FMD paves the way for a better understanding of the disease, which may hopefully lead to identification of accurate prognostic biomarkers and novel preventive and therapeutic options.

The association of FMD with the intronic variant in *PHACTR1* needs to be replicated in large and independent cohorts of patients with more balanced representation of renal and cervico-cephalic FMD, and tested in related conditions, such as SCAD. Whether this association is restricted to multifocal FMD or is also present in patients with focal FMD, and whether it is modulated by gender, ethnic background, tobacco exposure and/or other environmental factors remains to be demonstrated. More studies are also needed to better understand the dichotomic effect of the *PHACTR1* locus on FMD and related vascular diseases on one side, and on atherosclerotic coronary artery disease on the other side. Larger cohorts obtained through international collaborations should also facilitate identification of other susceptibility loci associated with FMD. Unveiling of genetic factors underlying FMD and their interactions with environmental factors may allow (i) to propose a more accurate classification of the disease; (ii) to identify subsets of patients with FMD associated with a higher risk of progression/complications; and (iii) to ascertain or not the dysplastic nature of arterial aneurysms and/or dissections occurring in patients without typical string-of-beads or focal FMD stenosis. Such advances in our understanding of the genetic basis of FMD are critically dependent on ongoing efforts to develop large scale registries including patients with FMD and related diseases, such as the US and European/International registries [3,39].

Conflicts of Interest: The authors declare no conflict of interest.

References

1. Persu, A.; Giavarini, A.; Touzé, E.; Januszewicz, A.; Sapoval, M.; Azizi, M.; Barral, X.; Jeunemaitre, X.; Morganti, A.; Plouin, P.F.; et al. European consensus on the diagnosis and management of fibromuscular dysplasia. *J. Hypertens.* **2014**, *32*, 1367–1378. [CrossRef] [PubMed]
2. Sharma, A.M.; Kline, B. The United States registry for fibromuscular dysplasia: New findings and breaking myths. *Tech. Vasc. Interv. Radiol.* **2014**, *17*, 258–263. [CrossRef] [PubMed]
3. Olin, J.W.; Froehlich, J.; Gu, X.; Bacharach, J.M.; Eagle, K.; Gray, B.H.; Jaff, M.R.; Kim, E.S.; Mace, P.; Matsumoto, A.H.; et al. The United States registry for fibromuscular dysplasia: Results in the first 447 patients. *Circulation* **2012**, *125*, 3182–3190. [CrossRef] [PubMed]
4. Plouin, P.-F.; Baguet, J.-P.; Thony, F.; Ormezzano, O.; Azarine, A.; Silhol, F.; Oppenheim, C.; Bouhanick, B.; Boyer, L.; Persu, A.; et al. High prevalence of multiple arterial bed lesions in patients with fibromuscular dysplasia: The ARCADIA registry (Assessment of Renal and Cervical Artery Dysplasia). *Hypertension* **2017**, *70*, 652–658. [CrossRef] [PubMed]
5. Olin, J.W.; Gornik, H.L.; Bacharach, J.M.; Biller, J.; Fine, L.J.; Gray, B.H.; Gray, W.A.; Gupta, R.; Hamburg, N.M.; Katzen, B.T.; et al. Fibromuscular dysplasia: State of the science and critical unanswered questions: A scientific statement from the American Heart Association. *Circulation* **2014**, *129*, 1048–1078. [CrossRef] [PubMed]
6. Cragg, A.H.; Smith, T.P.; Thompson, B.H.; Maroney, T.P.; Stanson, A.W.; Shaw, G.T.; Hunter, D.W.; Cochran, S.T. Incidental fibromuscular dysplasia in potential renal donors: Long-term clinical follow-up. *Radiology* **1989**, *172*, 145–147. [CrossRef] [PubMed]
7. Blondin, D.; Lanzman, R.; Schellhammer, F.; Oels, M.; Grotemeyer, D.; Baldus, S.E.; Rump, L.C.; Sandmann, W.; Voiculescu, A. Fibromuscular dysplasia in living renal donors: Still a challenge to computed tomographic angiography. *Eur. J. Radiol.* **2010**, *75*, 67–71. [CrossRef] [PubMed]
8. Saw, J.; Mancini, G.B.J.; Humphries, K.H. Contemporary review on spontaneous coronary artery dissection. *J. Am. Coll. Cardiol.* **2016**, *68*, 297–312. [CrossRef] [PubMed]
9. Kirton, A.; Crone, M.; Benseler, S.; Mineyko, A.; Armstrong, D.; Wade, A.; Sebire, G.; Crous-Tsanaclis, A.-M.; deVeber, G. Fibromuscular dysplasia and childhood stroke. *Brain J. Neurol.* **2013**, *136*, 1846–1856. [CrossRef] [PubMed]
10. Savard, S.; Steichen, O.; Azarine, A.; Azizi, M.; Jeunemaitre, X.; Plouin, P.-F. Association between 2 angiographic subtypes of renal artery fibromuscular dysplasia and clinical characteristics. *Circulation* **2012**, *126*, 3062–3069. [CrossRef] [PubMed]
11. Touzé, E.; Oppenheim, C.; Trystram, D.; Nokam, G.; Pasquini, M.; Alamowitch, S.; Hervé, D.; Garnier, P.; Mousseaux, E.; Plouin, P.-F. Fibromuscular dysplasia of cervical and intracranial arteries. *Int. J. Stroke Off. J. Int. Stroke Soc.* **2010**, *5*, 296–305. [CrossRef] [PubMed]
12. Stanley, J.C.; Gewertz, B.L.; Bove, E.L.; Sottiurai, V.; Fry, W.J. Arterial fibrodysplasia. Histopathologic character and current etiologic concepts. *Arch. Surg.* **1975**, *110*, 561–566. [CrossRef] [PubMed]
13. Ross, R.; Klebanoff, S.J. The smooth muscle cell. I. In vivo synthesis of connective tissue proteins. *J. Cell Biol.* **1971**, *50*, 159–171. [CrossRef] [PubMed]
14. Lüscher, T.F.; Lie, J.T.; Stanson, A.W.; Houser, O.W.; Hollier, L.H.; Sheps, S.G. Arterial fibromuscular dysplasia. *Mayo Clin. Proc.* **1987**, *62*, 931–952. [CrossRef]
15. Sang, C.N.; Whelton, P.K.; Hamper, U.M.; Connolly, M.; Kadir, S.; White, R.I.; Sanders, R.; Liang, K.-Y.; Bias, W. Etiologic factors in renovascular fibromuscular dysplasia. A case-control study. *Hypertension* **1989**, *14*, 472–479. [CrossRef] [PubMed]
16. Leung, D.Y.; Glagov, S.; Mathews, M.B. Cyclic stretching stimulates synthesis of matrix components by arterial smooth muscle cells in vitro. *Science* **1976**, *191*, 475–477. [CrossRef] [PubMed]

17. Kaufman, J.J.; Maxwell, M.H. Upright aortography in the study of nephroptosis, stenotic lesions of the renal artery, and hypertension. *Surgery* **1963**, *53*, 736–742. [PubMed]

18. Miller, D.J.; Marin, H.; Aho, T.; Schultz, L.; Katramados, A.; Mitsias, P. Fibromuscular dysplasia unraveled: The pulsation-induced microtrauma and reactive hyperplasia theory. *Med. Hypotheses.* **2014**, *83*, 21–24. [CrossRef] [PubMed]

19. Sottiurai, V.; Fry, W.J.; Stanley, J.C. Ultrastructural characteristics of experimental arterial medial fibroplasia induced by vasa vasorum occlusion. *J. Surg. Res.* **1978**, *24*, 167–177. [CrossRef]

20. Fievez, M.L. Fibromuscular dysplasia of arteries: A spastic phenomenon? *Med. Hypotheses.* **1984**, *13*, 341–349. [CrossRef]

21. Mackay, A.; Brown, J.J.; Cumming, A.M.; Isles, C.; Lever, A.F.; Robertson, J.I. Smoking and renal artery stenosis. *Br. Med. J.* **1979**, *2*, 770. [CrossRef] [PubMed]

22. Nicholson, J.P.; Teichman, S.L.; Alderman, M.H.; Sos, T.A.; Pickering, T.G.; Laragh, J.H. Cigarette smoking and renovascular hypertension. *Lancet* **1983**, *2*, 765–766. [CrossRef]

23. Boutouyrie, P.; Gimenez-Roqueplo, A.-P.; Fine, E.; Laloux, B.; Fiquet-Kempf, B.; Plouin, P.-F.; Jeunemaitre, X.; Laurent, S. Evidence for carotid and radial artery wall subclinical lesions in renal fibromuscular dysplasia. *J. Hypertens.* **2003**, *21*, 2287–2295. [CrossRef] [PubMed]

24. Savard, S.; Azarine, A.; Jeunemaitre, X.; Azizi, M.; Plouin, P.-F.; Steichen, O. Association of smoking with phenotype at diagnosis and vascular interventions in patients with renal artery fibromuscular dysplasia. *Hypertension* **2013**, *61*, 1227–1232. [CrossRef] [PubMed]

25. O'Connor, S.; Gornik, H.L.; Froehlich, J.B.; Gu, X.; Gray, B.H.; Mace, P.D.; Sharma, A.; Olin, J.W.; Kim, E.S.H. Smoking and Adverse Outcomes in Fibromuscular Dysplasia: U.S. Registry Report. *J. Am. Coll. Cardiol.* **2016**, *67*, 1750–1751. [CrossRef] [PubMed]

26. Halpern, M.M.; Sanford, H.S.; Viamonte, M. Renal-artery abnormalities in three hypertensive sisters. Probable familial fibromuscular hyperplasia. *JAMA* **1965**, *194*, 512–513. [CrossRef] [PubMed]

27. Hansen, J.; Holten, C.; Thorborg, J.V. Hypertension in two sisters caused by so-called fibromuscular hyperplasia of the renal arteries. *Acta Med. Scand.* **1965**, *178*, 461–474. [CrossRef] [PubMed]

28. Foster, J.H.; Oates, J.A.; Rhamy, R.K.; Klatte, E.C.; Burko, H.C.; Michelakis, A.M. Hypertension and fibromuscular dysplasia of the renal arteries. *Surgery* **1969**, *65*, 157–181. [PubMed]

29. Plagnol, P.; Gillet, J.M.; Cambuzat, J.M.; Broussin, J. Familial reno-vascular hypertension. *J. Radiol. Electrol. Med. Nucl.* **1975**, *56*, 173–174. [PubMed]

30. Morimoto, S.; Kuroda, M.; Uchida, K.; Funatsu, T.; Yamamoto, I.; Hashiba, T.; Kametani, T.; Takeda, R.; Matsubara, F. Occurrence of renovascular hypertension in two sisters. *Nephron* **1976**, *17*, 314–320. [CrossRef] [PubMed]

31. Kiando, S.R.; Barlassina, C.; Cusi, D.; Galan, P.; Lathrop, M.; Plouin, P.-F.; Jeunemaitre, X.; Bouatia-Naji, N. Exome sequencing in seven families and gene-based association studies indicate genetic heterogeneity and suggest possible candidates for fibromuscular dysplasia. *J. Hypertens.* **2015**, *33*, 1802–1810; discussion 1810. [CrossRef] [PubMed]

32. Pannier-Moreau, I.; Grimbert, P.; Fiquet-Kempf, B.; Vuagnat, A.; Jeunemaitre, X.; Corvol, P.; Plouin, P.-F. Possible familial origin of multifocal renal artery fibromuscular dysplasia. *J. Hypertens.* **1997**, *15*, 1797–1801. [CrossRef] [PubMed]

33. Green, R.; Gu, X.; Kline-Rogers, E.; Froehlich, J.; Mace, P.; Gray, B.; Katzen, B.; Olin, J.; Gornik, H.L.; Cahill, A.M.; et al. Differences between the pediatric and adult presentation of fibromuscular dysplasia: Results from the US Registry. *Pediatr. Nephrol.* **2016**, *31*, 641–650. [CrossRef] [PubMed]

34. Rushton, A.R. The genetics of fibromuscular dysplasia. *Arch. Intern. Med.* **1980**, *140*, 233–236. [CrossRef] [PubMed]

35. Perdu, J.; Boutouyrie, P.; Bourgain, C.; Stern, N.; Laloux, B.; Bozec, E.; Azizi, M.; Bonaiti-Pellié, C.; Plouin, P.-F.; Laurent, S.; et al. Inheritance of arterial lesions in renal fibromuscular dysplasia. *J. Hum. Hypertens.* **2007**, *21*, 393–400. [CrossRef] [PubMed]

36. Ganesh, S.K.; Morissette, R.; Xu, Z.; Schoenhoff, F.; Griswold, B.F.; Yang, J.; Tong, L.; Yang, M.-L.; Hunker, K.; Sloper, L.; et al. Clinical and biochemical profiles suggest fibromuscular dysplasia is a systemic disease with altered TGF-β expression and connective tissue features. *FASEB J.* **2014**, *28*, 3313–3324. [CrossRef] [PubMed]

37. O'Connor, S.; Kim, E.S.; Brinza, E.; Moran, R.; Fendrikova-Mahlay, N.; Wolski, K.; Gornik, H.L. Systemic connective tissue features in women with fibromuscular dysplasia. *Vasc. Med. Lond. Engl.* **2015**, *20*, 454–462. [CrossRef] [PubMed]

38. Poloskey, S.L.; Kim, E.S.; Sanghani, R.; Al-Quthami, A.H.; Arscott, P.; Moran, R.; Rigelsky, C.M.; Gornik, H.L. Low yield of genetic testing for known vascular connective tissue disorders in patients with fibromuscular dysplasia. *Vasc. Med. Lond Engl.* **2012**, *17*, 371–378. [CrossRef] [PubMed]

39. Persu, A.; van der Niepen, P.; Touzé, E.; Gevaert, S.; Berra, E.; Mace, P.; Plouin, P.-F.; Jeunemaitre, X. Revisiting Fibromuscular Dysplasia: Rationale of the European Fibromuscular Dysplasia Initiative. *Hypertension* **2016**, *68*, 832–839. [CrossRef] [PubMed]

40. Marais, L.; Boutouyrie, P.; Khettab, H.; Boulanger, C.; Lorthioir, A.; Franck, M.; Niarra, R.; Renard, J.; Chambon, Y.; Jeunemaitre, X.; et al. Structural and functional arterial abnormalities in fibromuscular dysplasia are in the continuum of hypertension: An imaging and biomechanical study. *Artery Res.* **2016**, *16*, 70–71. [CrossRef]

41. Bruno, R.M.; Lascio, N.D.; Guarino, D.; Vitali, S.; Rossi, P.; Caramella, D.; Taddei, S.; Ghiadoni, L.; Faita, F. Abstract P509: Identification of radial vascular wall abnormalities by very-high frequency ultrasound in patients with fibromuscular dysplasia: The fuchsia study. *Hypertension* **2017**, *70*, AP509.

42. Marks, S.D.; Gullett, A.M.; Brennan, E.; Tullus, K.; Jaureguiberry, G.; Klootwijk, E.; Stanescu, H.C.; Kleta, R.; Woolf, A.S. Renal FMD may not confer a familial hypertensive risk nor is it caused by *ACTA2* mutations. *Pediatr. Nephrol.* **2011**, *26*, 1857–1861. [CrossRef] [PubMed]

43. Lee, S.; Wu, M.C.; Lin, X. Optimal tests for rare variant effects in sequencing association studies. *Biostatistics* **2012**, *13*, 762–775. [CrossRef] [PubMed]

44. Guo, D.-C.; Duan, X.-Y.; Regalado, E.S.; Mellor-Crummey, L.; Kwartler, C.S.; Kim, D.; Lieberman, K.; de Vries, B.B.A.; Pfundt, R.; Schinzel, A.; et al. Loss-of-function mutations in *YY1AP1* lead to grange syndrome and a fibromuscular dysplasia-like vascular disease. *Am. J. Hum. Genet.* **2017**, *100*, 21–30. [CrossRef] [PubMed]

45. Grange, D.K.; Balfour, I.C.; Chen, S.C.; Wood, E.G. Familial syndrome of progressive arterial occlusive disease consistent with fibromuscular dysplasia, hypertension, congenital cardiac defects, bone fragility, brachysyndactyly, and learning disabilities. *Am. J. Med. Genet.* **1998**, *75*, 469–480. [CrossRef]

46. Kiando, S.R.; Tucker, N.R.; Castro-Vega, L.-J.; Katz, A.; D'Escamard, V.; Tréard, C.; Fraher, D.; Albuisson, J.; Kadian-Dodov, D.; Ye, Z.; et al. *PHACTR1* is a genetic susceptibility locus for fibromuscular dysplasia supporting its complex genetic pattern of inheritance. *PLoS Genet.* **2016**, *12*, e1006367. [CrossRef] [PubMed]

47. Debette, S.; Kamatani, Y.; Metso, T.M.; Kloss, M.; Chauhan, G.; Engelter, S.T.; Pezzini, A.; Thijs, V.; Markus, H.S.; Dichgans, M.; et al. Common variation in *PHACTR1* is associated with susceptibility to cervical artery dissection. *Nat. Genet.* **2015**, *47*, 78–83. [CrossRef] [PubMed]

48. Anttila, V.; Winsvold, B.S.; Gormley, P.; Kurth, T.; Bettella, F.; McMahon, G.; Kallela, M.; Malik, R.; de Vries, B.; Terwindt, G.; et al. Genome-wide meta-analysis identifies new susceptibility loci for migraine. *Nat. Genet.* **2013**, *45*, 912–917. [CrossRef] [PubMed]

49. Béjot, Y.; Aboa-Eboulé, C.; Debette, S.; Pezzini, A.; Tatlisumak, T.; Engelter, S.; Grond-Ginsbach, G.; Touzé, E.; Sessa, M.; Metso, T.; et al. Characteristics and outcomes of patients with multiple cervical artery dissection. *Stroke* **2014**, *45*, 37–41. [CrossRef] [PubMed]

50. Coronary Artery Disease (C4D) Genetics Consortium. A genome-wide association study in Europeans and South Asians identifies five new loci for coronary artery disease. *Nat. Genet.* **2011**, *43*, 339–344.

51. Gupta, R.M.; Hadaya, J.; Trehan, A.; Zekavat, S.M.; Roselli, C.; Klarin, D.; Emdin, C.A.; Hilvering, C.R.E.; Bianchi, V.; Mueller, C.; et al. A Genetic variant associated with five vascular diseases is a distal regulator of endothelin-1 gene expression. *Cell* **2017**, *170*, 522–533.e15. [CrossRef] [PubMed]

52. Perdu, J.; Gimenez-Roqueplo, A.-P.; Boutouyrie, P.; Beaujour, S.; Laloux, B.; Nau, V.; Fiquet-Kempf, B.; Emmerich, J.; Tichet, J.; Plouin, P.-F.; et al. Alpha1-antitrypsin gene polymorphisms are not associated with renal arterial fibromuscular dysplasia. *J. Hypertens.* **2006**, *24*, 705–710. [CrossRef] [PubMed]

53. Reschen, M.E.; Lin, D.; Chalisey, A.; Soilleux, E.J.; O'Callaghan, C.A. Genetic and environmental risk factors for atherosclerosis regulate transcription of phosphatase and actin regulating gene *PHACTR1*. *Atherosclerosis* **2016**, *250*, 95–105. [CrossRef] [PubMed]

54. Beaudoin, M.; Gupta, R.M.; Won, H.-H.; Lo, K.S.; Do, R.; Henderson, C.A.; Lavoie-St-Amour, C.; Langlois, S.; Rivas, D.; Lehoux, S.; et al. Myocardial infarction-associated SNP at 6p24 interferes with *MEF2* binding and associates with *PHACTR1* expression levels in human coronary arteries. *Arterioscler. Thromb. Vasc. Biol.* **2015**, *35*, 1472–1479. [CrossRef] [PubMed]
55. Bruno, R.M.; di Lascio, N.; Al Hussaini, A.; Guarino, D.; Vitali, S.; Rossi, P.; Caramella, D.; Cortese, B.; Faita, F.; Taddei, S.; et al. Disarray and remodeling of the radial artery in women with spontaneous coronary artery dissection: The FUCHSIA study. *Artery Res.* **2017**, *20*, 75–76. [CrossRef]

International Journal of
Molecular Sciences

Review

Kruppel-Like Factor 15 Is Critical for the Development of Left Ventricular Hypertrophy

Sheila K. Patel [1,*], **Jay Ramchand** [1,2], **Vincenzo Crocitti** [1] and **Louise M. Burrell** [1,2]

1   Department of Medicine, Austin Health, University of Melbourne, Melbourne, VIC 3084, Australia;
     jay.ramchand@unimelb.edu.au (J.R.); v.crocitti@student.unimelb.edu.au (V.C.);
     l.burrell@unimelb.edu.au (L.M.B.)
2   Department of Cardiology, Austin Health, Melbourne, VIC 3084, Australia
*   Correspondence: skpatel@unimelb.edu.au; Tel.: +61-3-9496-2362; Fax: +61-3-9457-5485

Received: 9 April 2018; Accepted: 24 April 2018; Published: 27 April 2018

Abstract: Left ventricular hypertrophy (LVH) is an independent risk factor for adverse cardiovascular events and is often present in patients with hypertension. Treatment to reduce blood pressure and regress LVH is key to improving health outcomes, but currently available drugs have only modest cardioprotective effects. Improved understanding of the molecular mechanisms involved in the development of LVH may lead to new therapeutic targets in the future. There is now compelling evidence that the transcription factor Kruppel-like factor 15 (KLF15) is an important negative regulator of cardiac hypertrophy in both experimental models and in man. Studies have reported that loss or suppression of KLF15 contributes to LVH, through lack of inhibition of pro-hypertrophic transcription factors and stimulation of trophic and fibrotic signaling pathways. This review provides a summary of the experimental and human studies that have investigated the role of KLF15 in the development of cardiac hypertrophy. It also discusses our recent paper that described the contribution of genetic variants in *KLF15* to the development of LVH and heart failure in high-risk patients.

Keywords: Kruppel-like factor 15; left ventricular hypertrophy; cardiac hypertrophy; heart failure; genetics of left ventricular hypertrophy

---

1. Introduction

Left ventricular hypertrophy (LVH) is an independent and potent risk factor for cardiovascular events [1] and is often present in patients with hypertension. LVH is diagnosed by electrocardiography, echocardiography or cardiac magnetic resonance imaging. The prevalence of LVH in the general population is 10% [1] but rises to 40–70% in those with risk factors such as hypertension, aortic stenosis, diabetes and chronic kidney disease [2–5]. As LVH per se causes no symptoms (preclinical), it is often only diagnosed after patients present with symptoms such as heart failure [1]. Treatment to regress LVH is key to improving health outcomes [6,7], but currently available drugs have only modest cardioprotective effects [8]. Improved understanding of the molecular mechanisms that contribute to the development of LVH provides an opportunity to identify pathways that may be new therapeutic targets.

Our group is interested in the function and regulation of transcription factors in the development of LVH. To date, studies have mainly focused on factors that induce hypertrophy such as GATA binding protein 4 (GATA4), myocyte enhancer factor 2 (MEF2) and serum response factor (SRF) [9] (Figure 1). However, there are also transcriptional repressors of hypertrophy such as Kruppel-like factor 15 (KLF15) [10], and compelling experimental evidence that suppression of KLF15 contributes to LVH through lack of inhibition of pro-hypertrophic transcription factors [9]. This review provides a summary of experimental and human studies that have investigated the role of KLF15 in the development of cardiac hypertrophy. It also discusses our recent paper that described the contribution of genetic variants in *KLF15* to the development of LVH and heart failure in high-risk patients.

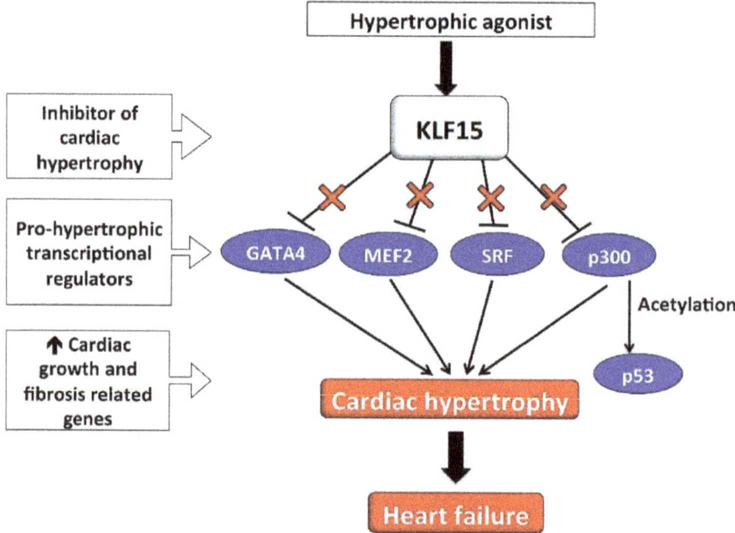

Figure 1. Schema illustrating how loss of the inhibitory effect of KLF15 on pro-hypertrophic cardiac transcriptional regulators may contribute to the development of LVH and heart failure. ↑ Increased; X in red font represents the inhibitory effect; GATA4, GATA binding protein 4; KLF15, Kruppel-like factor 15; MEF2, myocyte enhancer factor 2; SRF, serum response factor.

2. Kruppel-Like Factor 15 and Cardiac Hypertrophy

KLF15 is one member of a family of 18 Kruppel-like factors (KLF). The KLFs are a subclass of the zinc-finger family of DNA-binding transcriptional regulators, homologous to the *Drosophila* gap gene Kruppel [11]. KLF15 is expressed in myocytes and fibroblasts [12]. Both in vitro and in vivo studies support a role for KLF15 as a repressor of pathological cardiac hypertrophy and fibrosis [10,12,13]. Cardiac KLF15 expression is low during development and increases after birth to reach high levels in the adult rat and mouse heart [13,14]. The action of KLF15 to repress hypertrophy occurs through the inhibition of the activity of pro-hypertrophic transcriptional regulators, which affect atrial natriuretic peptide (ANP) and brain natriuretic peptide (BNP) promoter activity [13]. Expression of KLF15 is unchanged during physiologically induced LVH (i.e., exercise) but is decreased in pathological LVH [15].

Table 1 summarizes the results from in vivo studies that have explored KLF15 in experimental models of cardiac hypertrophy. Overall, regardless of the model used, the results have consistently shown that loss of KLF15 contributes to the development of cardiac hypertrophy. The approaches used to induce LVH include cardiac KLF15 null ($-/-$) mice, high blood pressure (angiotensin (Ang) II infusion [10], transgenic TGR(mRen2)27 rat (Ren-2) [15], high salt diet in Dahl salt-sensitive rats, isoproterenol infusion [16]) and surgical methods such as transaortic constriction [13], ascending aortic constriction [12,13], and aortic banding [17].

2.1. Mouse Models of KLF15 Gene Deletion, Overexpression and Cardiac Hypertrophy

KLF15 null mice are viable with increased heart cavity size and reduced left ventricular (LV) fractional shortening, but no increase in wall thickness [13]. Under conditions of pressure–overload due to ascending aortic constriction, KLF15 ($-/-$) null mice had marked LV cavity dilation, reduced systolic function, and increased cardiac mass compared to KLF15 wildtype (+/+) mice. Morphometric studies indicated that cardiomyocytes from KLF15 null mice were larger and longer compared to cardiomyocytes from wildtype mice with ascending aortic constriction. Others have reported that ascending aortic constriction in KLF15 null mice led to increased profibrotic factor connective tissue growth factor (CTGF) expression compared to sham operated KLF15 null mice [12]. Similar effects were seen in the study by Halder et al. [10] in which Ang II infusion led to impaired LV dysfunction and cavity dilation with increased cardiac mass in KLF15 null mice; the observed effects were not related to differences in blood pressure between wild type and null mice. Isoproterenol infusion also increased cardiac mass, cardiomyocyte cross-sectional area and led to significant fibrosis in KLF15 null mice compared to vehicle treated KLF15 null mice [16], with no effect in wildtype KLF15 mice [16]. Pressure-overload in mice with genetic deletion of KLF15 also led to a parallel increase in the cardiac expression of hypertrophic markers ANP, BNP and CTGF [10,12,13,16].

Halder et al. [10] examined the effect of restoring KLF15 levels on cardiac hypertrophy. The authors hypothesized that p53, a protein known to regulate the expression of genes involved in growth and apoptosis and activated by Ang II, played a role in cardiac decompensation. They reported that KLF15 null mice had a significant increase in p53 expression after Ang II infusion, and that KLF15 deficient hearts were rescued by p53 deletion or with curcumin, a potent p300 acetyltransferase inhibitor [10]. The p300 acetyltransferase inhibitor is an important regulator of p53 function and involved in acetylating GATA4, MEF2 and the Smads. Curcumin also ameliorated heart failure in KLF15 null mice, reduced cardiac mass, improved LV systolic function and decreased p53 abundance in heart tissue. The authors also investigated the effects of adenoviral KLF15 overexpression in neonatal rat ventricular cardiomyocytes (NRVM), which did not reduce p300 abundance or acetyltransferase activity [10].

The therapeutic potential of KLF15 overexpression on LVH in mice was examined using recombinant adenovirus (AAV9) to overexpress KLF15 and Ang II infusion to stimulate cardiac hypertrophy [18]. As expected Ang II infusion increased LV mass and cardiomyocyte size in AAV-green fluorescent protein (GFP, control vector) mice, with a blunted increase in LV mass and no change in cardiomyocyte size in Ang II infused mice infected with AAV9-KLF15. Ang II infusion increased interstitial fibrosis and LV mRNA expression of hypertrophic marker genes ANP and alpha skeletal actin (αSKA) in AAV9-GFP mice, but this effect was not modulated in the AAV9-KLF15 mice despite the reduction in LV mass. This result suggests that collagen deposition may occur through pathways independent of KLF15 and its effect to inhibit cardiac hypertrophy.

2.2. Rat Models of Cardiac Hypertrophy and KLF15

KLF15 expression was examined in 2 models of hypertension—the transgenic TGR(mRen2)27 rat (Ren-2) [15] and the high salt fed Dahl salt-sensitive rat [19]. The Ren-2 rat is characterized by severe hypertension, LVH, suppression of the kidney renin-angiotensin system and unchanged or suppressed levels of plasma renin, Ang I and II and angiotensinogen compared to wildtype rats [20]. Cardiac KLF15 expression was reduced at an early stage of LVH. In an elegant series of experiments, cardiac biopsies were taken from the rats with a similar degree of cardiac hypertrophy and normal systolic function, and followed over time. KLF15 mRNA expression decreased more in the hypertrophied hearts that progressed to heart failure compared to control and compensated rats, suggesting that expression of KLF15 may play a role in the prevention of heart failure.

Dahl salt-sensitive rats fed a high salt diet developed hypertension-induced LVH at 11 weeks, with LV dilatation and heart failure at 17 weeks [19]. Cardiac KLF15 mRNA was significantly reduced with LVH and decreased further with the development of heart failure. This change was associated with increased ANP gene expression with LVH and a further increase in levels with the development of heart failure [19].

Aortic stenosis is a common cause of LVH and KLF15 expression has been examined in two experimental models of aortic constriction—transaortic constriction and aortic banding. In adult

Sprague-Dawley rats, KLF15 mRNA expression was non-significantly reduced at 2 days post transaortic constriction with a significant reduction by 7 days post transaortic constriction [13].

Yu et al. [17] conducted studies in rats with pressure overload secondary to aortic banding and examined the effect of debanding and unloading pressure at 3 and 6-weeks. Aortic banding increased ejection fraction and LV pressure compared to the sham group, and both improved after debanding. At both 3 and 6-weeks after aortic banding, rats had decreased KLF15 mRNA, and increased CTGF, transforming growth factor-β (TGFβ) and myocardin-related transcription factor A (MRTF-A) with fibrosis and increased collagen I and III mRNA compared to sham-operated rats. KLF15 levels dropped further in the 6-week banded animals. Aortic debanding resulted in a decrease in collagen mRNA, increased KLF15 and reduced TGFβ mRNA.

2.3. In Vitro KLF15 and LVH Studies

Table 2 summarizes the results of *in vitro* studies of KLF15 which have been mainly conducted in NRVM, and include *KLF15* gene silencing [15], KLF15 overexpression [13,15], and cardiomyocyte hypertrophy induced with pro-hypertrophic factors [13,15]. Two studies were conducted in cardiac fibroblasts [12,17].

2.4. KLF15 Knock Down or Overexpression

In rat cardiomyocytes treated with KLF15 siRNAs, there were significant increases in cardiomyocyte size and ANP mRNA expression [15]. The loss of KLF15 alone was sufficient to induce cell hypertrophy [13]. Two studies explored overexpression of KLF15 in cardiomyocytes [13,15]. In contrast to the gene knock down study, overexpression of KLF15 reduced cardiomyocyte ANP and BNP mRNA [13,15] and led to reduced cardiomyocyte size in one study [13] but not the other [15].

Two studies examined KLF15 levels in cardiac fibroblasts. KLF15 inhibited the expression of CTGF in cardiac fibroblasts [12,17]. Adenoviral overexpression of KLF15 in neonatal rat ventricular fibroblasts (NRVF) inhibited both basal and TGFβ induced CTGF expression. Similar to TGFβ treatment in cardiomyocytes [15], TGFβ reduced KLF15 gene expression and induced CTGF in NRVF [12]. Co-immunoprecipitation and mobility shift experiments showed KLF15 inhibited recruitment of a co-activator P/CAF (a potent transcriptional co-activator of Smad3 target genes) to the CTGF promoter with no effect on Smad3 binding. KLF15 mediated repression of CTGF promoter was rescued by overexpressing P/CAF [12].

In primary neonate rat cardiac fibroblasts in which hypertrophy was also induced with TGFβ, silencing of the KLF15 gene with KLF15-shRNA significantly decreased KLF15 protein. In contrast, KLF15 overexpression with a recombinant adenovirus increased KLF15 protein expression. Fibroblasts stimulated with TGFβ alone showed increased collagen mRNA. TGFβ stimulated fibroblasts with KLF15 overexpression had less fibrosis and hypertrophy, reduced CTGF and myocardin related transcription factor A (MRFT-A) mRNA, all of which were increased in KLF15-shRNA infected fibroblasts [17].

Table 1. In vivo KLF15 experimental studies.

Experiment	Model/Animal Strain	Control Group	Summary of Main Findings	References
Genetic models	Ascending aortic constriction in KLF15 (−/−) and KLF15 (+/+) C57BL/6 mice	Sham-operation	KLF15 (−/−) vs. (+/+) - ↓ LV FS, ↑ cavity size; AAC in KLF15 (−/−) led to ↑ LV cavity dilation, ↓ systolic function,↑ cardiac mass, ↑ cardiomyocyte cell size, ↑ ANF and BNP mRNA	Fisch S et al., 2007 [13]
	Ascending aortic constriction in KLF15 (−/−) and KLF15 (+/+) C57BL/6 mice	Sham-operation	AAC led to ↑ cardiac CTGF mRNA in KLF15 (−/−); no change in controls	Wang B et al., 2008 [12]
	Ang II infusion in KLF15 (−/−) C57BL/6 mice	Saline infusion in KLF15 (+/+) mice	Ang II led to ↑ cardiac mass and cavity dilation, ↓ systolic function, ↓ LV KLF15 mRNA, ↑ ANF expression vs. control	Halder et al., 2010 [10]
Hypertension induced LVH	High salt diet in Dahl salt-sensitive rats; LVH at 11 weeks and heart failure at 17 weeks	Age-matched low salt diet	Further ↓ cardiac KLF15 mRNA as LVH progressed to heart failure; ↑ ANP with LVH and further ↓ with heart failure	Horie T et al., 2009 [19]
	Time course of cardiac biopsies in hypertensive transgenic TGR(mRen2)27 rats (Ren-2)	Wildtype littermates	Cardiac KLF15 mRNA ↓ with progression from LVH to heart failure	Leenders et al., 2010 [15]
	14-day Ang II infusion in C57BL/6 male mice.	Saline infusion	↓ KLF15 mRNA expression in ventricle	Halder et al., 2010 [10]
	Adenoviral (AAV9) KLF15 or GFP (control vector) over-expression in 8-week-old C57BL/6 mice + 28-day Ang II infusion	Saline infusion	↑ Interstitial fibrosis in both groups compared to controls; ↓ cardiac hypertrophy and cardiomyocyte area in AAV9-KLF15/Ang II vs. AAV9-GFP/Ang II	Leenders et al., 2012 [18]

Table 1. *Cont.*

Experiment	Model/Animal Strain	Control Group	Summary of Main Findings	References
	5-week isoproterenol or vehicle infusion in KLF15 (−/−) C57BL/6 mice	KLF15 (+/+) mice	↑ Cardiac mass, ↑ cardiomyocyte cross-sectional area, ↑ fibrosis in KLF15 (−/−) isoproterenol vs. vehicle; ↔ cardiac mass, cell size and fibrosis in KLF15 (+/+) isoproterenol vs. vehicle	Gao et al., 2017 [16]
Surgical induction of LVH	Pressure-overload hypertrophy induced by TAC in adult male Sprague-Dawley rats	No control group	↓ KLF15 LV mRNA expression at 2-days post-TAC and further ↓ at 7-days post-TAC	Fisch S et al., 2007 [13]
	Aortic banding in Sprague-Dawley rats with debanding at 3 and 6-weeks post-surgery	Time-matched sham-operated rats	↓ Cardiac KLF15 mRNA, ↑ interstitial fibrosis, ↑ CTGF and ↑ TGFβ mRNA at 3- and 6-week post-banding; debanding led to ↑ cardiac KLF15 and ↓ TGFβ mRNA	Yu et al., 2014 [17]

↑ Increased; ↓ decreased; ↔ no change. AAC, ascending aortic constriction; AAV9, adeno-associated virus 9; Ang II, angiotensin II; ANP, atrial natriuretic peptide; BNP, brain natriuretic peptide; CTGF connective tissue growth factor; FS, fractional shortening; GFP, green fluorescent protein; KLF15, Kruppel-like factor 15; LV, left ventricle; LVH, left ventricular hypertrophy; mRNA, messenger ribonucleic acid; TAC, transaortic constriction; TGFβ, transforming growth factor beta; TGR(mREN2)27, transgenic renin hypertensive model.

Table 2. In vitro KLF15 studies.

Experiment Type	Experimental Model/Cell	Control Group	Summary of Main Findings	References
KLF15 gene silencing	Neonatal rat (Lewis) and mouse (FBV mice) cardiomyocytes treated with 2 siRNAs against KLF15	Cells with non-targeted control siRNA	↑ Cardiomyocyte size, ↑ANP mRNA	Leender et al., 2010 [15]
	Neonatal Sprague-Dawley rat primary cardiac fibroblasts with viral KLF15 gene silencing. Hypertrophy induced with TGFβ	Fibroblasts either without TGFβ stimulation or control virus	↓ KLF15 mRNA	Yu et al., 2014 [17]

Table 2. *Cont.*

Experiment Type	Experimental Model/Cell	Control Group	Summary of Main Findings	References
KLF15 overexpression	Adenoviral overexpression of KLF15 in NRVM	GFP control vector	↓ ANP and BNP mRNA; ↓ cardiomyocyte size under basal and phenylephrine stimulated hypertrophy	Fisch S et al., 2007 [13]
	Lentiviral overexpression of KLF15 in neonatal rat (Lewis) cardiomyocytes	Cells with control vector	↓ ANP mRNA, ↔ cardiomyocyte size	Leender et al., 2010 [15]
	Neonatal Sprague-Dawley rat primary cardiac fibroblasts with adenoviral KLF15 overexpression. Hypertrophy induced with TGFβ.	Fibroblasts either without TGFβ stimulation or control virus	↑ KLF15 protein, ↓ fibrosis and hypertrophy, ↓ CTGF mRNA with TGFβ stimulation and KLF15 overexpression	Yu et al., 2014 [17]
	Adenoviral overexpression of KLF15 in NRVF	GFP control vector	Inhibits basal and TGFβ induced CTGF expression	Wang et al., 2008 [12]
Cardiac hypertrophy induced by pro-hypertrophic stimuli	Isolated NRVM. Stimulation of hypertrophy with phenylephrine, endothelin-1.	No controls	↓ KLF15 and ↑ANP and BNP mRNA expression with pro-hypertrophic stimuli	Fisch S et al., 2007 [13]
	Isolated NRVF 2 day old Sprague-Dawley rats stimulated with TGFβ	NRVF under basal conditions	↓ KLF15 and ↓ CTGF mRNA post TGFβ1 stimulation	Wang et al., 2008 [12]
	Neonatal rat (Lewis) LV cardiomyocytes. Stimulation of hypertrophy (phenylephrine, endothelin-1, TGFβ) or by stimuli known to stimulate physiological hypertrophy (insulin, IGF-1, IGF-2)	Control cells	↓ KLF15 expression with all hypertrophic stimuli. ↔ KLF15 expression with physiological growth. TGFβ knockdown abolished ↓ KLF15	Leender et al., 2010 [15]
	Neonatal rat (Lewis) and mouse (FBV mice) cardiomyocytes, treated with two p38 MAPK inhibitors and TGFβ stimulation	Cells without p38 MAPK inhibitor or TGFβ stimulation.	TGFβ – induced ↓ KLF15 expression abolished by p38 MAPK inhibitors	Leender et al. 2010 [15]

↑ Increased; ↓ decreased; ↔ no change. Ang II, angiotensin II; ANP, atrial natriuretic peptide; BNP, brain natriuretic peptide; CTGF, connective tissue growth factor; GFP, green fluorescent protein; IGF-1, insulin-like growth factor 1; IGF-2, insulin-like growth factor 2; KLF15, Kruppel-like factor 15; LV, left ventricle; MAPK, mitogen-activated protein kinase; mRNA, messenger ribonucleic acid; NRVF, neonatal rat ventricular fibroblasts; NRVM, neonatal rat ventricular myocytes; TGFβ, transforming growth factor beta.

2.5. Hypertrophy Induced by Stimulation with Pro-Hypertrophic Factors

KLF15 mRNA expression is low in NRVM but induced with serum starvation with an expression pattern that is antiparallel to that of ANP and BNP gene expression [13]. Treatment of NRVM with the pro-hypertrophic stimuli, endothelin-1 and phenylephrine, decreased KLF15 mRNA expression and induced ANP and BNP expression. Similarly, in cultured neonatal rat cardiomyocytes, stimulation with phenylephrine, endothelin-1 and TGFβ significantly reduced KLF15 expression levels [15]. However KLF15 expression was unchanged when cardiomyocyte growth was induced with physiological hypertrophy stimuli (i.e., insulin, insulin-like growth factor 1 or 2). These results support a role for the involvement of KLF15 in pathological hypertrophy but not in physiological hypertrophy. TGFβ-mediated down regulation of KLF15 was abolished with knockdown of the TGFβ receptor 1 [15]. TGFβ-mediated activation of p38 mitogen-activated protein kinase was also necessary and sufficient to decrease KLF15 expression. Adenoviral overexpression of an upstream p38 kinase (MKK6) that induced increased p38 phosphorylation led to an 80% decrease in KLF15 mRNA and the induction of BNP. No other studies have explored TGFβ stimulated hypertrophy and KLF15 levels in cardiomyocytes.

3. Human Studies of KLF15

3.1. KLF15 Expression in Human Cardiac Tissue

Consistent with the observation in experimental studies, there is evidence that loss of cardiac KLF15 expression may contribute to pathological LVH in humans (Table 3). Three separate studies investigated KLF15 expression in LV tissue obtained from patients undergoing cardiac surgery. In patients with LVH secondary to aortic stenosis, KLF15 protein was reduced in myocardial needle biopsy samples taken from the anterior LV of patients undergoing open heart surgery ($n = 8$) compared to patients undergoing coronary bypass grafting ($n = 6$) [13]. These patients were selected from a larger group of patients from an earlier study [21]. The characteristics of the subgroup of patients used in the paper by Fisch et al. [13] were not provided. KLF15 mRNA was also significantly reduced in LV samples of patients with non-ischaemic cardiomyopathy ($n = 36$) compared to control tissue from non-failing hearts deemed unsuitable for transplantation ($n = 30$) [10].

In patients with end-stage heart failure undergoing implantation and explantation of a left ventricular assist device (LVAD) as a bridge to transplantation, pre-device implantation LV KLF15 mRNA expression was reduced in the failing heart compared to control non-failing hearts, with significant recovery of KLF15 expression after mechanical unloading [14]. The latter study used 3–4 samples per group, collected as part of a larger study (36 pre-LVAD, 30 post-LVAD) [22].

3.2. Genetic Studies of KLF15 in Patients with LVH

Conventional risk factors do not explain all the variability in LVH [23], and a heritable component underlying LVH is recognized. We recently explored the genetic association of *KLF15* single nucleotide polymorphisms (SNP) in patients with type 2 diabetes and an echocardiographic assessment of LV mass [2]. The intronic *KLF15* SNP rs9838915 A allele was significantly associated with LV mass and the association was replicated in an independent cohort of >5000 type 2 diabetes patients (Table 3). The *KLF15* rs9838915 A allele predicted the first hospitalization with heart failure [24]. No other genetic studies have been conducted in this area. Analysis of the GTEx data, [25], identified 84 expression quantitative trait loci downstream of *KLF15* as highly significant (p values 0.00011 to 3.8^{10-8}, false discovery rate <5%, alternative allele effect sizes from 0.12–0.16) modifiers of KLF15 gene expression in transformed fibroblasts. Although the rs9838915 SNP was not examined in GTEx, the data suggest that *KLF15* SNPs have the functional potential to influence KLF15 expression. Recently, Ferreira et al. [26] conducted a bioinformatics analysis focused on non-synonymous variants in KLF genes using algorithms to predict the effect of variants on the genes structure and function. Two *KLF15* intronic variants, one located in the conserved Zinc-finger domain and the other in a non-conserved region was predicted to affect DNA binding or protein destabilization and hence may lead to disrupted KLF15 protein function. The frequency of these variants is <0.0001 in the population and therefore unlikely to contribute to LVH in the general population.

Table 3. Human studies of KLF15.

Population Group and Sample Size (*n*)	Summary of Main Findings	References
LV tissue from patients with LVH due to aortic stenosis (*n* = 8) and control subjects undergoing CABG without LVH (*n* = 6).	KLF15 protein detected in nuclei of myocytes from control patients. ↓ cardiac KLF15 protein in LVH due to aortic stenosis	Fisch S et al. 2007 [13]
LV tissue from heart transplant patients with systolic heart failure (non-ischemic cardiomyopathy, *n* = 36) and non-failing control hearts (*n* = 30).	↓ cardiac KLF15 mRNA in failing hearts	Halder et al. 2010 [10]
LV tissue taken pre- and post-LVAD implantation/explantation in end-stage heart failure patients and non-failing hearts. (*n* = 3–4/group)	↓ cardiac KLF15 mRNA pre-LVAD compared to control non-failing hearts. ↑ cardiac KLF15 mRNA expression post LVAD	Prosdocimo et al. 2014 [14]
Type 2 diabetes patients (*n* = 318) with LVH and heart failure outcomes; replication LVH cohort (*n* = 5631)	↑ LV mass and LVH with *KLF15* SNP rs9838915 A allele; finding replicated in independent cohort. *KLF15* rs9838915 A allele predicted development of first hospitalization with heart failure	Patel et al. 2017 [24]

↑ Increased; ↓ decreased. CABG, coronary artery bypass graft; mRNA, messenger ribonucleic acid; LV, left ventricle; LVAD, left ventricular assisted device; LVH, left ventricular hypertrophy; KLF15, Kruppel-like factor 15; SNP, single nucleotide polymorphism.

4. Proposed Mechanisms: KLF15 and Cardiac Hypertrophy

A number of mechanisms by which KLF15 contributes to cardiac hypertrophy have been proposed. Gene reporter studies show that KLF15 inhibits MEF2 and GATA4 DNA-binding transcriptional activation and that KLF15 negatively regulates cardiac hypertrophy by preventing DNA-binding of MEF2 and GATA4 to their transcriptional targets [13]. KLF15 activation represses the acetyltransferase p300 mediated acetylation of p53 to maintain normal heart function [10]. Conversely, KLF15 deficiency leads to hyperacetylation of p53 in the mice hearts and in human hearts. This action of KLF15 is important as p300 acetylates several pro-hypertrophic markers involved in pathogenic remodeling including GATA4, MEF2 and Smad. Furthermore, p300 is a direct transcriptional target of TGFβ in fibroblasts and its levels are elevated in experimental models of fibrosis [27]. The KLF15 protein domain map shows highly conserved regions between residues 140–160 in the putative p300-interacting transactivation domain and conserved sequences in this region between species [14]. In the same study, KLF15 deficient mice were rescued by p53 deletion or p300 inhibition by curcumin. The authors propose that loss of KLF15 is not a response to heart failure but that loss of KLF15 contributes to progression to heart failure by removing the ability to repress key cardiac transcription factors that enable growth [10].

KLF15 competitively inhibits SRF binding to the transcriptional co-activator myocardin thus preventing expression of cardiotrophic genes such as ANP [15]. GST-pulldown assays revealed no direct interaction between KLF15 and SRF or MEF2 [15] suggesting previous interactions reported may be indirect [13]. KLF15 also competitively inhibits binding of MRTF-A and MRTF-B to SRF, repressing SRF-dependent cardiotrophic gene expression [18]. CTGF, a key mediator of fibrosis in pathological cardiac hypertrophy [28], is negatively regulated by KLF15 and increased TGFβ levels activate p38-MAPK signaling which downregulates KLF15 leading myocardin to bind to SRF and to activation of cardiotrophic genes [15]. Therefore, KLF15 acts as a negative repressor of cardiac hypertrophy via inhibitory competitive binding to myocardin [15,18].

Tandler et al. [29] postulated that KLF15 deficiency could impair the mitochondrial division process, creating enlarged mitochondria that contribute to hypertrophic cardiomyocytes. KLF15 deficient mice had sparse dispersions of megamitochondria within LV tissue on electron microscopy [29]. Although normal morphological mitochondria were more frequently observed, megamitochondria were up to three times wider and significantly longer in length with morphological features suggestive of inability to undergo fission. Others reported no differences in mitochondrial ultrastructure in wild-type and knock-out KLF15 mice on electron microscopy [14]. A role for KLF15 in cardiac metabolism has also been suggested. KLF15 positively influences the energetics of cardiac metabolism and KLF15 null mice have impaired mitochondrial fatty acid [14].

5. Summary

In the past decade, there has been increasing evidence from both experimental and human studies that KLF15 is an important regulator of cardiac hypertrophy. The studies clearly support a pathological role of KLF15 in the development of cardiac hypertrophy and identify KLF15 as part of a transcriptional pathway regulating the cardiac response to hypertrophic stimuli. Thus, loss of KLF15 leads to transcription repression and stimulates fibrotic signaling pathways. Overexpression studies suggest KLF15 inhibits hypertrophy and reduces cardiomyocyte size and pro-hypertrophic and pro-fibrotic markers such as ANP, BNP and CTGF. Genetic variants in *KLF15* may also contribute to the development of LVH and predict heart failure outcomes in those with LVH. The genetic findings complement the human studies which reported reduced KLF15 expression in the hearts of patients with LVH and heart failure. The only suggested therapeutic pathway for preventing loss of KLF15 and transition to heart failure thus far has been inhibition of the p53 and p300 pathway.

Conflicts of Interest: The authors declare no conflict of interest.

References

1. Levy, D.; Garrison, R.J.; Savage, D.D.; Kannel, W.B.; Castelli, W.P. Prognostic implications of echocardiographically determined left ventricular mass in the Framingham Heart Study. *N. Engl. J. Med.* **1990**, *322*, 1561–1566. [CrossRef] [PubMed]
2. Srivastava, P.M.; Calafiore, P.; Macisaac, R.J.; Patel, S.K.; Thomas, M.C.; Jerums, G.; Burrell, L.M. Prevalence and predictors of cardiac hypertrophy and dysfunction in patients with Type 2 diabetes. *Clin. Sci.* **2008**, *114*, 313–320. [CrossRef] [PubMed]
3. Modin, D.; Biering-Sørensen, S.R.; Mogelvang, R.; Landler, N.; Jensen, J.S.; Biering-Sørensen, T. Prognostic Value of Echocardiography in Hypertensive Versus Nonhypertensive Participants From the General Population. *Hypertension* **2018**, *71*, 742–751. [CrossRef] [PubMed]
4. Kearney, L.; Ord, M.; Buxton, B.; Matalanis, G.; Patel, S.; Burrell, L.; Srivastava, P. Usefulness of the Charlson co-morbidity index to predict outcomes in patients >60 years old with aortic stenosis during 18 years of follow-up. *Am. J. Cardiol.* **2012**, *110*, 695–701. [CrossRef] [PubMed]
5. Bansal, N.M.; Keane, P.; Delafontaine, D.; Dries, E.; Foster, C.A.; Gadegbeku, A.S.; Go, L.L.; Hamm, J.W.; Kusek, A.O.; Ojo, M.; et al. A longitudinal study of left ventricular function and structure from CKD to ESRD: The CRIC study. *Clin. J. Am. Soc. Nephrol.* **2013**, *8*, 355–362. [CrossRef] [PubMed]
6. Devereux, R.B.; Roman, M.J.; De Simone, G.; O'Grady, M.J.; Paranicas, M.; Yeh, J.; Fabsitz, R.R.; Howard, B.V. Relations of left ventricular mass to demographic and hemodynamic variables in American Indians: The Strong Heart Study. *Circulation* **1997**, *96*, 1416–1423. [CrossRef] [PubMed]
7. Dahlof, B.R.B.; Devereux, S.E.; Kjeldsen, S.; Julius, G.; Beevers, U.; de Faire, F.; Fyhrquist, H.; Ibsen, K.; Kristiansson, O.; Lederballe-Pedersen, L.H.; et al. Cardiovascular morbidity and mortality in the Losartan Intervention For Endpoint reduction in hypertension study (LIFE): A randomised trial against atenolol. *Lancet* **2002**, *359*, 995–1003. [CrossRef]

8.   Diez, J.B.; Lopez, A.; Gonzalez, A.; Querejeta, R. Clinical aspects of hypertensive myocardial fibrosis. *Curr. Opin. Cardiol.* **2001**, *16*, 328–335. [CrossRef] [PubMed]

9.   Heineke, J.; Molkentin, J.D. Regulation of cardiac hypertrophy by intracellular signalling pathways. *Nat. Rev. Mol. Cell Biol.* **2006**, *7*, 589–600. [CrossRef] [PubMed]

10.   Haldar, S.M.Y.; Lu, D.; Jeyaraj, D.; Kawanami, Y.; Cui, S.J.; Eapen, C.; Hao, Y.; Li, Y.Q.; Doughman, M.; Watanabe, K.; et al. KLF15 deficiency is a molecular link between heart failure and aortic aneurysm formation. *Sci. Transl. Med.* **2010**, *2*. [CrossRef] [PubMed]

11.   Stanojevic, D.T.; Hoey, T.; Levine, M. Sequence-specific DNA-binding activities of the gap proteins encoded by hunchback and Kruppel in Drosophila. *Nature* **1989**, *341*, 331–335. [CrossRef] [PubMed]

12.   Wang, B.S.M.; Haldar, Y.; Lu, O.A.; Ibrahim, S.; Fisch, S.; Gray, A.; Leask, A.; Jain, M.K. The Kruppel-like factor KLF15 inhibits connective tissue growth factor (CTGF) expression in cardiac fibroblasts. *J. Mol. Cell Cardiol.* **2008**, *45*, 193–197. [CrossRef] [PubMed]

13.   Fisch, S.S.; Gray, S.; Heymans, S.M.; Haldar, B.; Wang, O.; Pfister, L.; Cui, A.; Kumar, Z.; Lin, S.; Sen-Banerjee, H.; et al. Kruppel-like factor 15 is a regulator of cardiomyocyte hypertrophy. *Proc. Natl. Acad. Sci. USA* **2007**, *104*, 7074–7079. [CrossRef] [PubMed]

14.   Prosdocimo, D.A.P.; Anand, X.; Liao, H.; Zhu, S.; Shelkay, P.; Artero-Calderon, L.; Zhang, J.; Kirsh, D.; Moore, M.G.; Rosca, E.; et al. Kruppel-like factor 15 is a critical regulator of cardiac lipid metabolism. *J. Biol. Chem.* **2014**, *289*, 5914–5924. [CrossRef] [PubMed]

15.   Leenders, J.J.W.J.; Wijnen, M.; Hiller, I.; van der Made, V.; Lentink, R.E.; van Leeuwen, V.; Herias, S.; Pokharel, S.; Heymans, L.J.; de Windt, M.A.; et al. Regulation of cardiac gene expression by KLF15, a repressor of myocardin activity. *J. Biol. Chem.* **2010**, *285*, 27449–27456. [CrossRef] [PubMed]

16.   Gao, L.Y.; Guo, X.; Liu, D.; Shang, D.; Du, Y. KLF15 protects against isoproterenol-induced cardiac hypertrophy via regulation of cell death and inhibition of Akt/mTOR signaling. *Biochem. Biophys. Res. Commun.* **2017**, *487*, 22–27. [CrossRef] [PubMed]

17.   Yu, Y.; Ma, J.; Xiao, Y.; Yang, Q.; Kang, H.; Zhen, J.; Chen, L. KLF15 is an essential negative regulatory factor for the cardiac remodeling response to pressure overload. *Cardiology* **2015**, *130*, 143–152. [CrossRef] [PubMed]

18.   Leenders, J.J.; Wijnen, W.J.; van Der Made, I.; Hiller, M.; Swinnen, M.; Vandendriessche, T.; Chuah, M.; Pinto, Y.M.; Creemers, E.E. Repression of cardiac hypertrophy by KLF15: Underlying mechanisms and therapeutic implications. *PLoS ONE* **2012**, *7*, e36754. [CrossRef] [PubMed]

19.   Horie, T.; Ono, K.; Nishi, H.; Iwanaga, Y.; Nagao, K.; Kinoshita, M.; Kuwabara, Y.; Takanabe, R.; Hasegawa, K.; Kita, T.; et al. MicroRNA-133 regulates the expression of GLUT4 by targeting KLF15 and is involved in metabolic control in cardiac myocytes. *Biochem. Biophys. Res. Commun.* **2009**, *389*, 315–320. [CrossRef] [PubMed]

20.   Langheinrich, M.; Ae Lee, M.; Böhm, M.; Pinto, Y.M.; Ganten, D.; Paul, M. The hypertensive Ren-2 transgenic rat TGR (mREN2)27 in hypertension research. Characteristics and functional aspects. *Am. J. Hypertens.* **1996**, *9*, 506–512. [CrossRef]

21.   Heymans, S.B.; Schroen, P.; Vermeersch, H.; Milting, F.; Gao, A.; Kassner, H.; Gillijns, P.; Herijgers, W.; Flameng, P.; Carmeliet, F.; et al. Increased cardiac expression of tissue inhibitor of metalloproteinase-1 and tissue inhibitor of metalloproteinase-2 is related to cardiac fibrosis and dysfunction in the chronic pressure-overloaded human heart. *Circulation* **2005**, *112*, 1136–1144. [CrossRef] [PubMed]

22.   Chokshi, A.K.; Drosatos, F.H.; Cheema, R.; Ji, T.; Khawaja, S.; Yu, T.; Kato, R.; Khan, H.; Takayama, R.; Knoll, H.; et al. Ventricular assist device implantation corrects myocardial lipotoxicity, reverses insulin resistance, and normalizes cardiac metabolism in patients with advanced heart failure. *Circulation* **2012**, *125*, 2844–2853. [CrossRef] [PubMed]

23.   Bella, J.N.; Goring, H.H. Genetic epidemiology of left ventricular hypertrophy. *Am. J. Cardiovasc. Dis.* **2012**, *2*, 267–278. [PubMed]

24.   Patel, S.K.; Wai, B.; Lang, C.C.; Levin, D.; Palmer, C.N.; Parry, H.M.; Velkoska, E.; Harrap, S.B.; Srivastava, P.M.; Burrell, L.M. Genetic variation in Kruppel like factor 15 is associated with left ventricular hypertrophy in patients with type 2 diabetes: Discovery and replication cohorts. *EBioMedicine* **2017**, *18*, 171–178. [CrossRef] [PubMed]

25.   GTEx Consortium. The Genotype-Tissue Expression (GTEx) pilot analysis: Multitissue gene regulation in humans. *Science* **2015**, *348*, 648–660.

26. Ferreira, K.K.; de Morais Gomes, E.R.; de Lima Filho, J.L.; Castelletti, C.H.; Martins, D.B. Bioinformatics analysis of non-synonymous variants in the KLF genes related to cardiac diseases. *Gene* **2018**, *650*, 68–76. [CrossRef] [PubMed]

27. Ghosh, A.K.; Varga, J. The transcriptional coactivator and acetyltransferase p300 in fibroblast biology and fibrosis. *J. Cell. Physiol.* **2007**, *213*, 663–671. [CrossRef] [PubMed]

28. Candido, R.; Forbes, J.M.; Thomas, M.C.; Thallas, V.; Dean, R.G.; Burns, W.C.; Tikellis, C.; Ritchie, R.H.; Twigg, S.M.; Cooper, M.E.; et al. A breaker of advanced glycation end products attenuates diabetes-induced myocardial structural changes. *Circ. Res.* **2003**, *92*, 785–792. [CrossRef] [PubMed]

29. Tandler, B.; Fujioka, H.; Hoppel, C.L.; Haldar, S.M.; Jain, M.K. Megamitochondria in Cardiomyocytes of a Knockout (KLF15$^{-/-}$) Mouse. *Ultrastruct. Pathol.* **2015**, *39*, 336–339. [CrossRef] [PubMed]

International Journal of
Molecular Sciences

Review

DNA Methylation and Histone Modification in Hypertension

Shaunrick Stoll [1], Charles Wang [2] and Hongyu Qiu [1,*

[1] Division of Pharmacology and Physiology, Department of Basic Sciences, School of Medicine, Loma Linda University, Loma Linda, CA 92350, USA, sstoll@llu.edu

[2] Center for Genomics, Department of Basic Sciences, School of Medicine, Loma Linda University, Loma Linda, CA 92350, USA; chwang@llu.edu

* Correspondence: hqiu@llu.edu; Tel.: +1-909-651-5894; Fax: +1-909-558-0119

Received: 10 February 2018; Accepted: 9 April 2018; Published: 12 April 2018

Abstract: Systemic hypertension, which eventually results in heart failure, renal failure or stroke, is a common chronic human disorder that particularly affects elders. Although many signaling pathways involved in the development of hypertension have been reported over the past decades, which has led to the implementation of a wide variety of anti-hypertensive therapies, one half of all hypertensive patients still do not have their blood pressure controlled. The frontier in understanding the molecular mechanisms underlying hypertension has now advanced to the level of epigenomics. Particularly, increasing evidence is emerging that DNA methylation and histone modifications play an important role in gene regulation and are involved in alteration of the phenotype and function of vascular cells in response to environmental stresses. This review seeks to highlight the recent advances in our knowledge of the epigenetic regulations and mechanisms of hypertension, focusing on the role of DNA methylation and histone modification in the vascular wall. A better understanding of the epigenomic regulation in the hypertensive vessel may lead to the identification of novel target molecules that, in turn, may lead to novel drug discoveries for the treatment of hypertension.

Keywords: DNA methylation; histone modifications; vascular smooth muscle cells; endothelial cells; hypertension

---

1. Introduction

Systemic hypertension (or Hypertension in this review) refers to a condition of high blood pressure (BP) in the systemic arteries, the vessels that carry blood from the heart to the body's tissues, which distinguishes the condition from a local high BP such as in pulmonary (lung) hypertension. Hypertension has been defined as a BP reading of 140/90 mm Hg or higher in adults for many years. However, the latest report (2017) from the American College of Cardiology/American Heart Association (ACC/AHA) provides an updated guideline that classifies hypertension as a BP reading of 130/80 mm Hg or higher [1]. Hypertension remains a major risk factor for myocardial infarction, heart failure, end-stage renal disease, and stroke despite extensive research for decades and many therapeutic approaches [2]. Hypertension affects nearly one third of all adults in the US and only one half of the hypertensive patients have their BP under control [3]. The ACC/AHA report also notes that males are more likely to develop hypertension compared to women at the pre-menopausal ages, suggesting a role for hormone signaling in the management of BP [1]. Although the causes of hypertension have not been fully outlined, certain chronic conditions may increase the risk of the development of hypertension which include general risk factors such as aging, smoking, low socioeconomic and educational status, overweight/obesity, unhealthy diet, and physical inactivity as well as other secondary disorders such as chronic kidney disease (CKD), genetic family history, diabetes mellitus, obstructive sleep apnea and psychosocial stress [1]. Current nonpharmacological therapy focuses on maintaining a healthy diet, exercise, and other lifestyle changes, while pharmacological treatment includes: thiazide or thiazide-type diuretics, angiotensin-converting enzyme (ACE) inhibitors, dihydropyridines, aldosterone receptor blockers (spironolactone and derivatives), beta blockers, and vasodilators, direct renin inhibitors, and alpha-1 blockers [1]. Despite the plethora of treatment options, the high BP of 16.1 million hypertensive patients remains uncontrolled [3].

While many signaling pathways involved in the development of hypertension have been discovered, the molecular mechanisms related to the epigenomic regulation underlying vascular dysfunction are still not well understood. Epigenomics refers to the genome-wide study of gene regulation that alters gene activity without changing the DNA sequence [2]. Epigenomic regulation provides a link between the genotype and phenotype and is essential for many normal cellular functions, and inappropriate regulation at the epigenomic level may lead to some major adverse effects on cellular function of a tissue or organ which may lead to the development of diseases. Different epigenetic regulations have been identified in hypertension, including methylation, acetylation, phosphorylation, ubiquitination, and sumolyation.

It has been well-known that the renin-angiotensin-aldosterone system (RAAS), a hormone system that is integral to the physiological regulation of BP, plays a crucial role in the development of hypertension, thus, the epigenomic alterations of RAAS-regulated genes and their effects have been extensively tested in hypertensive models [4]. Increasing evidence is emerging that epigenomics also plays an important role in gene expression which may alter the phenotype and function of the vascular wall or cells in response to environmental stresses [5], underscoring our need to consider the contribution of epigenomic regulation of the vascular walls to hypertension. This review seeks to highlight recent advances in our knowledge of the changes in DNA methylation and histone modification which occur during the development of hypertension, as summarized in Table 1. Despite the well-known importance of RAAS in the development of hypertension, we have only briefly summarized them in this present manuscript since the relative epigenomic alterations of RAAS have been extensively discussed in a recent review [4]. To concentrate on the updated information, our present review focuses on the contribution of alterations to the vasculature in the development of systemic hypertension. An increased understanding of the epigenomic regulation of these vascular cells will lead to the identification of novel target molecules that may, in turn, lead to novel drug discoveries for the treatment of hypertension.

Table 1. DNA methylation and histone modification associated with hypertension.

Genes	Mark	Status	Species	Models	Tissues/Cells	Function	Ref
DNA methylation							
Atgr1α	5mC	Hypo	Rat	SHR	Aorta and mesentery artery	Increased expression of receptor and effect of RAAS	[6]
Atgr1β	5mC	Hypo	Rat	Maternal low protein rat		RAAS	[7]
Ace-1	5mC	Hypo	Mice	Maternal protein deficient mice		RAAS	[8]
ACE-1	5mC	Hyper	Human		Human PBMCs; cell culture (HepG2, HT29, HMEC-1, SUT)	RAAS	[9]
HSD11B2	5mC	Hyper	Human	Glucocorticoid treatment	Human PBMCs	Renal sodium balance	[10]
Sslc12a2 (NKCC1)	5mC	Hypo	Rat	SHR	Aorta and heart	Ionic balance	[11]

Table 1. *Cont.*

Genes	Mark	Status	Species	Models	Tissues/Cells	Function	Ref
ESR1 (ERα)	5mC	Hyper	Sheep		Uterine artery	vasodilation	[12]
SRF, MYOCD, MYH11	5mC	Hyper	Human		Human coronary artery SMCs	contraction phenotype	[13]
ADD1	5mC	Hypo	Human		Human PBMCs	Ionic balance	[14]
SCNN1A	5mC	Hyper	Human		Human PBMCs	Ionic balance	[15]
SCNN1B	5mC	CpG1 Hyper, CpG2 Hypo	Human		Human PBMCs	Ionic balance	[16]
TLR2	5mC	Hypo	Human		Human PBMCs	Chronic inflammation	[17]
EHMT2	5mC	Hypo	Human		Human PBMCs	Chronic inflammation	[18]
Histone modification							
Ace1	H3Ac, H3K4me3, H3K9me2	Hyper, Hyper, Hypo	Rat	SHR	Heart, kidney	RAAS	[19]
SM22	H3Ac	Hyper	Mouse		10T1/2 cells	Contractile phenotype	[20]
Nlrp3	H3K9Ac	Hyper	Rat	SHR	VSMCs	Chronic inflammation	[21]
NOS3 (eNOS)	H3K9Ac, H4K12 H3K4 me2, H3K4me3	Hyper	Human		Cell culture; HUVEC, HMVEC, VSMC, HEPG2, HeLa, JEG-3	Vasodilation in endothelial cells	[22]
Slc12a2 (NKCC1)	H3Ac, H3K27me3	Hyper, Hypo	Rat	Angiotensin II delivery	Aorta	Ionic balance	[23]

Abbreviations: 5mc—5-methylcytosine, RAAS—Renin-angiotensin-aldosterone system, PBMC—peripheral blood mononuclear cell, SHR—spontaneously hypertensive rats, VSMC—vascular smooth muscle cell, HUVEC—human umbilical vein endothelial cell, HMVEC—human dermal microvascular endothelial cells, H3Ac—Histone 3 acetylation, H3K4me2—dimethylation of histone 3 lysine 4, H3K4me3—trimethylation of histone 3 lysine 4.

2. Discovery, Development, and Detection

2.1. DNA Methylation

DNA methylation involves the addition of methyl groups to the cytosine residues of DNA [24]. Initial studies of X-chromosome inactivation in mice provided the evidence that DNA could be silenced without changes in the DNA sequence itself [25,26]. Later studies showed that DNA methylation can act as a mechanism through which this silencing may occur [27,28]. Importantly, DNA methylation marks can be copied from the parent strand to the daughter strand. Another notable finding is the identification of the main methylation target: the sequence of CpG, shorthand for 5′-C-phosphate-G-3′, that is, cytosine and guanine separated by only one phosphate group [27,28]. These findings eventually led to the development of methylation-sensitive restriction enzymes, which became important tools in epigenetic analyses [29,30].

DNA methlytransferases are responsible for moving methyl groups from S-adenosyl methionine (SAM) to CpG islands on DNA. For example, DNA methlytransferases 1, 3A, and 3B (DNMT1/3A/3B) add methyl groups to the carbon at position 5 of cytosine resides which are adjacent to guanine residues to produce 5-methylcytosine (5mC) [31,32]. While CpG is the main target, methylation may occur on CpHpG, where H may be A, T, or C [31]. The methylation mark 5mC is generally associated with gene repression within the gene promoter [31,33]. Alternatively, actively transcribed genes may be methylated within the transcriptional region which includes parts actually translated into protein (exons) and parts that will be cut out of the mRNA and are not translated into protein (introns) [31]. The methylation pattern of a cell may vary in response to stress, as different genes are turned on and off. For example, it was shown that in the nuclei of mouse cardiomyocytes, 127 genes gained methylation and 313 lost methylation of their transcriptional region during the postnatal period [34]. While methylation at transcription start sites represses transcription, it is yet unclear whether methylation within exons serve clear functions, as high methylation seems to be positively correlated with transcription. One suggestion is that DNA methylation within exons may affect alternative splicing [35].

Methylation patterns can also change under stress and disease, as shown in the methylation of promoters of tumor suppressor genes in cancer [36]. Importantly, DNA methylation is reversible, with the removal reaction catalyzed by histone lysine demethylases (KDM) [37], providing therapeutic potential to prevent or change the pathological DNA methylation that may be related to the diseases.

Initial methylation detection assays utilized restriction endonucleases with known methylation sensitive CpG sites and their methylation insensitive isoschizomers [30]. Genomic DNA was then subjected to cleavage by both sets of restriction enzymes, and the differences between the fragments would outline where the methylated cytosines were, as revealed by Southern blot analysis, 2-dimensional gel electrophoresis, polymerase chain reaction (PCR), and, more recently, next-generation sequencing [31]. Alternatively, methylated cytosines may be targeted using specific antibodies by a process known as methylated DNA immunoprecipitation (MeDIP) or affinity enrichment [38]. One of the most common current methods of detecting epigenetic modifications is bisulfite sequencing. Unmethylated cytosine residues are converted to uracil using sodium bisulfite and alkaline treatment, while methylated cytosines are left intact. The bisulfite-treated DNA is then sequenced to reveal the methylated cytosines [31].

In addition to DNA methylation (5mC), other DNA modifications include hydroxymethylation (5hmC) and formylcytosine (5fC) [39]. Ten eleven translocation (TET) enzymes oxidize 5mCs to 5hmCs, and further catalyze the conversion to 5fC and 5-carboxylcytosine (5-caC) which can then be replaced by unmethylated cytosine after undergoing thymine DNA glycosylase (TDG)-mediated base excision and DNA base excision repair [33,40]. The TET enzymes may be significant to future therapies as they allow exploitation of the oxidation of 5mCs as a means of reversing gene repression by DNA methylation, should the repression of signature genes be identified.

2.2. Histone Modification

Histones are important proteins responsible for maintaining the structure of chromatin and play a role in the dynamic and long term regulation of genes. The *N*-terminal tail of histone 3 (H3), one of the five histones found in eukaryotic nuclei, is subject to methylation or acetylation of lysine and arginine residues as well as phosphorylation of serine and threonine residues [41]. Histones are acetylated by histone acetyltransferases (HAT) and deacytelated by histone deacetylases (HDAC). These modifications may have opposing effects based on which residue is modified or which moiety is added. For example, methylation of H3 lysine 4 (H3K4) is a hallmark of actively transcribed DNA while methylation of H3 lysine 9 (H3K9) is associated with repressed gene expression [42,43]. Acetylation of histones is associated with transcriptionally active DNA, as the *N*-terminal of the histone tails become neutralized and their affinity for DNA reduced, so loosening the conformation of the chromatin [44]. Interestingly, while methylation at H3K9 allows for the binding of heterochromodoman protein 1 (HP1), leading to the repression of transcription, methylation at H3K4 blocks the binding of transcriptional repressor, nucleosome remodeling and deacetylase (NuRD), leading to transcription. Acetylation of lysines allows for the binding of bromodomain proteins, such as histone acetyltransferase, GCN5 [43].

The most common histone modification assay is the chromatin immunoprecipitation (ChIP), which uses antibodies against site-specific epigenetic marks to identify histone-DNA complexes with that mark [45]. Following fragmentation, histone-DNA complexes that showcase specific modifications are immunoprecipated using an antibody against that modification. Enriched and purified DNA fragments can then be detected by quantitative PCR (ChIP-qPCR), microarray analysis (ChIP-on-chip) or deep sequencing (ChIP-seq) [46]. ChIP-seq uses next-generation sequencing instead of amplifying purified DNA that was hybridized to a DNA microarray [47]. Importantly, ChIP-seq allows for mapping of the newly sequenced DNA to the reference genome, enabling researchers to determine the genome-wide distribution of particular modifications [46].

Conventional epigenomic analyses require large amounts of starting material due to the low DNA recovery efficiency after bisulfite conversion and enzyme digestion [48]. However, using large sample sizes masks cell to cell heterogeneity which may be important to the disease state. As molecular technology progresses to the single-cell level, many new approaches are being developed to study the single-cell epigenome, albeit with limited efficiency and scalability [39,48], including: single-cell bisulfite sequencing [49], single-cell reduced representation bisulfite sequencing [50], DNase I sequencing [51], single-cell DamID sequencing [52] and single-cell ATAC-seq [53]. Drawbacks of these methods include low coverage of CpG islands as compared to bulk methods, but the methodology is expected to grow to meet these challenges [54]. One such recent development is single cell combinatorial indexing for methylation analysis (sci-MET) which boasts of a 69% alignment rate to bulk methods [55].

3. Epigenomic Regulation in Hypertensive Vasculature

While numerous studies have outlined the epigenetic mechanisms in pulmonary hypertension (reviewed in [56–59]), the epigenomic regulation in systemic hypertension remains largely undescribed. Due to the well-known involvement of the renin-angiotensin-aldosterone system (RAAS) system on arterial pressure regulation, the effects of epigenomic regulation of the RAAS system have been extensively tested in animal models of systemic hypertension [4]. For example, it has been shown that the angiotensin 1α receptor (AT1aR), encoded by *Atgr1α*, is significantly increased in spontaneously hypertensive rats (SHR) compared to its counterpart, the Wistar-Kyoto rats (WKY), which may be responsible for the increased BP in SHR [6]. Bisulfite sequencing further revealed that the *Atgr1α* promoter is hypomethylated at its CpG islands in SHR at 20 weeks compared to WKY rats, implicating that the methylation of *Atgr1α* CpG islands reduces its expression, leading to normotensive BP [6]. In addition, hypomethylation of the promoter regions of the angiotensin II type 1β receptor (AT1bR) gene, (*Atgr1β*), in the adrenal glands of the maternal low protein rat exhibited hypertension in response to salt intake [7]. A similar effect was seen in another study in mice where maternal protein deficiency during pregnancy reduced methylation of promoter regions of the angiotensin I converting enzyme gene (*Ace-1*), which is responsible for converting angiotensin I to the active angiotensin II, eventually leading to hypertension in offspring [8]. In human cell lines, the luciferase activity of the *ACE-1* promoter/reporter constructs of somatic ACE (sACE) was inhibited by DNA methylation and subsequent inhibition of DNA methylation and/or histone deacetylation by 5-aza-cytidine injections in rats restored sACE expression in the lung and liver, highlighting the epigenetic regulation of sACE in hypertension [9].

The hydroxysteroid dehydrogenase-11β2 enzyme (HSD11B2) is responsible for degrading cortisol to biologically inert cortisone. Cortisol can be found in the blood at concentrations 2-3 orders of magnitude higher than aldosterone, the key mineralocorticoid in the RAAS [60]. Although cortisol and aldosterone bind mineralocorticoid receptors with similar affinity, degradation of cortisol to cortisone by the HSD11B2 enzyme in mineralocorticoid target tissues ensures that aldosterone is able to bind to the mineralocorticoid receptors [61]. In normal situations, aldosterone regulates sodium reabsorption due to the inactivation of cortisol to cortisone by the 11β HSD enzyme. Thus cortisol only has effects in the absence of this enzyme. Hypermethylation of the *HSD11B2* gene promoter impairs HSD11B2-mediated degradation of cortisol to cortisone, leading to an altered tetrahydrocortisol (THF) to tetrahydrocortisone (THE) ratio [10,62]. A high concentration of cortisol in mineralocorticoid target tissues and enables cortisol to regulate sodium reabsorption by the kidney, and ultimately arterial pressure. The hypermethylation of the HSD11B2 promoter may also contribute to the development of apparent mineralocorticoid excess (AME) [63]. This presents a case where both genetic and epigenetic regulation may be present.

Interestingly, in addition to the reports of the regulation of DNA methylation on RAAS, SHR aortas also showed higher enrichment of H3Ac and H3K4me3, while enrichment of H3K9me2 was reduced on the angiotensin-converting enzyme 1 (*ACE1*) promoter [19]. Thus, both the regulation of DNA methylation and histone modification exist on RAAS system.

Despite the importance of epigenetic regulation in the RAAS system on arterial pressure regulation, the blood vessels are the end-effect targets of RAAS and the function of blood vessels, particular the arteries, play a determinant role in the control of BP. Thus, in addition to the epigenetic regulation in the RAAS system described above, we will outline the recent studies on the epigenomic regulation in vascular tissue and cells, focusing on DNA methylation and histone modification in hypertension.

3.1. Epigenomic Regulation in the Whole Vessel

Several epigenomic studies have been conducted at the level of the whole vessel in animals. The Na^+–K^+–$2Cl^-$ cotransporter 1 (NKCC1), which is encoded by *Slc12a2*, regulates the exchange of sodium, potassium, and chlorine ions across cells of various types, including VSMCs and endothelial cells, regulating ionic balance and cell volume [64]. It has been showed that NKCC1$^{-/-}$ mice displayed a >15 mmHg reduction in systolic BP compared to wild-type and reduced ability to maintain vascular tone [65]. Pharmacological inhibition of NKCC1 by bumetanide led to an immediate 5% reduction in BP, highlighting its importance in BP regulation [66]. Combined bisulfite restriction assay and bisulfite sequencing revealed that the *Slc12a2* promoter was hypomethylated in SHR aorta and heart compared to WKY, which resulted in an upregulation of NKCC1 in both mRNA and protein levels detected by PCR and western blots in SHR vs WKY [11].

Furthermore, DNA methylation has been also linked to the estrogen receptor-mediated vascular regulation. Estrogen induces vasodilation as well as inhibits the response of blood vessels to injury, by interacting directly with the vasculature [67]. Two estrogen receptors have been identified: estrogen receptor α (ERα) and estrogen receptor β (ERβ). In uterine arteries of pregnant sheep, hypermethylation of the ERα (*ESR1*) promoter during hypoxia reduced its expression, leading to preeclampsia and impaired cardiovascular homeostasis [12]. Promoter hypermethylation inhibited transcription factor binding and promoter activity, leading to gene repression [12]. Although these mechanisms need to be confirmed in the human, this evidence suggests that the protective nature of estrogen signaling may be regulated through epigenetic mechanisms, which may also contribute to the gender disparity in hypertensive patients.

Moreover, histone modification has been found to be involved in the regulation of vascular function in hypertension. Additionally, NKCC1 can also be regulated by histone modification. Angiotensin II (Ang II) was delivered in rats, in vivo, to increase BP and the changes in NKCC1 mRNA, protein and epigenetic modifications at the *slc12a2* promoter were measured in the aorta [23]. Real-time PCR and western blot revealed a progressive increase in NKCC1 expression over the period of Ang II delivery. Interestingly, H3Ac levels were consistently increased in Ang II infused rats whereas H3K27me3 (a repressive histone code) levels were decreased as compared to sham [23]. These results together suggest that NKCC1 might be regulated by both DNA methylation and histone modification [4].

Overexpression of SIRT1, a histone deacetylase, reduced angiotensin-II induced hypertrophy, in vitro, and vascular remodeling and hypertension, in vivo [68,69]. Among the affected parameters were reduced reactive oxygen species (ROS) generation, vascular inflammation, and collagen synthesis in arterial walls. SIRT1 overexpression also decreased the association between nuclear factors on specific binding sites on TGF [69].

In addition to these studies on the whole vascular wall in hypertensive animal models, many observations on the epigenomic regulation have been done on specific cells of the vascular wall or the cells interacting with vessels such as the blood cells circulating around the body. We next outline these investigations.

3.2. Epigenomic Regulation in Vascular Smooth Muscle Cells (VSMCs)

Vascular smooth muscle cells (VSMCs) play an integral role in the regulation of peripheral resistance by modulating vascular tone. Autonomic nervous activity as well humoral agents are able to regulate contraction of VSMCs, leading to changes in BP [70]. It has been noted that VSMCs play a key role in the pathophysiology of hypertension due to their remarkable ability to dedifferentiate, allowing them to switch between contractile and synthetic states, in response to environmental cues or stress [71]. Studies have shown that this phenotypic switch between contractile and synthetic states in VSMCs could be governed by epigenetic modifications [71,72]. In addition, our recent studies also showed the alteration of intrinsic VSMC stiffness contributes to the development of hypertension [68,69]. A thorough understanding of epigenetic regulatory mechanisms in the VSMCs is mandatory as we develop new antihypertensive therapies.

Ten-eleven translocation-2 (TET2), a key enzyme in the DNA demethylation pathway, is a major governor of SMC plasticity and is highly expressed in VSMCs [13]. TET2 knockdown leads to a decreased expression of contractile markers such as myocardin (MYOCD), serum response factor (SRF), and myosin heavy chain 11(MYH11) and an increase in proliferative markers such as kruppel like factor 4 (KLF4) and myosin heavy chain 10 (MYH10), while TET2 overexpression restored contractile markers and inhibited synthetic genes, suggesting that TET2 serves as an important switch in VSMCs [13]. Importantly, it was shown that TET2 binds to the promoters of MYOCD, SRF, and MYH11, implicating its role in demethylating contractile genes [13]. Stimulation of the TET proteins may offer therapeutic potential in hypertension and atherosclerosis where pathological VSMC switching have been described, and re-differentiation of VSMCs is needed.

In addition to DNA methylation, histone modifications are also found to be able to modulate gene expression in VSMCs. Serum response factor (SRF), a key mediator of SMC transcription, binds to highly conserved domains near gene promoters known as the SRF binding sites or CArG boxes. Recruitment of SRF to the CArG boxes of SMC marker genes has been implicated to be associated with hyperacetylation of H3 and H4 in SMC-differentiated cells [20]. SRF complexes with cAMP-response element-binding protein (CREB)-binding protein on the hyperacetylated *SM22* gene promoter, leading to its expression [73]. Moreover, p300 mediated-acetylation of myocardin is critical for its dissociation from the inhibitory effects of histone deacetylase 5 (HDAC5) [74,75]. Myocardin's acetylation enhances its binding of SRF and the CArG boxes and is required for VSMC gene transcription [76]. Interestingly, our recent studies showed that upregulation of SRF/myocardin in VSMCs mediates the intrinsic VSMC and aortic stiffness in hypertension [77,78], indicating the key role of this signaling pathway in the regulation of BP. The further investigation of the epigenetic regulation of SRF/myocardin signaling may be a promising target for future studies.

It is generally accepted that oxidative stress and mild chronic vascular inflammation contribute to the pathophysiology of hypertension [79]. Activation of the nucleotide-binding oligomerization domain-like receptor protein 3 (NLRP3) inflammasome, a cytosolic complex for early inflammatory responses, generates proinflammatory cytokines such as interleukin 1 β (IL-1β) and interleukin 18 (IL-18) through the activation of caspase-1 [80]. These cytokines have been implicated in hypertension and consideration should be given to their therapeutic potential [81]. Recently, the *NLRP3* gene promoter was shown to have increased acetylation at lysine 9 of histone 3 and HAT expression in SHR VSMCs [21]. Importantly, inhibition of NFκβ and HAT by curcumin prevented the activation of the NLRP3 inflammasome, VSMC phenotypic switching, and proliferation in VSMCs of SHR [21]. These findings imply that inhibition of the inflammasome, IL-1β, IL-18, and HAT by reducing histone acetylation at the *NLRP3* gene promoter may prove effective in controlling chronic inflammation in hypertension and reducing the pathology of the disease.

3.3. Epigenomic Regulation in Endothelial Cell Dysregulation

While VSMCs comprise a major part of the medial layer of vascular wall, the endothelial cells which line the walls are in direct contact with circulating stimuli and their contribution to the BP control has been extensively studied. It has been widely accepted that endothelial dysfunction plays an important role in systemic hypertension. Both DNA methylation and histone modification have been implicated in the regulation of endothelial cells in hypertension.

The endothelium releases various vasoactive factors, some of which are vasodilatory, e.g., nitric oxide (NO), prostacyclin (PGI$_2$), and endothelium derived hyperpolarizing factor (EDHF); while some of which are vasoconstrictive, e.g., thromboxane (TXA$_2$) and endothelin-1 (ET-1) [82]. The endothelial nitric oxide synthase (eNOS), which is expressed primarily in endothelial cells, is the primary mechanism by which NO is produced in the vessel. Disturbances in the NO pathway have been linked to the predominance of vasoconstrictors which feed vicious cycles to maintain high BP [83]. Notably, the expression of eNOS may be controlled by cell-specific histone modifications as acetylated histone H3 lysine 9, histone H4 lysine 12, and di- and tri-methylated lysine 4 of histone H3 are all present in the *NOS3* gene promoter in human umbilical vein endothelial cells (HUVEC) and human dermal microvascular endothelial cells (HMVEC) but are absent in non-eNOS expressing cells like VSMCs and HeLa cells [22]. This mechanism was further explained by the discovery of HDAC1 selectively bound to the *NOS3* gene promoter in VSMCs, thus reducing the histone acetylation and transcription of eNOS in SMC [22]. Epigenetic regulation of eNOS expression is key to the tissue specificity observed in NO and dysregulation of epigenetic signaling during disease may contribute to increased vascular tone as a result of decreased NO synthesis in vessels.

In addition, an experiment in the deoxycorticosterone acetate (DOCA) salt-sensitive hypertension model of the Wistar rat, showed that animals treated with resveratrol, a known antihypertensive agent, increased the total antioxidant capacity and hydrogen sulfide levels which are independent of the change NO levels in circulation [84]. It also showed that resveratrol altered the staining of the H3K27me3 pattern of the aorta and renal artery sections, suggesting that its protective effects may be due to its effects of epigenetic modifications of the vessels [84].

4. Clinical Application of Epigenomic Studies in Human Systemic Hypertension

Although VSMCs and ECs are the main components of the vascular wall, inflammatory cells and circulating blood cells also play essential roles in the development of hypertension, and their epigenetic regulation should be reported. As human studies increase and biomedical technology advances, many researchers have opted to use peripheral blood to isolate DNA and measure epigenetic programming. While one can speculate that there may be cell to cell or tissue to tissue variation within the human, these studies offer a powerful tool in exploring differences in epigenetic patterns from patient to patient and identifying potential risk factors and key marks. One of the first questions that researchers asked was whether there was a difference in the DNA methylation of hypertensive patients from normotensive individuals that can be detected in blood cells. Interestingly, the total 5mC level was shown lower in the peripheral blood cells of patients with essential hypertension, and even lower in patients with Stage 1 hypertension [85]. This finding indicates that some epigenomic signatures in hypertensive patients can be detected in the peripheral blood at an early stage of hypertension, thus providing a predictive biomarker for the development of hypertension. While initial DNA methylation studies focused on the global level of 5mC, subsequent research has shown that the epigenetic regulation of specific DNA sequences in peripheral blood cells can contribute to the hypertensive state.

Since hypertension is a multifactorial disease which may be mediated by alterations in multiple biological pathways, research continues to identify possible targets whose manipulation may have effects on systemic BP. It has been shown that DNA variation is correlated with DNA methylation. Genome-wide linkage and association studies (GWAS) have identified single nucleotide polymorphisms (SNPs), which influence BP. SNPs that create CpG sites may be targets for epigenetic modifications, just as loss of these sites will prevent DNA methylation. For example, a recent trans-ancestry genome-wide association study identified genetic variants at 12 new loci which have been differently DNA methylated in hypertension, four of these genes encoded proteins that participate in the regulation of vascular tone and VSMC plasticity, including: insulin like growth factor binding brotein 3 (IGFBP3), potassium two pore domain channel subfamily K member 3 (KCNK3), phosphodiesterase 3A (PDE3A), and PR domain-containing protein 6 (PRDM 6) [86]. It is remarkable that a two-fold enrichment between sentinel BP single nucleotide polymorphisms (SNPs) and DNA methylation was observed, highlighting the importance of genetic variation and providing an additional hypothesis for the racial disparity observed in the incidence of hypertension [86]. While additional studies are needed to confirm whether DNA methylation of these specific genes is enough to cause changes in BP, it is exciting to note that DNA methylation explains part of the relationship between sentinel SNPs and BP ($r = 0.52$; $p = 0.005$) [86].

In a study of the DNA methylation levels of 5 CpG dinucleotides in 62 patients, reduced promoter methylation of the α-adducin (ADD1) gene was been shown to be linked to an increased essential hypertension risk [14]. Females were shown to have higher methylation of the ADD1 promoter, with a sharp decline in post-menopausal women, suggesting that estrogen signaling may also influence epigenetic regulation of the ADD1 gene [14]. Although α-adducin is a cytoskeletal protein, an ADD1 variant has been implicated in stimulation of renal sodium reabsorption and, subsequently, hypertension [87].

Epithelial sodium channel (ENaC), a nonvoltage-gated sodium channel, comprises three homologous subunits (α, β, and γ), with the alpha subunit (SCNN1A) being indispensable for channel activity [15]. The ENaC plays a critical role in maintaining extracellular fluid volume and BP. Hypermethylation of SCNN1A at CpG islands within the transcriptional region was associated with increased risk of developing hypertension. Importantly, the methylation level was regulated by age, gender, and antihypertensive therapy [15]. Hypermethylation of CpG islands within exons is typical of expressed genes, suggesting that SCNN1A gene is highly expressed in hypertension. Furthermore, CpG1 hypermethylation and CpG2 hypomethylation of the gene promoter of the amiloride-senstive sodium channel beta subunit (SCNN1B) were linked to hypertension, with the tested antihypertensive therapy only affecting CpG1 levels [16].

Toll-like receptors (TLRs) contribute to chronic inflammation, a mechanism that has be shown to play a role in the development of hypertension [17]. Hypomethylation of the TLR2 promoter at CpG6 was shown to be associated with an increased risk of systemic hypertension [17]. Notably, an important link between environmental cues, such as alcohol consumption and smoking, and specific CpG islands were identified as the risk factors of hypertension [17]. The identification of methylation patterns as a result of behavior is significant since hypertension has often been described as a lifestyle disease but precise mechanisms by which chronic inflammation through TLRs contribute to hypertension are still unclear.

Euchromatic histone-lysine in methyltransferase 2 (EHMT2; also called G9a) is a lysine histone methytransferase found in leukocytes which regulates the expression of IL-17 through methylation pattern H3K9me2 [88,89]. In a longitudinal genome wide methylation study of Roux-en-Y bypass patients, the EHMT2 promoter was found to be hypomethylated in hypertension, suggesting that epigenetic regulation of EHMT2 contributes to vascular pathophysiology [18]. It remains to be explored whether IL-17 has effects in human hypertension and whether epigenetic regulation of EHMT2 is alter the hypertensive state.

Together, all of the evidence indicates that, although epigenetic regulations of systemic hypertension in the vasculature are far from fully understood, the study of epigenetic regulation in systemic hypertension holds valuable potential for the identification of biomarkers and may provide for useful therapeutic targets. The discovery of new drugs based on the epigenomic regulation of the hypertensive vasculature, both synthetic and natural, offers new hope to protect ourselves against the silent killer.

5. Mechanisms Underlying Epigenetic Alterations in Hypertension

It is well known that hypertension can be influenced by genetic and environmental factors. Although the mechanisms underlying the epigenetic alterations during the development of hypertension have not been completely elucidated, epigenetic programming caused by adverse fetal environments, in utero, has been shown to be strongly correlated with the hypertension in adult offspring [90]. For example, intrauterine exposure to a maternal low-protein diet in the rat results in the development of hypertension in adult offspring rats which is associated with hypomethylated *Atgr1β* gene promoters along with increased adrenal expression of AT1bR [7,91–93]. In addition, prenatal inflammation exposure due to the maternal inflammatory diseases is highly associated with adult hypertension in the offspring [94–97]. It has been also showed that prenatal exposure to lipopolysaccharide (LPS), a nonspecific immune-inflammatory stimulant, led to hypertension in adult offspring rats, correlating with the augmentation of histone H3 acetylation (H3AC) on the angiotensin-converting enzyme 1 (ACE1) promoter, which induced the upregulation of the (*ACE1*) expression in renal cortex tissues [98–100]. Importantly, prenatal anti-inflammatory treatment is able to prevent offspring from fetal programming hypertension [100]. These results together indicate that epigenetic alterations induced during the fetal development are crucial contributors to the development of hypertension.

In addition to these "congenital" effects, the epigenetic alterations can be acquired during the early period of life due to the increased vulnerability of the reprogramming cells to environmental stress, such as malnutrition, toxic chemicals, infections and mental stress. This concept is further supported by the epidemiological investigations on human population which showed that social and environmental stresses during the early period of life influence epigenetic processes that contribute to the adult race-based US health disparities in diseases including hypertension [101]. Furthermore, several lines of evidence suggest that environmentally-induced epigenomic alterations can be transmitted to subsequent generations with disease phenotypes [102]. These observations, together, indicate the interaction of the genetic and environmental factors on the epigenomic alterations in the development of hypertension. On the other hand, it has been shown that increased hemodynamic forces are able to induce DNA methylation and histone modification in VSMCs and ECs [103]. In accordance with these understandings, it is important to note that while the epigenomic alterations are a part of the mechanisms involved in blood pressure elevation, they can also be a consequence of blood pressure alterations. Additional studies are therefore needed to determine causative agents of human epigenomic manipulation and their roles in development of hypertension.

Despite this evidence, the mechanisms by which DNA demethylation or histone modification are involved in the gene specific regulation of hypertension have not been fully elucidated. One mechanism suggests that the cell specific DNA demethylation is linked to a unique response to hormone stimulation [104]. Several studies suggest that interactions with sequence-specific DNA binding proteins and co-repressor complexes can target certain proteins to histones in a gene-specific manner [105,106]. Furthermore, other studies have shown that in addition to histone methylases, there are multiple histone demethylases, such lysine-specific demethylase 1-(LSD1), which removes mono- or di-methyl groups from H3K4 and the Jumonji C-(JmjC) domain-containing demethylases 5, which removes the tri-methylated modification and demethylates histones in a gene-specific manner by interactions between demethylases and DNA sequence specific nuclear factor complexes. Moreover, recent studies have shown that specific histone demethylases may regulate androgen-mediated transcriptional responses [107–109]. However, further studies are needed to confirm these mechanisms in the hypertensive individuals.

6. Conclusions

The root cause of hypertension is still to be elucidated, but there is little doubt that epigenomic changes in the vessels make a contribution to the disease. As illustrated in Figure 1, epigenetic regulation participates in the development of hypertension through a comprehensive mechanism which targets different levels of complexity including the RAAS system, the vascular wall and specific cell types within the vessels. Epigenetic regulation may also affect the circulating blood cells that interact with vessels. While regulators of the RAAS system are a current key focal point, more focus should be given to the components of blood vessels themselves and their regulation.

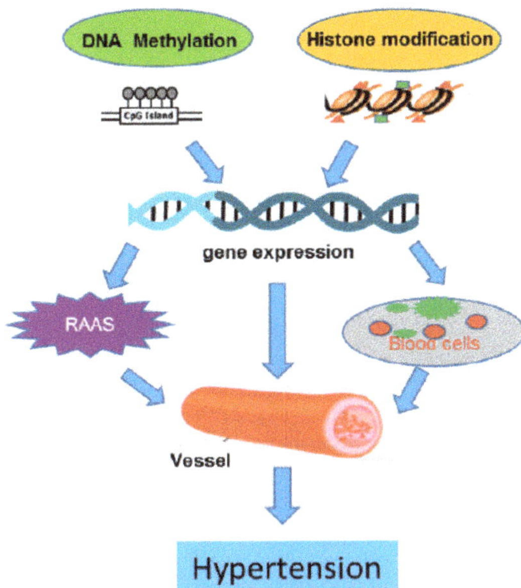

Figure 1. The illustration of the regulation DNA methylation and histone modification in hypertension.

Acknowledgments: This work is supported by NIH grants HL115195-01 (H.Q.), HL137962 (H.Q.) and National natural Science Foundation of China grant 31330029 (N.D.). Besides, we thank Nianguo Dong for his contribution of consultant and support on this paper.

Author Contributions: Shaunrick Stoll, Charles Wang and Hongyu Qiu have written and revised the manuscript and have approved the final version of the manuscript.

Conflicts of Interest: The authors declare no conflict of interest.

References

1. Whelton, P.K.; Carey, R.M.; Aronow, W.S.; Casey, D.E., Jr.; Collins, K.J.; Dennison Himmelfarb, C.; DePalma, S.M.; Gidding, S.; Jamerson, K.A.; Jones, D.W.; et al. 2017 acc/aha/aapa/abc/acpm/ags/apha/ash/aspc/nma/pcna guideline for the prevention, detection, evaluation, and management of high blood pressure in adults: Executive summary: A report of the american college of cardiology/american heart association task force on clinical practice guidelines. *Hypertension* **2017**. [CrossRef]
2. Liang, M.; Cowley, A.W., Jr.; Mattson, D.L.; Kotchen, T.A.; Liu, Y. Epigenomics of hypertension. *Semin. Nephrol.* **2013**, *33*, 392–399. [CrossRef] [PubMed]
3. Merai, R.; Siegel, C.; Rakotz, M.; Basch, P.; Wright, J.; Wong, B.; Dhsc; Thorpe, P. Cdc grand rounds: A public health approach to detect and control hypertension. *Morb. Mortal. Wkly. Rep.* **2016**, *65*, 1261–1264. [CrossRef] [PubMed]

4.   Wise, I.A.; Charchar, F.J. Epigenetic modifications in essential hypertension. *Int. J. Mol. Sci.* **2016**, *17*, 451. [CrossRef] [PubMed]

5.   Alexander, M.R.; Owens, G.K. Epigenetic control of smooth muscle cell differentiation and phenotypic switching in vascular development and disease. *Annu. Rev. Physiol.* **2012**, *74*, 13–40. [CrossRef] [PubMed]

6.   Pei, F.; Wang, X.; Yue, R.; Chen, C.; Huang, J.; Huang, J.; Li, X.; Zeng, C. Differential expression and DNA methylation of angiotensin type 1a receptors in vascular tissues during genetic hypertension development. *Mol. Cell. Biochem.* **2015**, *402*, 1–8. [CrossRef] [PubMed]

7.   Bogdarina, I.; Welham, S.; King, P.J.; Burns, S.P.; Clark, A.J. Epigenetic modification of the renin-angiotensin system in the fetal programming of hypertension. *Circ. Res.* **2007**, *100*, 520–526. [CrossRef] [PubMed]

8.   Goyal, R.; Goyal, D.; Leitzke, A.; Gheorghe, C.P.; Longo, L.D. Brain renin-angiotensin system: Fetal epigenetic programming by maternal protein restriction during pregnancy. *Reprod. Sci.* **2010**, *17*, 227–238. [CrossRef] [PubMed]

9.   Rivière, G.; Lienhard, D.; Andrieu, T.; Vieau, D.; Frey, B.M.; Frey, F.J. Epigenetic regulation of somatic angiotensin-converting enzyme by DNA methylation and histone acetylation. *Epigenetics* **2014**, *6*, 478–489. [CrossRef]

10.   Friso, S.; Pizzolo, F.; Choi, S.W.; Guarini, P.; Castagna, A.; Ravagnani, V.; Carletto, A.; Pattini, P.; Corrocher, R.; Olivieri, O. Epigenetic control of 11 beta-hydroxysteroid dehydrogenase 2 gene promoter is related to human hypertension. *Atherosclerosis* **2008**, *199*, 323–327. [CrossRef] [PubMed]

11.   Lee, H.A.; Baek, I.; Seok, Y.M.; Yang, E.; Cho, H.M.; Lee, D.Y.; Hong, S.H.; Kim, I.K. Promoter hypomethylation upregulates na+-k+-2cl-cotransporter 1 in spontaneously hypertensive rats. *Biochem. Biophys. Res. Commun.* **2010**, *396*, 252–257. [CrossRef] [PubMed]

12.   Dasgupta, C.; Chen, M.; Zhang, H.; Yang, S.; Zhang, L. Chronic hypoxia during gestation causes epigenetic repression of the estrogen receptor-α gene in ovine uterine arteries via heightened promoter methylation. *Hypertension* **2012**, *60*, 697–704. [CrossRef] [PubMed]

13.   Liu, R.; Jin, Y.; Tang, W.H.; Qin, L.; Zhang, X.; Tellides, G.; Hwa, J.; Yu, J.; Martin, K.A. Ten-eleven translocation-2 (tet2) is a master regulator of smooth muscle cell plasticity. *Circulation* **2013**, *128*, 2047–2057. [CrossRef] [PubMed]

14.   Zhang, L.N.; Liu, P.P.; Wang, L.; Yuan, F.; Xu, L.; Xin, Y.; Fei, L.J.; Zhong, Q.L.; Huang, Y.; Xu, L.; et al. Lower add1 gene promoter DNA methylation increases the risk of essential hypertension. *PLoS ONE* **2013**, *8*, e63455. [CrossRef] [PubMed]

15.   Mao, S.Q.; Fan, R.; Gu, T.L.; Zhong, Q.L.; Gong, M.L.; Dong, C.Z.; Hao, L.M.; Zhang, L.N. Hypermethylation of scnn1a gene-body increases the risk of essential hypertension. *Int. J. Clin. Exp. Pathol.* **2016**, *9*, 8047–8056.

16.   Zhong, Q.; Liu, C.; Fan, R.; Duan, S.; Xu, X.; Zhao, J.; Mao, S.; Zhu, W.; Hao, L.; Yin, F.; et al. Association of scnn1b promoter methylation with essential hypertension. *Mol. Med. Rep.* **2016**, *14*, 5422–5428. [CrossRef] [PubMed]

17.   Mao, S.; Gu, T.; Zhong, F.; Fan, R.; Zhu, F.; Ren, P.; Yin, F.; Zhang, L. Hypomethylation of the toll-like receptor-2 gene increases the risk of essential hypertension. *Mol. Med. Rep.* **2017**, *16*, 964–970. [CrossRef] [PubMed]

18.   Bostrom, A.E.; Mwinyi, J.; Voisin, S.; Wu, W.; Schultes, B.; Zhang, K.; Schioth, H.B. Longitudinal genome-wide methylation study of roux-en-y gastric bypass patients reveals novel CpG sites associated with essential hypertension. *BMC Med. Genom.* **2016**, *9*, 20. [CrossRef] [PubMed]

19.   Lee, H.A.; Cho, H.M.; Lee, D.Y.; Kim, K.C.; Han, H.S.; Kim, I.K. Tissue-specific upregulation of angiotensin-converting enzyme 1 in spontaneously hypertensive rats through histone code modifications. *Hypertension* **2012**, *59*, 621–626. [CrossRef] [PubMed]

20.   Manabe, I.; Owens, G.K. Recruitment of serum response factor and hyperacetylation of histones at smooth muscle-specific regulatory regions during differentiation of a novel p19-derived in vitro smooth muscle differentiation system. *Circ. Res.* **2001**, *88*, 1127–1134. [CrossRef] [PubMed]

21.   Sun, H.J.; Ren, X.S.; Xiong, X.Q.; Chen, Y.Z.; Zhao, M.X.; Wang, J.J.; Zhou, Y.B.; Han, Y.; Chen, Q.; Li, Y.H.; et al. Nlrp3 inflammasome activation contributes to vsmc phenotypic transformation and proliferation in hypertension. *Cell Death Dis.* **2017**, *8*, e3074. [CrossRef] [PubMed]

22.   Fish, J.E.; Matouk, C.C.; Rachlis, A.; Lin, S.; Tai, S.C.; D'Abreo, C.; Marsden, P.A. The expression of endothelial nitric-oxide synthase is controlled by a cell-specific histone code. *J. Biol. Chem.* **2005**, *280*, 24824–24838. [CrossRef] [PubMed]

23. Cho, H.M.; Lee, D.Y.; Kim, H.Y.; Lee, H.A.; Seok, Y.M.; Kim, I.K. Upregulation of the na(+)-k(+)-2cl(-) cotransporter 1 via histone modification in the aortas of angiotensin ii-induced hypertensive rats. *Hypertens. Res.* **2012**, *35*, 819–824. [CrossRef] [PubMed]

24. Felsenfeld, G. A brief history of epigenetics. *Cold Spring Harb. Perspect. Biol.* **2014**, *6*, a018200. [CrossRef] [PubMed]

25. Ohno, S.; Kaplan, W.D.; Kinosita, R. Formation of the sex chromatin by a single x-chromosome in liver cells of rattus norvegicus. *Exp. Cell Res.* **1959**, *18*, 415–418. [CrossRef]

26. Lyon, M.F. Gene action in the x-chromosome of the mouse (*Mus musculus* L.). *Nature* **1961**, *190*, 372–373. [CrossRef] [PubMed]

27. Riggs, A.D. X inactivation, differentiation, and DNA methylation. *Cytogenet. Genome Res.* **1975**, *14*, 9–25. [CrossRef] [PubMed]

28. Holliday, R.; Pugh, J.E. DNA modification mechanisms and gene activity during development. *Science* **1975**, *187*, 226–232. [CrossRef] [PubMed]

29. Doskočil, J.; Šorm, F. Distribution of 5-methylcytosine in pyrimidine sequences of deoxyribonucleic acids. *Biochim. Biophys. Acta* **1962**, *55*, 953–959. [CrossRef]

30. Bird, A.P.; Southern, E.M. Use of restriction enzymes to study eukaryotic DNA methylation. *J. Mol. Biol.* **1978**, *118*, 27–47. [CrossRef]

31. Laird, P.W. Principles and challenges of genomewide DNA methylation analysis. *Nat. Rev. Genet.* **2010**, *11*, 191–203. [CrossRef] [PubMed]

32. Kietzmann, T.; Petry, A.; Shvetsova, A.; Gerhold, J.M.; Gorlach, A. The epigenetic landscape related to reactive oxygen species formation in the cardiovascular system. *Br. J. Pharmacol.* **2017**, *174*, 1533–1554. [CrossRef] [PubMed]

33. Branco, M.R.; Ficz, G.; Reik, W. Uncovering the role of 5-hydroxymethylcytosine in the epigenome. *Nat. Rev. Genet.* **2011**, *13*, 7–13. [CrossRef] [PubMed]

34. Gilsbach, R.; Preissl, S.; Gruning, B.A.; Schnick, T.; Burger, L.; Benes, V.; Wurch, A.; Bonisch, U.; Gunther, S.; Backofen, R.; et al. Dynamic DNA methylation orchestrates cardiomyocyte development, maturation and disease. *Nat. Commun.* **2014**, *5*, 5288. [CrossRef] [PubMed]

35. Jones, P.A. Functions of DNA methylation: Islands, start sites, gene bodies and beyond. *Nat. Rev. Genet.* **2012**, *13*, 484–492. [CrossRef] [PubMed]

36. Schubeler, D. Function and information content of DNA methylation. *Nature* **2015**, *517*, 321–326. [CrossRef] [PubMed]

37. Allis, C.D.; Jenuwein, T. The molecular hallmarks of epigenetic control. *Nat. Rev. Genet.* **2016**, *17*, 487–500. [CrossRef] [PubMed]

38. Weber, M.; Davies, J.J.; Wittig, D.; Oakeley, E.J.; Haase, M.; Lam, W.L.; Schubeler, D. Chromosome-wide and promoter-specific analyses identify sites of differential DNA methylation in normal and transformed human cells. *Nat. Genet.* **2005**, *37*, 853–862. [CrossRef] [PubMed]

39. Kelsey, G.; Stegle, O.; Reik, W. Single-cell epigenomics: Recording the past and predicting the future. *Science* **2017**, *358*, 69–75. [CrossRef] [PubMed]

40. Rasmussen, K.D.; Helin, K. Role of tet enzymes in DNA methylation, development, and cancer. *Genes Dev.* **2016**, *30*, 733–750. [CrossRef] [PubMed]

41. Millis, R.M. Epigenetics and hypertension. *Curr. Hypertens. Rep.* **2011**, *13*, 21–28. [CrossRef] [PubMed]

42. Lachner, M.; O'Carroll, D.; Rea, S.; Mechtler, K.; Jenuwein, T. Methylation of histone h3 lysine 9 creates a binding site for hp1 proteins. *Nature* **2001**, *410*, 116–120. [CrossRef] [PubMed]

43. Zegerman, P.; Canas, B.; Pappin, D.; Kouzarides, T. Histone h3 lysine 4 methylation disrupts binding of nucleosome remodeling and deacetylase (nurd) repressor complex. *J. Biol. Chem.* **2002**, *277*, 11621–11624. [CrossRef] [PubMed]

44. Struhl, K. Histone acetylation and transcriptional regulatory mechanisms. *Genes Dev.* **1998**, *12*, 599–606. [CrossRef] [PubMed]

45. Collas, P. The current state of chromatin immunoprecipitation. *Mol. Biotechnol.* **2010**, *45*, 87–100. [CrossRef] [PubMed]

46. Kimura, H. Histone modifications for human epigenome analysis. *J. Hum. Genet.* **2013**, *58*, 439–445. [CrossRef] [PubMed]

47. Barski, A.; Cuddapah, S.; Cui, K.; Roh, T.Y.; Schones, D.E.; Wang, Z.; Wei, G.; Chepelev, I.; Zhao, K. High-resolution profiling of histone methylations in the human genome. *Cell* **2007**, *129*, 823–837. [CrossRef] [PubMed]

48. Liu, T.; Wu, H.; Wu, S.; Wang, C. Single-cell sequencing technologies for cardiac stem cell studies. *Stem Cells Dev.* **2017**, *26*, 1540–1551. [CrossRef] [PubMed]

49. Smallwood, S.A.; Lee, H.J.; Angermueller, C.; Krueger, F.; Saadeh, H.; Peat, J.; Andrews, S.R.; Stegle, O.; Reik, W.; Kelsey, G. Single-cell genome-wide bisulfite sequencing for assessing epigenetic heterogeneity. *Nat. Methods* **2014**, *11*, 817–820. [CrossRef] [PubMed]

50. Guo, H.; Zhu, P.; Wu, X.; Li, X.; Wen, L.; Tang, F. Single-cell methylome landscapes of mouse embryonic stem cells and early embryos analyzed using reduced representation bisulfite sequencing. *Genome Res.* **2013**, *23*, 2126–2135. [CrossRef] [PubMed]

51. Jin, W.; Tang, Q.; Wan, M.; Cui, K.; Zhang, Y.; Ren, G.; Ni, B.; Sklar, J.; Przytycka, T.M.; Childs, R.; et al. Genome-wide detection of dnase i hypersensitive sites in single cells and ffpe tissue samples. *Nature* **2015**, *528*, 142–146. [CrossRef] [PubMed]

52. Kind, J.; Pagie, L.; de Vries, S.S.; Nahidiazar, L.; Dey, S.S.; Bienko, M.; Zhan, Y.; Lajoie, B.; de Graaf, C.A.; Amendola, M.; et al. Genome-wide maps of nuclear lamina interactions in single human cells. *Cell* **2015**, *163*, 134–147. [CrossRef] [PubMed]

53. Buenrostro, J.D.; Wu, B.; Litzenburger, U.M.; Ruff, D.; Gonzales, M.L.; Snyder, M.P.; Chang, H.Y.; Greenleaf, W.J. Single-cell chromatin accessibility reveals principles of regulatory variation. *Nature* **2015**, *523*, 486–490. [CrossRef] [PubMed]

54. Wu, H.; Wang, C.; Wu, S. Single-cell sequencing for drug discovery and drug development. *Curr. Top. Med. Chem.* **2017**, *17*, 1769–1777. [CrossRef] [PubMed]

55. Mulqueen, R.M.; Pokholok, D.; Norberg, S.; Fields, A.J.; Sun, D.; Torkenczy, K.A.; Shendure, J.; Trapnell, C.; O'Roak, B.J.; Xia, Z.; et al. Scalable and efficient single-cell DNA methylation sequencing by combinatorial indexing. *bioRxiv* **2017**. [CrossRef]

56. Kim, J.D.; Lee, A.; Choi, J.; Park, Y.; Kang, H.; Chang, W.; Lee, M.S.; Kim, J. Epigenetic modulation as a therapeutic approach for pulmonary arterial hypertension. *Exp. Mol. Med.* **2015**, *47*, e175. [CrossRef] [PubMed]

57. Chelladurai, P.; Seeger, W.; Pullamsetti, S.S. Epigenetic mechanisms in pulmonary arterial hypertension: The need for global perspectives. *Eur. Respir. Rev.* **2016**, *25*, 135–140. [CrossRef] [PubMed]

58. Saco, T.V.; Parthasarathy, P.T.; Cho, Y.; Lockey, R.F.; Kolliputi, N. Role of epigenetics in pulmonary hypertension. *Am. J. Physiol. Cell Physiol.* **2014**, *306*, C1101–C1105. [CrossRef] [PubMed]

59. Kim, G.H.; Ryan, J.J.; Marsboom, G.; Archer, S.L. Epigenetic mechanisms of pulmonary hypertension. *Pulm. Circ.* **2011**, *1*, 347–356. [CrossRef] [PubMed]

60. Raftopoulos, L.; Katsi, V.; Makris, T.; Tousoulis, D.; Stefanadis, C.; Kallikazaros, I. Epigenetics, the missing link in hypertension. *Life Sci.* **2015**, *129*, 22–26. [CrossRef] [PubMed]

61. Funder, J.; Pearce, P.; Smith, R.; Smith, A. Mineralocorticoid action: Target tissue specificity is enzyme, not receptor, mediated. *Science* **1988**, *242*, 583–585. [CrossRef] [PubMed]

62. Ferrari, P.; Sansonnens, A.; Dick, B.; Frey, F.J. In vivo 11 -hsd-2 activity: Variability, salt-sensitivity, and effect of licorice. *Hypertension* **2001**, *38*, 1330–1336. [CrossRef] [PubMed]

63. Pizzolo, F.; Friso, S.; Morandini, F.; Antoniazzi, F.; Zaltron, C.; Udali, S.; Gandini, A.; Cavarzere, P.; Salvagno, G.; Giorgetti, A.; et al. Apparent mineralocorticoid excess by a novel mutation and epigenetic modulation by HSD11B2 promoter methylation. *J. Clin. Endocrinol. Metab.* **2015**, *100*, E1234–E1241. [CrossRef] [PubMed]

64. Xu, J.C.; Lytle, C.; Zhu, T.T.; Payne, J.A.; Benz, E., Jr.; Forbush, B., 3rd. Molecular cloning and functional expression of the bumetanide-sensitive Na-K-Cl cotransporter. *Proc. Natl. Acad. Sci. USA* **1994**, *91*, 2201–2205. [CrossRef] [PubMed]

65. Meyer, J.W.; Flagella, M.; Sutliff, R.L.; Lorenz, J.N.; Nieman, M.L.; Weber, C.S.; Paul, R.J.; Shull, G.E. Decreased blood pressure and vascular smooth muscle tone in mice lacking basolateral na(+)-k(+)-2cl(-) cotransporter. *Am. J. Physiol. Heart Circ. Physiol.* **2002**, *283*, H1846–H1855. [CrossRef] [PubMed]

66. Garg, P.; Martin, C.F.; Elms, S.C.; Gordon, F.J.; Wall, S.M.; Garland, C.J.; Sutliff, R.L.; O'Neill, W.C. Effect of the Na-K-2Cl cotransporter NKCC1 on systemic blood pressure and smooth muscle tone. *Am. J. Physiol. Heart Circ. Physiol.* **2007**, *292*, H2100–H2105. [CrossRef] [PubMed]

67. Mendelsohn, M.E.; Karas, R.H. The protective effects of estrogen on the cardiovascular system. *N. Engl. J. Med.* **1999**, *340*, 1801–1811. [CrossRef] [PubMed]

68. Li, L.; Gao, P.; Zhang, H.; Chen, H.; Zheng, W.; Lv, X.; Xu, T.; Wei, Y.; Liu, D.; Liang, C. Sirt1 inhibits angiotensin ii-induced vascular smooth muscle cell hypertrophy. *Acta Biochim. Biophys. Sin.* **2011**, *43*, 103–109. [CrossRef] [PubMed]

69. Gao, P.; Xu, T.T.; Lu, J.; Li, L.; Xu, J.; Hao, D.L.; Chen, H.Z.; Liu, D.P. Overexpression of sirt1 in vascular smooth muscle cells attenuates angiotensin ii-induced vascular remodeling and hypertension in mice. *J. Mol. Med.* **2014**, *92*, 347–357. [CrossRef] [PubMed]

70. Brozovich, F.V.; Nicholson, C.J.; Degen, C.V.; Gao, Y.Z.; Aggarwal, M.; Morgan, K.G. Mechanisms of vascular smooth muscle contraction and the basis for pharmacologic treatment of smooth muscle disorders. *Pharmacol. Rev.* **2016**, *68*, 476–532. [CrossRef] [PubMed]

71. Liu, R.; Leslie, K.L.; Martin, K.A. Epigenetic regulation of smooth muscle cell plasticity. *Biochim. Biophys. Acta* **2015**, *1849*, 448–453. [CrossRef] [PubMed]

72. Levy, E.; Spahis, S.; Bigras, J.L.; Delvin, E.; Borys, J.M. The epigenetic machinery in vascular dysfunction and hypertension. *Curr. Hypertens. Rep.* **2017**, *19*, 52. [CrossRef] [PubMed]

73. Qiu, P.; Li, L. Histone acetylation and recruitment of serum responsive factor and creb-binding protein onto sm22 promoter during sm22 gene expression. *Circ. Res.* **2002**, *90*, 858–865. [CrossRef] [PubMed]

74. Cao, D.; Wang, C.; Tang, R.; Chen, H.; Zhang, Z.; Tatsuguchi, M.; Wang, D.Z. Acetylation of myocardin is required for the activation of cardiac and smooth muscle genes. *J. Biol. Chem.* **2012**, *287*, 38495–38504. [CrossRef] [PubMed]

75. Cao, D.; Wang, Z.; Zhang, C.L.; Oh, J.; Xing, W.; Li, S.; Richardson, J.A.; Wang, D.Z.; Olson, E.N. Modulation of smooth muscle gene expression by association of histone acetyltransferases and deacetylases with myocardin. *Mol. Cell. Biol.* **2005**, *25*, 364–376. [CrossRef] [PubMed]

76. Bauer, A.J.; Martin, K.A. Coordinating regulation of gene expression in cardiovascular disease: Interactions between chromatin modifiers and transcription factors. *Front. Cardiovasc. Med.* **2017**, *4*, 19. [CrossRef] [PubMed]

77. Zhou, N.; Lee, J.J.; Stoll, S.; Ma, B.; Wiener, R.; Wang, C.; Costa, K.D.; Qiu, H. Inhibition of SRF/myocardin reduces aortic stiffness by targeting vascular smooth muscle cell stiffening in hypertension. *Cardiovasc. Res.* **2017**, *113*, 171–182. [CrossRef] [PubMed]

78. Zhou, N.; Lee, J.J.; Stoll, S.; Ma, B.; Costa, K.D.; Qiu, H. Rho kinase regulates aortic vascular smooth muscle cell stiffness via actin/SRF/myocardin in hypertension. *Cell. Physiol. Biochem.* **2017**, *44*, 701–715. [CrossRef] [PubMed]

79. Siti, H.N.; Kamisah, Y.; Kamsiah, J. The role of oxidative stress, antioxidants and vascular inflammation in cardiovascular disease (a review). *Vasc. Pharmacol.* **2015**, *71*, 40–56. [CrossRef] [PubMed]

80. Kang, T.B.; Yang, S.H.; Toth, B.; Kovalenko, A.; Wallach, D. Activation of the NLRP3 inflammasome by proteins that signal for necroptosis. *Methods Enzymol.* **2014**, *545*, 67–81. [PubMed]

81. Krishnan, S.M.; Sobey, C.G.; Latz, E.; Mansell, A.; Drummond, G.R. Il-1β and il-18: Inflammatory markers or mediators of hypertension? *Br. J. Pharmacol.* **2014**, *171*, 5589–5602. [CrossRef] [PubMed]

82. Sandoo, A.; van Zanten, J.J.; Metsios, G.S.; Carroll, D.; Kitas, G.D. The endothelium and its role in regulating vascular tone. *Open Cardiovasc. Med. J.* **2010**, *4*, 302–312. [CrossRef] [PubMed]

83. Nadar, S.; Blann, A.; Lip, G. Endothelial dysfunction: Methods of assessment and application to hypertension. *Curr. Pharm. Des.* **2004**, *10*, 3591–3605. [CrossRef] [PubMed]

84. Han, S.; Uludag, M.O.; Usanmaz, S.E.; Ayaloglu-Butun, F.; Akcali, K.C.; Demirel-Yilmaz, E. Resveratrol affects histone 3 lysine 27 methylation of vessels and blood biomarkers in doca salt-induced hypertension. *Mol. Biol. Rep.* **2015**, *42*, 35–42. [CrossRef] [PubMed]

85. Smolarek, I.; Wyszko, E.; Barciszewska, A.M.; Nowak, S.; Gawronska, I.; Jablecka, A.; Barciszewska, M.Z. Global DNA methylation changes in blood of patients with essential hypertension. *Med. Sci. Monit.* **2010**, *16*, CR149–CR155. [PubMed]

86. Kato, N.; Loh, M.; Takeuchi, F.; Verweij, N.; Wang, X.; Zhang, W.; Kelly, T.N.; Saleheen, D.; Lehne, B.; Leach, I.M.; et al. Trans-ancestry genome-wide association study identifies 12 genetic loci influencing blood pressure and implicates a role for DNA methylation. *Nat. Genet.* **2015**, *47*, 1282–1293. [CrossRef] [PubMed]

87. Staessen, J.A.; Bianchi, G. Adducin and hypertension. *Pharmacogenomics* **2005**, *6*, 665–669. [CrossRef] [PubMed]

88. Brown, S.E.; Campbell, R.D.; Sanderson, C.M. Novel NG36/G9a gene products encoded within the human and mouse mhc class iii regions. *Mamm. Genome* **2014**, *12*, 916–924. [CrossRef]

89. Lehnertz, B.; Northrop, J.P.; Antignano, F.; Burrows, K.; Hadidi, S.; Mullaly, S.C.; Rossi, F.M.; Zaph, C. Activating and inhibitory functions for the histone lysine methyltransferase G9a in T helper cell differentiation and function. *J. Exp. Med.* **2010**, *207*, 915–922. [CrossRef] [PubMed]

90. Handy, D.E.; Castro, R.; Loscalzo, J. Epigenetic modifications: Basic mechanisms and role in cardiovascular disease. *Circulation* **2011**, *123*, 2145–2156. [CrossRef] [PubMed]

91. Sherman, R.C.; Langley-Evans, S.C. Early administration of angiotensin-converting enzyme inhibitor captopril, prevents the development of hypertension programmed by intrauterine exposure to a maternal low-protein diet in the rat. *Clin. Sci.* **1998**, *94*, 373–381. [CrossRef] [PubMed]

92. Jackson, A.A.; Dunn, R.L.; Marchand, M.C.; Langley-Evans, S.C. Increased systolic blood pressure in rats induced by a maternal low-protein diet is reversed by dietary supplementation with glycine. *Clin. Sci.* **2002**, *103*, 633–639. [CrossRef] [PubMed]

93. Lillycrop, K.A.; Slater-Jefferies, J.L.; Hanson, M.A.; Godfrey, K.M.; Jackson, A.A.; Burdge, G.C. Induction of altered epigenetic regulation of the hepatic glucocorticoid receptor in the offspring of rats fed a protein-restricted diet during pregnancy suggests that reduced DNA methyltransferase-1 expression is involved in impaired DNA methylation and changes in histone modifications. *Br. J. Nutr.* **2007**, *97*, 1064–1073. [PubMed]

94. Alexander, B.T.; Ojeda, N.B. Prenatal inflammation and the early origins of hypertension. *Clin. Exp. Pharmacol. Physiol.* **2008**, *35*, 1403–1404. [CrossRef] [PubMed]

95. Samuelsson, A.M.; Alexanderson, C.; Molne, J.; Haraldsson, B.; Hansell, P.; Holmang, A. Prenatal exposure to interleukin-6 results in hypertension and alterations in the renin-angiotensin system of the rat. *J. Physiol.* **2006**, *575*, 855–867. [CrossRef] [PubMed]

96. Hao, X.Q.; Zhang, H.G.; Li, S.H.; Jia, Y.; Liu, Y.; Zhou, J.Z.; Wei, Y.L.; Hao, L.Y.; Tang, Y.; Su, M.; et al. Prenatal exposure to inflammation induced by zymosan results in activation of intrarenal renin-angiotensin system in adult offspring rats. *Inflammation* **2010**, *33*, 408–414. [CrossRef] [PubMed]

97. Liao, W.; Wei, Y.; Yu, C.; Zhou, J.; Li, S.; Pang, Y.; Li, G.; Li, X. Prenatal exposure to zymosan results in hypertension in adult offspring rats. *Clin. Exp. Pharmacol. Physiol.* **2008**, *35*, 1413–1418. [CrossRef] [PubMed]

98. Hao, X.Q.; Zhang, H.G.; Yuan, Z.B.; Yang, D.L.; Hao, L.Y.; Li, X.H. Prenatal exposure to lipopolysaccharide alters the intrarenal renin-angiotensin system and renal damage in offspring rats. *Hypertens. Res.* **2010**, *33*, 76–82. [CrossRef] [PubMed]

99. Wei, Y.L.; Li, X.H.; Zhou, J.Z. Prenatal exposure to lipopolysaccharide results in increases in blood pressure and body weight in rats. *Acta Pharmacol. Sin.* **2007**, *28*, 651–656. [CrossRef] [PubMed]

100. Wang, J.; Yin, N.; Deng, Y.; Wei, Y.; Huang, Y.; Pu, X.; Li, L.; Zheng, Y.; Guo, J.; Yu, J.; et al. Ascorbic acid protects against hypertension through downregulation of ace1 gene expression mediated by histone deacetylation in prenatal inflammation-induced offspring. *Sci. Rep.* **2016**, *6*, 39469. [CrossRef] [PubMed]

101. Kuzawa, C.W.; Sweet, E. Epigenetics and the embodiment of race: Developmental origins of us racial disparities in cardiovascular health. *Am. J. Hum. Biol.* **2009**, *21*, 2–15. [CrossRef] [PubMed]

102. Jirtle, R.L.; Skinner, M.K. Environmental epigenomics and disease susceptibility. *Nat. Rev. Genet.* **2007**, *8*, 253–262. [CrossRef] [PubMed]

103. Chen, M.; Zhang, L. Epigenetic mechanisms in developmental programming of adult disease. *Drug Discov. Today* **2011**, *16*, 1007–1018. [CrossRef] [PubMed]

104. Kim, M.S.; Kondo, T.; Takada, I.; Youn, M.Y.; Yamamoto, Y.; Takahashi, S.; Matsumoto, T.; Fujiyama, S.; Shirode, Y.; Yamaoka, I.; et al. DNA demethylation in hormone-induced transcriptional derepression. *Nature* **2009**, *461*, 1007–1012. [CrossRef] [PubMed]

105. Li, J.; Wang, J.; Wang, J.; Nawaz, Z.; Liu, J.M.; Qin, J.; Wong, J. Both corepressor proteins smrt and n-cor exist in large protein complexes containing hdac3. *EMBO J.* **2000**, *19*, 4342–4350. [CrossRef] [PubMed]

106. Kao, H.Y.; Downes, M.; Ordentlich, P.; Evans, R.M. Isolation of a novel histone deacetylase reveals that class i and class ii deacetylases promote smrt-mediated repression. *Gene Dev.* **2000**, *14*, 55–66. [PubMed]

107. Tsukada, Y.; Fang, J.; Erdjument-Bromage, H.; Warren, M.E.; Borchers, C.H.; Tempst, P.; Zhang, Y. Histone demethylation by a family of jmjc domain-containing proteins. *Nature* **2006**, *439*, 811–816. [CrossRef] [PubMed]

108. Metzger, E.; Wissmann, M.; Yin, N.; Muller, J.M.; Schneider, R.; Peters, A.H.; Gunther, T.; Buettner, R.; Schule, R. LSD1 demethylates repressive histone marks to promote androgen-receptor-dependent transcription. *Nature* **2005**, *437*, 436–439. [CrossRef] [PubMed]

109. Yamane, K.; Toumazou, C.; Tsukada, Y.; Erdjument-Bromage, H.; Tempst, P.; Wong, J.; Zhang, Y. JHDM2A, a JmjC-containing H3K9 demethylase, facilitates transcription activation by androgen receptor. *Cell* **2006**, *125*, 483–495. [CrossRef] [PubMed]

International Journal of
Molecular Sciences

Review

Comparative Genomics and Transcriptome Profiling in Primary Aldosteronism

Elke Tatjana Aristizabal Prada [1], Isabella Castellano [2], Eva Sušnik [1], Yuhong Yang [1], Lucie S. Meyer [1], Martina Tetti [3], Felix Beuschlein [1,4], Martin Reincke [1] and Tracy A. Williams [1,3,*]

[1]   Medizinische Klinik und Poliklinik IV, Klinikum der Universität, Ludwig-Maximilians-Universität München, 80336 Munich, Germany; elke_tatjana_aristizabal.prada@med.uni-muenchen.de (E.T.A.P.); Eva.Susnik@med.uni-muenchen.de (E.S.); Yuhong.Yang@med.uni-muenchen.de (Y.Y.); Lucie.Meyer@med.uni-muenchen.de (L.S.M.); Felix.Beuschlein@med.uni-muenchen.de (F.B.); Martin.Reincke@med.uni-muenchen.de (M.R.)

[2]   Division of Pathology, Department of Medical Sciences, University of Torino, 10124 Torino, Italy; isabella.castellano@unito.it

[3]   Division of Internal Medicine and Hypertension, Department of Medical Sciences, University of Torino, 10126 Torino, Italy; marty_queen@alice.it

[4]   Klinik für Endokrinologie, Diabetologie und Klinische Ernährung, UniversitätsSpital Zürich, CH-8091 Zurich, Switzerland

*   Correspondence: Tracy.Williams@med.uni-muenchen.de; Tel.: +49-089-440-052-941

Received: 15 February 2018; Accepted: 6 April 2018; Published: 9 April 2018

Abstract: Primary aldosteronism is the most common form of endocrine hypertension with a prevalence of 6% in the general population with hypertension. The genetic basis of the four familial forms of primary aldosteronism (familial hyperaldosteronism FH types I–IV) and the majority of sporadic unilateral aldosterone-producing adenomas has now been resolved. Familial forms of hyperaldosteronism are, however, rare. The sporadic forms of the disease prevail and these are usually caused by either a unilateral aldosterone-producing adenoma or bilateral adrenal hyperplasia. Aldosterone-producing adenomas frequently carry a causative somatic mutation in either of a number of genes with the *KCNJ5* gene, encoding an inwardly rectifying potassium channel, a recurrent target harboring mutations at a prevalence of more than 40% worldwide. Other than genetic variations, gene expression profiling of aldosterone-producing adenomas has shed light on the genes and intracellular signalling pathways that may play a role in the pathogenesis and pathophysiology of these tumors.

Keywords: primary aldosteronism; aldosterone; aldosterone-producing adenoma; transcriptome profiing

1. Introduction

Primary aldosteronism (PA) is the most common potentially curable form of hypertension with a prevalence of 5–10% in patients with hypertension [1–3] and is characterized by the excessive production of aldosterone. Aldosterone excess has detrimental effects independent of blood pressure control as demonstrated by the increased risk of cardiovascular and cerebrovascular events and target organ damage in patients with PA relative to matched patients with essential hypertension. PA is mainly caused by either an aldosterone-producing adenoma (APA) or bilateral adrenal hyperplasia (BAH). The diagnosis of an APA is appealing because hypertension can be cured or markedly improved by unilateral adrenalectomy and resolve the aldosterone excess in the majority of cases [4]. In contrast, patients with BAH are usually treated with a mineralocorticoid receptor antagonist but plasma renin levels should be monitored as a therapeutic response (as well as blood pressure) because suppressed renin, independent of blood pressure control, is associated with an increased risk of cardiometabolic events and death relative to patients with essential hypertension [5]. A landmark in PA research was the discovery of germline and somatic mutations that drive the aldosterone overproduction in patients with PA, discoveries that were made possible by the application of next-generation sequencing [6–12]. The genetic basis of four familial forms of PA has now been described as well as somatic mutations in APAs and these are discussed in more detail below.

Herein, we outline novel contributions and major discoveries that have been made in the field of PA research over the last few years. We describe the genetic basis of familial forms of hyperaldosteronism and the identification of somatic mutations that lead to excess aldosterone production. Differential gene expression profiles between APAs and reference tissues are highlighted, as well as key signalling pathways and molecular mechanisms that may drive cell proliferation and constitutive aldosterone production in APAs. Additionally, the possible influence of genetics and genomics on surgical outcome and the potential application of next-generation sequencing methods and transcriptome profiling as possible prognostic tools are described.

2. Synthesis of Aldosterone

The primary function of aldosterone is to maintain fluid and electrolyte balance for the control of blood pressure. The main physiological regulators of aldosterone synthesis are angiotensin II, potassium and adrenocorticotropic hormone. Aldosterone synthesis is restricted to the *zona glomerulosa* (ZG), the outer layer of the adrenal cortex (Figure 1) where aldosterone synthase converts deoxycorticosterone to aldosterone by three successive steps of 11β-hydroxylation, 18-hydroxylation and 18-oxidation by a single enzyme, aldosterone synthase (encoded by *CYP11B2*). *CYP11B2* displays a high level of intron and exon sequence homology to the *CYP11B1* gene localized in the *zona fasciculata* (ZF) that encodes 11β-hydroxylase that catalyses the final step in the conversion of 11-deoxycortisol to cortisol (Figure 1). Angiotensin II and potassium regulate aldosterone production via Ca^{2+} signalling which also plays a key role in the aldosterone excess in PA due to the somatic and germline mutations in ion channels and transporters.

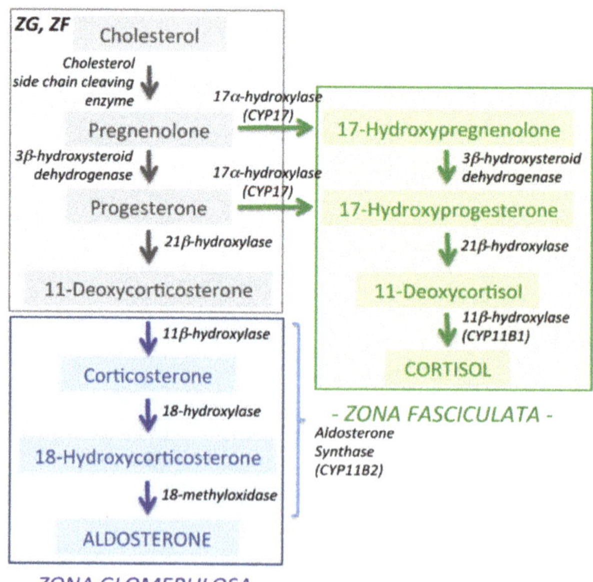

Figure 1. Aldosterone synthesis in the adrenal cortex. Aldosterone is synthesized in the *zona glomerulosa* (ZG) distinct from the synthesis of cortisol in the *zona fasciculata* (ZF). Aldosterone synthase encoded by *CYP11B2* performs all three enzymatic steps in the conversion of deoxycorticosterone to aldosterone.

3. Familial Forms of Hyperaldosteronism

There are currently 4 recognised forms of familial hyperaldosteronism (FH types I–IV) and the genetic basis of each type is summarized in Table 1.

Table 1. Familial forms of hyperaldosteronism.

Subtype of Primary Aldosteronism	Genetic Variant	Encoded Protein	Brief Description
FH Type I	*CYP11B1/CYP11B2* hybrid gene	CYP11B2	Ectopic expression in ZF; regulated by ACTH
FH Type II	*CLCN2* mutations	CIC-2	Chloride voltage-gated channel 2
FH Type III	*KCNJ5* mutations	GIRK4	Potassium Voltage-Gated Channel Subfamily J Member 5
FH Type IV	*CACNA1H* mutations	Cav3.2	Calcium Voltage-Gated Channel Subunit α1H

3.1. Familial Hyperaldosteronism Type I

Familial hyperaldosteronism type I (FH type I or GRA, glucocorticoid remediable aldosteronism) is caused by a hybrid *CYP11B1/CYP11B2* gene inherited as an autosomal dominant trait. The hybrid gene results from an asymmetrical crossing over between the highly homologous *CYP11B1* (encoding 11β-hydroxylase) and *CYP11B2* (encoding aldosterone synthase) genes and comprises 5′ sequences of *CYP11B1* (including the promoter region) and 3′ sequences of *CYP11B2* (including the coding region of aldosterone synthase). Thus, in FH type I, aldosterone synthase is ectopically expressed in the ZF under the control of adrenocorticotropic hormone (ACTH) rather than restricted to the ZG under the control of angiotensin II [13,14].

3.2. Familial Hyperaldosteronism Type II

Familial hyperaldosteronism type II (FH type II) was first described by Stowasser et al. [15] in a kindred with an autosomal dominant form of PA. Other kindreds were subsequently described and a linkage with a locus on chromosome 7p22 was reported in some but not all families but sequencing the entire linked locus did not identify the genetic cause [16]. In 2018, Scholl et al. [17] identified the genetic variant responsible in the original kindred with FH type II described by Michael Stowasser as a heterozygous variant of the *CLCN2* gene that caused early-onset primary aldosteronism and hypertension often with hypokalaemia. *CLCN2* encodes CIC-2, a homodimer voltage-gated chloride channel expressed in the adrenal gland predominantly in the ZG [17]. In the original family with FH type II, eight individuals were carriers of the *CLCN2* mutation (resulting in the CIC-2 p.Arg172Gln substitution) and of these, seven tested positive with a screening test for primary aldosteronism (elevated aldosterone-to-renin ratio). One carrier for the CIC-2 p.Arg172Gln variant had a normal aldosterone-to-renin ratio, and therefore did not have primary aldosteronism, indicating an incomplete penetrance of the allele. Scholl et al. found the p.Arg172Gln substitution in an additional kindred and two further cases of p.Arg172Gln mutations (1 occurring de novo) in 2 unrelated patients with early-onset PA [17] as well as other *CLCN2* variants encoding 4 different mutations in CIC-2 (a de novo p.Met22Lys mutation, p.Tyr26Asn, p.Ser865Arg and p.Lys362del). At the same time, Fernandes-Rosa et al., reported a de novo heterozygous p.Gly24Asp mutation in the CIC-2 chloride channel associated with PA [18]. Electrophysiological recordings showed that the mutated CIC-2 channels display modified gating resulting in increased chloride efflux compared with wild-type CIC-2 channels. The increased chloride efflux leads to depolarization of adrenocortical cells, activation of voltage-dependent Ca^{2+} channels, Ca^{2+} influx, increased *CYP11B2* gene expression and aldosterone production.

3.3. Familial Hyperaldosteronism Type III

Choi et al. identified the genetic basis of familial hyperaldosteronism type III (FH type III) in 2011 by next-generation sequencing [6]. A gain-of-function mutation in the *KCNJ5* gene (encoding the G-protein-coupled inwardly rectifying potassium channel GIRK4) was identified in the male index case and his two daughters. The mutation results in the substitution of a threonine residue (p.Thr158Ala) located just above the selectivity filter of the channel pore which interferes with the Thr158-Pro128 hydrogen bonding [6]. Patch clamp recordings of human embryonic kidney cells expressing the mutated GIRK4 p.Thr158Ala channel showed that the mutation results in a loss of selectivity for K^+ and permissively allows the passage of Na^+ resulting in membrane depolarization [6]. In adrenal cells, membrane depolarization leads to the opening of voltage gated Ca^{2+} channels and Ca^{2+} influx activating the Ca^{2+} signalling pathway and aldosterone production. Expression of GIRK4 p.Thr158Ala in the human adrenocortical carcinoma cell line (HAC15) caused a marked increase in aldosterone secretion that was dependent on membrane depolarization and Na^+ and Ca^{2+} influx [19]. Until 2017, 22 patients with FH type III were described in the literature from 12 families [20]. Notable is that over half of the patients described with FH type III (14 of 22 cases occurring in 7 of 12 families) carried mutations of the Gly151 residue (p.Gly151Glu or p.Gly151Arg) in the GlyTyrGly motif implicated in K^+ selectivity [21].

3.4. Familial Hyperaldosteronism Type IV

Familial hyperaldosteronism type IV (FH type IV) is caused by gain-of-function mutations in Cav3.2, a T type Ca^{2+} channel encoded by *CACNA1H*. FH type IV was first identified in 2015 by exome sequencing of 40 unrelated subjects with early-onset hyperaldosteronism and hypertension (<10 years of age) [22]. Scholl et al. identified five subjects with the same heterozygous mutation in *CACNA1H* encoding the Ca^{2+} voltage gated channel ($Ca_v3.2$) resulting in a $Ca_v3.2$ p.Met1549Val substitution [16]. Comparisons of whole cell patch clamp recordings of $Ca_v3.2$ p.Met1549Val and wild-type $Ca_v3.2$ expressed in human embryonic kidney cells showed that the p.Met1549Val mutation causes an impairment of channel activation and inactivation. The mutant channel displayed slightly slower activation and much slower inactivation time constants compared with the wild-type channel as well as a tail current indicating that a proportion of the mutated channels remain non-inactivated. These properties would lead to an increase in Ca^{2+} influx in adrenal ZG cells and signal an increase in aldosterone production. Validation of this concept was subsequently demonstrated by the same group by expression of the $Ca_v3.2$ p.Met1549Val mutation in human adrenocortical (HAC15) cells which resulted in an increase in *CYP11B2* gene transcription and aldosterone secretion relative to cells expressing the wild-type channel [23]. Following the discovery by Scholl, additional mutations in *CACNA1H* were described involving a substitution of the same Met1549 residue (Met1549Ile) or other amino acid residues (Ser196Leu, Val1951Glu and Pro2083Leu) [24].

4. Somatic Mutations in Aldosterone-Producing Adenomas

The most frequent genetic variation in APA is a somatic mutation of the *KCNJ5* gene [25]. First identified by Choi et al. in 2011 by exome sequencing, mutations in *KCNJ5* were identified in 8 of 22 APA resulting in GIRK p.Gly151Arg or p.Leu168Arg mutations [6]. Both mutations were demonstrated to interfere with the selectivity filter of the channel pore and result in membrane depolarization causing the opening of voltage gated Ca^{2+} channels in adrenal glomerulosa cells and Ca^{2+} influx [6]. Somatic *KCNJ5* mutations are found at a prevalence of 40–50% [26–30] although a higher prevalence has been reported in populations from Japan and China [31,32]. Following the description of the *KCNJ5* mutations, the application of next generation sequencing rapidly identified additional somatic mutations associated with aldosterone overproduction in sporadic PA. These include heterozygous gain-of-function mutations in $Ca_v1.3$ (the α1D subunit of the L-type voltage-dependent calcium channel) encoded by *CACNA1D* [8,9] and the ion transporters,

Na$^+$/K$^+$-ATPase (encoded by *ATP1A1*) and Ca^{2+}-ATPase (encoded by *ATP2B3*) [7,8]. These mutations result in an increase in intracellular Ca^{2+} concentration thereby causing an increase in transcription of the *CYP11B2* gene that encodes aldosterone synthase. Activating mutations in exon 3 of *CTNNB1* that encodes β-catenin have been identified in APAs as well as in other adrenal tumours [33]. Despite these major advances, the mechanisms underlying the deregulated cell growth of APAs are probably not explained by somatic mutations and the GIRK4 Thr158Ala mutation does not enhance proliferation of adrenal cells in vitro [19]. Herein we discuss the transcriptome studies that have identified genes and signalling pathways that may function in the pathophysiology and pathogenesis of APA.

5. Gene Expression Profiling

Gene expression studies have identified genes with a potential role in the pathogenesis and pathophysiology of APAs (Table 2). Despite inter-study heterogeneity of gene expression data, which may be accounted for by the use of different reference tissues (adjacent cortex or normal adrenal tissue or, in some cases, non-functioning adrenocortical adenomas) and different diagnostic criteria [34–38], such studies have proven valuable in the identification of genes and signalling pathways with a potential role in the pathogenesis and pathophysiology of APAs.

Gene expression studies employing microarrays have shown a higher expression of *CYP11B2* in APA compared with normal adrenals [39–41] and by SAGE (serial analysis of gene expression) in an APA compared with adjacent cortical tissue [42]. However, another study of APA transcriptomes reported two distinct and opposing expression profiles for genes encoding steroidogenic enzymes with *CYP11B2* displaying increased or decreased expression levels with respect to normal adrenal tissue [43]. This apparent paradoxical decrease of the gene expressing aldosterone synthase in a tumour that overexpresses aldosterone may be accounted for by sampling areas of the normal adrenal reference tissue. In fact, these may contain aldosterone-producing cell clusters (APCC) that express high levels of *CYP11B2* with somatic *CACNA1D*, *ATP1A1* or *ATP2B3* mutations [44–46]. Conversely large APA with low expression of *CYP11B2* that give rise to inappropriate aldosterone production might occur. Another possibility is that non-APA nodules were used in the study due to non-selective diagnostic criteria.

Many studies have described an association of somatic APA mutations with histological phenotype. APAs carrying *KCNJ5* mutations have been widely reported to comprise predominantly large lipid-rich ZF-like cells (Figure 2) [8,47–50]. Some studies have also described a predominance of small compact ZG-like cells in APA harbouring *CACNA1D*, *ATP1A1* or *ATP2B3* mutations [8,27,49,51] and somatic APA genotype is associated with plasma steroid profiles [52]. Such genotype-phenotype associations indicate that APA genotype may influence transcriptome signatures. Histological differences between large lipid-rich ZF-like cells and small compact ZG-like cells in APA are shown in Figure 2.

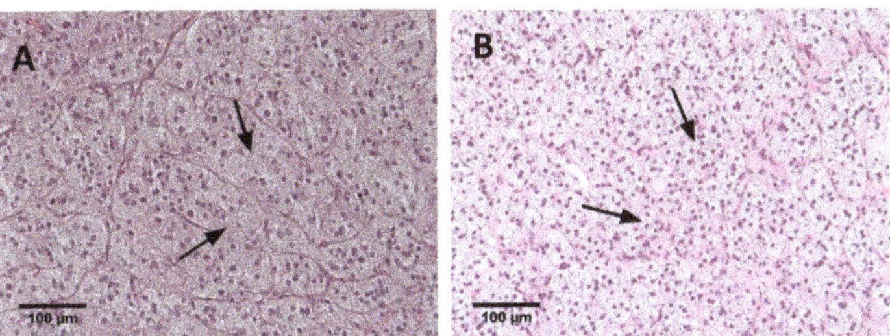

Figure 2. Histopathological phenotype of aldosterone-producing adenomas. Haematoxylin and eosin staining of an aldosterone-producing adenoma showing large lipid-rich cells of the ZF type (indicated with arrows) (panel **A**) or a predominance of smaller compact cells of the ZG type (indicated with arrows) (panel **B**).

No differences in the transcriptome profiles of APA with and without *KCNJ5* mutations were initially described [26]. However, later studies reported distinct expression profiles of APA with *KCNJ5* mutations compared with APA without *KCNJ5* mutations (with higher *CYP11B2* expression in the tumours with *KCNJ5* mutations) [53]. Different expression profiles were reported in APA with *ATP1A1* and *ATP2B3* mutations relative to APA with *KCNJ5* mutations (with higher *CYP11B2* expression in the tumors with *ATP1A1* and *ATP2B3* mutations) [54]. Azizan et al. [35] demonstrated marked differences in *CYP17A1* gene expression from microarrays, validated by real-time PCR, in APA with a ZF phenotype compared to those APA with a ZG phenotype [47]. If *CYP17A1* and *CYP11B2* are expressed in the same cell then cortisol can be metabolized further to produce the hybrid steroids 18-hydroxycortisol and 18-oxocortisol [55]. Higher levels of these hybrid steroids are associated with FH type I and FH type III (although not in all cases) and in patients with an APA with a *KCNJ5* mutation [12,52].

CYP17 expression in APA has been shown to be associated with APA phenotype with marked upregulation in adenomas comprising predominantly ZF-type cells [47]. *NURR1* (*NR4A2*, encoding Nur-related factor 1) and *NGFIB* (*NR4A1*, encoding nerve growth factor IB), genes that encode transcription factors playing a key role in the regulation of *CYP11B2* gene transcription [56], are upregulated in APA. Also, genes encoding the nuclear receptor transcription factors SF-1 (*NR5A1*) and DAX1 (*NR0B1*) that are essential for adrenal development and steroidogenesis, are upregulated in APA [39]. Although low DAX1 expression in adrenocortical tumours is associated with aldosterone production [57]. A target gene of SF-1, *VSNL1* [58], is upregulated in APA and *VSNL1* in vitro overexpression in the NCI H295R cell line results in an increase in aldosterone production under both basal and angiotensin II-stimulated conditions [59].

Several genes encoding G-protein-coupled receptors are among the genes upregulated in APA, including those encoding the luteinizing hormone receptor (LH-R encoded by *LHCGR*), gonadotropin releasing hormone receptor (GnRHR encoded by *GNRHR*), serotonin receptor 4 (HTR4), melanocortin 2 receptor (MC2R), and the angiotensin II type 1 receptor (AGTR1) [60]. Overexpression of LH-R in the adrenocortical carcinoma NCI H295R cell line causes a concentration-dependent increase in *CYP11B2* expression after stimulation with luteinizing hormone [40]. Accordingly, the expression of *LH-R* and *GnRHR* in APAs has been proposed to be related to increased aldosterone production during pregnancy [61]. Therefore, the presence of activating APA *CTNNB1* mutations might contribute to an abnormal receptor activation [60].

NEFM, encoding the medium neurofilament protein, is highly upregulated only in APAs without *KCNJ5* mutations and is selectively expressed in the ZG and in APA comprised of predominantly ZG cells [62,63]. Dopamine regulates aldosterone production via activation of its G-protein-coupled receptor (GPCR) subtypes and silencing of *NEFM* amplified aldosterone stimulation by a DR1 (dopamine receptor subunit 1) agonist and aldosterone secretion in response to the DR1 agonist was greater in primary cultures of APAs composed of primarily ZF cells compared with cultures of APAs with ZG cells. These data indicate a role for *NEFM* in aldosterone production and cell proliferation [63].

Analysis of the methylome of APAs demonstrated hypomethylation of GPCR genes and a strong association of promoter hypomethylation of the HTR4 and PTGER1 genes with the upregulation of mRNA levels, validated by real-time PCR, was demonstrated in APAs compared with non-functioning adrenocortical adenomas [64]. Methylation of HTR4 and PTGER1 was significantly inversely correlated with their respective mRNA expression levels [64]. The most hypomethylated promoter in APA is the *PCP4* (encoding purkinje cell protein 4) promoter with demethylation associated with enhanced gene transcription [65]. *CYP11B2* was also extensively hypomethylated in APAs [64] but although hypomethylation was not associated with gene expression levels in this study it could facilitate gene transcription [64]. In contrast, Howard et al., reported hypomethylation of APAs with hypomethylation and overexpression of *CYP11B2* [64].

Calcium is a key intracellular messenger for aldosterone production and the intracellular Ca^{2+} signaling pathway is independent of the renin–angiotensin–aldosterone system in APAs [66]. A number of genes involved in Ca^{2+} signaling or Ca^{2+} sequestration have been reported as upregulated in APAs and are described in more detail below. *VSNL1* that encodes a Ca^{2+}-sensor protein and a target of the nuclear receptor SF-1 [58] was one of several upregulated genes in APAs by microarray analysis compared with normal adrenals validated by real-time PCR [41]. In NCI H295R adrenal cells, overexpression of *VSNL1* resulted in an upregulation of *CYP11B2* gene expression under both basal and angiotensin II-stimulated conditions thereby implicating a role for *VSNL1* in aldosterone production. Analysis of a larger sample set of tumours showed that *VSNL1* was overexpressed in APAs carrying a *KCNJ5* mutation compared with those APA without a *KCNJ5* mutation. A potential role for the calcium sensor in the protection of cells in an adenoma via Ca^{2+}-related anti-apoptotic cell death mechanisms was hypothesized [59]. The expression of the VSNL1 protein in an APA (carrying a *KCNJ5* mutation) that displays strong CYP11B2 immunostaining is shown in Figure 3.

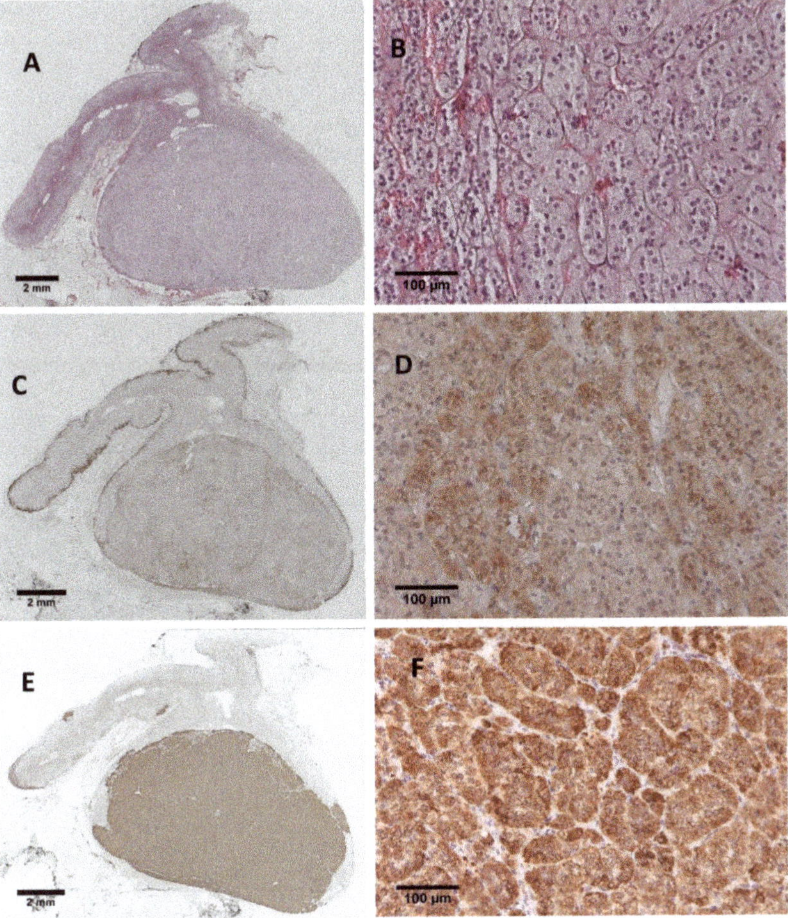

Figure 3. VSNL1 and CYP11B2 immunohistochemistry. An aldosterone-producing adenoma with a *KCNJ5* mutation stained with haematoxylin and eosin panels (**A,B**); immunostained for VSNL1, panels (**C,D**); and for CYP11B2 panels (**E,F**). The VSNL1 antibody was from Merck and the CYP11B2 was a kind gift from Prof Celso Gomez-Sanchez, University of Mississippi, Oxford, MS, USA.

The *CALN1* gene, that encodes the Ca^{2+} binding protein calneuron 1, has been reported as upregulated in APA in two transcriptome studies [41,67]. CALN1 was shown to potentiate aldosterone production and silencing *CALN1* led to a decrease in Ca^{2+} storage in the endoplasmic reticulum and abrogated angiotensin II-mediated aldosterone secretion in an adrenocortical carcinoma cell line [41].

CALM2 encoding calmodulin 2 is a Ca^{2+}-binding protein expressed in a wide-range of tissues involved in signalling, cell cycle progression and proliferation. *CALM2* was highly upregulated in a transcriptome comparison of APAs with the adjacent ZG [42]. The increased expression of *PCP4* in APA cells is likely to play a role in APA pathophysiology because PCP4 modulates Ca^{2+}-binding by calmodulin and activates the calcium-calmodulin cascade leading to an increased expression of *CYP11B2* [68].

GSTA1 (encoding glutathione-S-transferase, an enzyme that protects cells from reactive oxygen species, ROS) gene expression is inversely correlated with the level of aldosterone production in APAs with a *KCNJ5* mutation and appears to regulate aldosterone secretion via ROS and Ca^{2+} signalling [69]. *GSTA1* overexpression suppressed aldosterone biosynthesis, while silencing of *GSTA1* increased aldosterone production through increasing ROS, superoxide, H2O2 levels, Ca2+ influx and the expression of *CAMK1* (encoding Ca^{2+}/calmodulin dependent protein kinase 1) and the transcription factors *NR4A1* (also called *NGFIB*) and *NR4A2* that regulate *CYP11B2* gene expression [69].

The epidermal growth factor-like teratocarcinoma-derived growth factor-1 gene (*TDGF1*) was identified as the most highly expressed gene in APAs compared with normal adrenals by microarray analysis [41]. *TDGF1* was also identified as upregulated in an APA relative to the paired adjacent cortex by serial analysis of gene expression [41]. Overexpression of *TDGF1* in NCI H295R adrenal cells activated the PI3K-Akt signalling pathway and led to an increase in aldosterone production, indicating a potential role in APA pathophysiology [41]. The activation of PI3K/Akt mTOR signalling, a pathway with a known role in cell proliferation, was also reported in patients with PA [41].

Wnt plays a key role in the development of the adrenal cortex and the dysregulation of this signalling pathway is associated with tumorigenesis [70]. The Wnt/β-catenin pathway is constitutively activated in around 70% of APAs [71] with the decreased expression on the Wnt inhibitor *SFRP2* (encoding secreted frizzled related protein 2) likely playing a role in the deregulated Wnt/β-catenin signalling [72]. *SFRP2* was also four-fold down-regulated in APAs compared with normal adrenals in an oligonucleotide microarray [72]. Mice with an ablation of *Sfrp2* display enhanced aldosterone production [72]. β-catenin appears to mediate aldosterone production by increasing the transcription of several genes including AT_1R, *CYP21* and *CYP11B2* as well as upregulating expression of transcription factors NURR1 (*NR4A2*) and NUR77 (*NR4A1*) [72].

NPNT (nephronectin), a secreted matrix protein, was most highly expressed in APAs with a ZG phenotype with *CTNNB1* mutations. Thereby it may represent a potential biomarker to recognize a subtype of APAs and indicates a further mechanism by which the Wnt/β-catenin signalling pathway may upregulate aldosterone production [72]. These studies show that aberrant Wnt/β-catenin pathway activation is associated with APA development and suggests that the Wnt/β-catenin signalling mediates aldosterone production at multiple levels [71].

Table 2. Differentially expressed genes in aldosterone-producing adenomas and their reference tissues used in transcriptome studies.

Gene	Encoded Protein and Description	Reference Tissue	Ref.
Upregulated Genes			
CYP11B2	Aldosterone synthase- steroid hydroxylase cytochrome P450 enzyme with 11β-hydroxylase, 18-hydroxylase and 18-oxidase activities	AAC; NLA	[34,39–41]
Calcium Signaling			
VSNL1	Visinin-like 1, calcium sensor protein of visinin/recoverin subfamily	NLA	[59]
CALN1	Calneuron 1, calcium-binding protein with high similarity to calmodulin family	NLA	[41,67]
CALM2	Calmodulin 2, calcium-binding protein of calmodulin family.	Adjacent ZG	[42]
PCP4	Purkinje cell protein 4, regulates calmodulin activity by modulating calcium binding by calmodulin	NFA	[68]
Nuclear receptor Transcription Factors			
NR4A1	Nuclear receptor subfamily 4 group A member 1; steroid-thyroid hormone-retinoid receptor superfamily.	WT-*KCNJ5*-APAs	[56]
NR4A2	Nuclear receptor subfamily 4 group A member 2; steroid-thyroid hormone-retinoid receptor superfamily.	WT-*KCNJ5*-APAs	[56]
NR5A1	Nuclear receptor subfamily 5 group A member 1 (SF1); transcriptional activator of sex determination.	AAC	[39]
NR0B1	Nuclear receptor subfamily 0 group B member 1 (DAX1); functions in proper formation of adult adrenal gland formation.	AAC	[39]
G-protein-coupled Receptors			
LHCGR	Luteinizing hormone/choriogonadotropin receptor	NLA	[60]
GNRHR	Gonadotropin releasing hormone receptor	NLA	[60]
HTR4	5-hydroxytryptamine receptor 4	NLA; NFA	[60,64]
PTGER1	Prostaglandin E receptor 1	NFA	[64]
MC2R	Melanocortin 2 receptor	NLA	[60]
AGTR1	Angiotensin II receptor type I	NLA	[60]
Others			
NEFM	Medium neurofilament protein- biomarker of neuronal damage	*KCNJ5*-mut APAs; ZF-like APAs	[62,63]
TDGF1	Teratocarcinoma-derived growth factor 1- signaling protein that functions in development and tumor growth	NLA	[41]
NPNT	Nephronectin, a secreted matrix protein	NLA	[72]
Downregulated Genes			
GSTA1	Glutathione S-transferase alpha 1- member of a family of enzymes that protect cells from reactive oxygen species	WT-*KCNJ5*-APAs; NLA	[69]
SFPR2	Secreted frizzled related protein 2- agonist of Wnt signaling	NLA	[72]

AAC: adjacent adrenal cortex; APAs: aldosterone-producing adenomas; *KCNJ5*-mut APAs: APAs with *KCNJ5* mutations; NFA: non-functioning adrenocortical adenomas; NLA: normal adrenals; WT-*KCNJ5* APAs: APAs with wild type *KCNJ5* gene; ZF: *zona fasciculata*; ZG: *zona glomerulosa*.

6. Single Nucleotide Polymorphisms

Single nucleotide polymorphisms (SNPs) are the most frequent genetic variation in the human genome. A rare nonsynonymous SNP (rs7102584) resulting in a GIRK4.p.Glu282Gln substitution was identified in 12 of 251 patients (5%) with sporadic PA (9 with bilateral and 3 with unilateral PA) compared with a prevalence of 2% in the 1000 genomes cohort [73]. Five common SNPs of the *KCNJ5* gene (rs6590357, rs4937391, rs3740835, rs2604204, and rs11221497) were found in patients with sporadic PA and essential hypertension and a significant association of the rs2604204 variant with sporadic PA in Chinese males was found indicating a potential role for this polymorphism in the pathogenesis of sporadic PA in this specific subgroup of patients [74].

7. Influence of Genetics and Genomics on Surgical Outcome

High blood pressure may persist after adrenalectomy due to contributory factors other than PA and the surgical cure rate of patients with APAs varies widely. Comparison of the transcriptomes of APAs with normal adrenals identified two subgroups of APAs based on their expression profiles (low *versus* high mRNA levels) of genes encoding steroidogenic enzymes. APAs with a low level of

CYP11B2 gene transcription are associated with a longer known duration of hypertension and a lower rate of long-term cure [43].

Microarray analysis identified differentially expressed genes in a comparison of the transcriptomes of APAs from patients with persistent hypertension after adrenalectomy with that of patients with APA who were cured by surgery [75]. The differentially expressed genes were associated with five different pathways that included lipid metabolism and cell differentiation and indicate the possibility of using genomic approaches to identify drug targets and prognostic markers [75].

A number of studies have investigated the effect of *KCNJ5* mutational status as a marker for surgical outcome. In a prospective study by the TAIPAI (Taiwan Primary Aldosteronsim Investigation study group) of 108 patients that were divided into *KCNJ5* mutated and non-mutated groups, patients with an APA carrying a *KCNJ5* mutation aged between 37 and 60 years may have an advantage in blood pressure response to surgery but mutation status is not associated with an improvement in arterial stiffness [76].

Cardiovascular complications before and after unilateral adrenalectomy in patients harboring APAs with and without *KCNJ5* gene mutations were evaluated in a Japanese population. The *KCNJ5*-mutated group displayed a significant improvement in left ventricular mass index which was independently associated with the presence of APA *KCNJ5* mutations whereas the group without *KCNJ5* mutations had no such improvement [77]. A higher left ventricular mass index and plasma aldosterone concentration in patients with APA *KCNJ5* mutations relative to those without *KCNJ5* mutations has also been reported [35]. Despite the increased cardiac damage, patients with *KCNJ5* mutations exhibited a decrease of blood pressure and plasma aldosterone concentrations and a regression of left ventricular mass index similar to the *KCNJ5* wild type group after adrenalectomy [35]. Another study reported an association of APA *KCNJ5* mutations with lower blood pressure and the higher likelihood of cure of PA by adrenalectomy relative to patients with APA without *KCNJ5* mutations [36].

8. Conclusions and Perspectives

Major discoveries have been made in the field of PA research over the last few years mainly due to the application of next-generation sequencing methods. Four familial forms of hyperaldosteronism are now recognized with the genetic basis of three of these uncovered by exome sequencing. Somatic mutations have been identified in ion channels and transporters that alter intracellular ion homeostasis and drive the constitutive aldosterone production in over half of aldosterone-producing adenomas. Differential gene expression studies have further highlighted key signalling pathways and molecular mechanisms that may drive cell proliferation and aldosterone overproduction in aldosterone-producing adenomas. Transcriptome analysis methods may have a future application in the identification of prognostic markers to identify post-operative cardiovascular events.

Acknowledgments: This work was supported by the European Research Council (ERC) under the European Union's Horizon 2020 research and innovation programme (grant agreement No. 694913 to Martin Reincke) and by the Deutsche Forschungsgemeinschaft (DFG) (within the CRC/Transregio 205/1 "The Adrenal: Central Relay in Health and Disease" to Felix Beuschlein, Martin Reincke and Tracy A. Williams; and grants RE 752/20-1 to Martin Reincke and grants BE 2177/13-1 and BE 2177/18-1 to Felix Beuschlein) and the Else Kröner-Fresenius Stiftung in support of the German Conns Registry-Else-Kröner Hyperaldosteronism Registry (2013_A182 and 2015_A171 to Martin Reincke).

Author Contributions: All the authors contributed substantially to the work presented in this manuscript. Tracy A. Williams conceived the work; Felix Beuschlein, Martin Reincke and Isabella Castellano contributed materials; Eva Sušnik performed immunohistochemistry; Lucie S. Meyer, Yuhong Yang and Martina Tetti performed the literature search and prepared art work and tables; Elke Tatjana Aristizabal Prada wrote the first draft of the manuscript; Tracy A. Williams, Felix Beuschlein and Martin Reincke revised the manuscript; all authors critically revised the final version.

Conflicts of Interest: The authors declare no conflict of interest.

References

1. Monticone, S.; D'Ascenzo, F.; Moretti, C.; Williams, T.A.; Veglio, F.; Gaita, F.; Mulatero, P. Cardiovascular events and target organ damage in primary aldosteronism compared with essential hypertension: A systematic review and meta-analysis. *Lancet Diabetes Endocrinol.* **2018**, *6*, 41–50. [CrossRef]
2. Kayser, S.C.; Dekkers, T.; Groenewoud, H.J.; van der Wilt, G.J.; Carel Bakx, J.; van der Wel, M.C.; Hermus, A.R.; Lenders, J.W.; Deinum, J. Study Heterogeneity and Estimation of Prevalence of Primary Aldosteronism: A Systematic Review and Meta-Regression Analysis. *J. Clin. Endocrinol. Metab.* **2016**, *101*, 2826–2835. [CrossRef] [PubMed]
3. Monticone, S.; Burrello, J.; Tizzani, D.; Bertello, C.; Viola, A.; Buffolo, F.; Gabetti, L.; Mengozzi, G.; Williams, T.A.; Rabbia, F.; et al. Prevalence and Clinical Manifestations of Primary Aldosteronism Encountered in Primary Care Practice. *J. Am. Coll. Cardiol.* **2017**, *69*, 1811–1820. [CrossRef] [PubMed]
4. Williams, T.A.; Lenders, J.W.M.; Mulatero, P.; Burrello, J.; Rottenkolber, M.; Adolf, C.; Satoh, F.; Amar, L.; Quinkler, M.; Deinum, J.; et al. Outcomes after adrenalectomy for unilateral primary aldosteronism: An international consensus on outcome measures and analysis of remission rates in an international cohort. *Lancet Diabetes Endocrinol.* **2017**, *5*, 689–699. [CrossRef]
5. Hundemer, G.L.; Curhan, G.C.; Yozamp, N.; Wang, M.; Vaidya, A. Cardiometabolic outcomes and mortality in medically treated primary aldosteronism: A retrospective cohort study. *Lancet Diabetes Endocrinol.* **2018**, *6*, 51–59. [CrossRef]
6. Choi, M.; Scholl, U.I.; Yue, P.; Bjorklund, P.; Zhao, B.; Nelson-Williams, C.; Ji, W.; Cho, Y.; Patel, A.; Men, C.J.; et al. K^+ channel mutations in adrenal aldosterone-producing adenomas and hereditary hypertension. *Science* **2011**, *331*, 768–772. [CrossRef] [PubMed]
7. Beuschlein, F.; Boulkroun, S.; Osswald, A.; Wieland, T.; Nielsen, H.N.; Lichtenauer, U.D.; Penton, D.; Schack, V.R.; Amar, L.; Fischer, E.; et al. Somatic mutations in *ATP1A1* and *ATP2B3* lead to aldosterone-producing adenomas and secondary hypertension. *Nat. Genet.* **2013**, *45*, 440–444. [CrossRef] [PubMed]
8. Azizan, E.A.; Poulsen, H.; Tuluc, P.; Zhou, J.; Clausen, M.V.; Lieb, A.; Maniero, C.; Garg, S.; Bochukova, E.G.; Zhao, W.; et al. Somatic mutations in ATP1A1 and CACNA1D underlie a common subtype of adrenal hypertension. *Nat. Genet.* **2013**, *45*, 1055–1060. [CrossRef] [PubMed]
9. Scholl, U.I.; Goh, G.; Stolting, G.; de Oliveira, R.C.; Choi, M.; Overton, J.D.; Fonseca, A.L.; Korah, R.; Starker, L.F.; Kunstman, J.W.; et al. Somatic and germline CACNA1D calcium channel mutations in aldosterone-producing adenomas and primary aldosteronism. *Nat. Genet.* **2013**, *45*, 1050–1054. [CrossRef] [PubMed]
10. Akerstrom, T.; Maharjan, R.; Sven Willenberg, H.; Cupisti, K.; Ip, J.; Moser, A.; Stalberg, P.; Robinson, B.; Alexander Iwen, K.; Dralle, H.; et al. Activating mutations in CTNNB1 in aldosterone producing adenomas. *Sci. Rep.* **2016**, *6*, 19546. [CrossRef] [PubMed]
11. Asbach, E.; Williams, T.A.; Reincke, M. Recent Developments in Primary Aldosteronism. *Exp. Clin. Endocrinol. Diabetes* **2016**, *124*, 335–341. [CrossRef] [PubMed]
12. Prada, E.T.A.; Burrello, J.; Reincke, M.; Williams, T.A. Old and New Concepts in the Molecular Pathogenesis of Primary Aldosteronism. *Hypertension* **2017**, *70*, 875–881. [CrossRef] [PubMed]
13. Lifton, R.P.; Dluhy, R.G.; Powers, M.; Rich, G.M.; Cook, S.; Ulick, S.; Lalouel, J.M. A chimaeric 11 beta-hydroxylase/aldosterone synthase gene causes glucocorticoid-remediable aldosteronism and human hypertension. *Nature* **1992**, *355*, 262–265. [CrossRef] [PubMed]
14. Lifton, R.P.; Dluhy, R.G.; Powers, M.; Rich, G.M.; Gutkin, M.; Fallo, F.; Gill, J.R., Jr.; Feld, L.; Ganguly, A.; Laidlaw, J.C.; et al. Hereditary hypertension caused by chimaeric gene duplications and ectopic expression of aldosterone synthase. *Nat. Genet.* **1992**, *2*, 66–74. [CrossRef] [PubMed]
15. Stowasser, M.; Gordon, R.D.; Tunny, T.J.; Klemm, S.A.; Finn, W.L.; Krek, A.L. Familial hyperaldosteronism type II: Five families with a new variety of primary aldosteronism. *Clin. Exp. Pharmacol. Physiol.* **1992**, *19*, 319–322. [CrossRef] [PubMed]
16. So, A.; Duffy, D.L.; Gordon, R.D.; Jeske, Y.W.; Lin-Su, K.; New, M.I.; Stowasser, M. Familial hyperaldosteronism type II is linked to the chromosome 7p22 region but also shows predicted heterogeneity. *J. Hypertens.* **2005**, *23*, 1477–1484. [CrossRef] [PubMed]

17. Scholl, U.I.; Stolting, G.; Schewe, J.; Thiel, A.; Tan, H.; Nelson-Williams, C.; Vichot, A.A.; Jin, S.C.; Loring, E.; Untiet, V.; et al. CLCN2 chloride channel mutations in familial hyperaldosteronism type II. *Nat. Genet.* **2018**, *50*, 349. [CrossRef] [PubMed]

18. Fernandes-Rosa, F.L.; Daniil, G.; Orozco, I.J.; Goppner, C.; El Zein, R.; Jain, V.; Boulkroun, S.; Jeunemaitre, X.; Amar, L.; Lefebvre, H.; et al. A gain-of-function mutation in the CLCN2 chloride channel gene causes primary aldosteronism. *Nat. Genet.* **2018**, *50*, 355. [CrossRef] [PubMed]

19. Oki, K.; Plonczynski, M.W.; Luis Lam, M.; Gomez-Sanchez, E.P.; Gomez-Sanchez, C.E. Potassium channel mutant KCNJ5 T158A expression in HAC-15 cells increases aldosterone synthesis. *Endocrinology* **2012**, *153*, 1774–1782. [CrossRef] [PubMed]

20. Monticone, S.; Tetti, M.; Burrello, J.; Buffolo, F.; De Giovanni, R.; Veglio, F.; Williams, T.A.; Mulatero, P. Familial hyperaldosteronism type III. *J. Hum. Hypertens.* **2017**, *31*, 776–781. [CrossRef] [PubMed]

21. Dibb, K.M.; Rose, T.; Makary, S.Y.; Claydon, T.W.; Enkvetchakul, D.; Leach, R.; Nichols, C.G.; Boyett, M.R. Molecular basis of ion selectivity, block, and rectification of the inward rectifier Kir3.1/Kir3.4 K(+) channel. *J. Biol. Chem.* **2003**, *278*, 49537–49548. [CrossRef] [PubMed]

22. Scholl, U.I.; Stolting, G.; Nelson-Williams, C.; Vichot, A.A.; Choi, M.; Loring, E.; Prasad, M.L.; Goh, G.; Carling, T.; Juhlin, C.C.; et al. Recurrent gain of function mutation in calcium channel CACNA1H causes early-onset hypertension with primary aldosteronism. *eLife* **2015**, *4*, e06315. [CrossRef] [PubMed]

23. Reimer, E.N.; Walenda, G.; Seidel, E.; Scholl, U.I. CACNA1H(M1549V) Mutant Calcium Channel Causes Autonomous Aldosterone Production in HAC15 Cells and Is Inhibited by Mibefradil. *Endocrinology* **2016**, *157*, 3016–3022. [CrossRef] [PubMed]

24. Daniil, G.; Fernandes-Rosa, F.L.; Chemin, J.; Blesneac, I.; Beltrand, J.; Polak, M.; Jeunemaitre, X.; Boulkroun, S.; Amar, L.; Strom, T.M.; et al. CACNA1H Mutations Are Associated With Different Forms of Primary Aldosteronism. *EBioMedicine* **2016**, *13*, 225–236. [CrossRef] [PubMed]

25. Williams, T.A.; Monticone, S.; Mulatero, P. KCNJ5 mutations are the most frequent genetic alteration in primary aldosteronism. *Hypertension* **2015**, *65*, 507–509. [CrossRef] [PubMed]

26. Boulkroun, S.; Beuschlein, F.; Rossi, G.P.; Golib-Dzib, J.F.; Fischer, E.; Amar, L.; Mulatero, P.; Samson-Couterie, B.; Hahner, S.; Quinkler, M.; et al. Prevalence, clinical, and molecular correlates of KCNJ5 mutations in primary aldosteronism. *Hypertension* **2012**, *59*, 592–598. [CrossRef] [PubMed]

27. Akerstrom, T.; Crona, J.; Delgado Verdugo, A.; Starker, L.F.; Cupisti, K.; Willenberg, H.S.; Knoefel, W.T.; Saeger, W.; Feller, A.; Ip, J.; et al. Comprehensive re-sequencing of adrenal aldosterone producing lesions reveal three somatic mutations near the KCNJ5 potassium channel selectivity filter. *PLoS ONE* **2012**, *7*, e41926. [CrossRef] [PubMed]

28. Mulatero, P.; Monticone, S.; Rainey, W.E.; Veglio, F.; Williams, T.A. Role of KCNJ5 in familial and sporadic primary aldosteronism. *Nat. Rev. Endocrinol.* **2013**, *9*, 104–112. [CrossRef] [PubMed]

29. Williams, T.A.; Monticone, S.; Schack, V.R.; Stindl, J.; Burrello, J.; Buffolo, F.; Annaratone, L.; Castellano, I.; Beuschlein, F.; Reincke, M.; et al. Somatic ATP1A1, ATP2B3, and KCNJ5 mutations in aldosterone-producing adenomas. *Hypertension* **2014**, *63*, 188–195. [CrossRef] [PubMed]

30. Fernandes-Rosa, F.L.; Williams, T.A.; Riester, A.; Steichen, O.; Beuschlein, F.; Boulkroun, S.; Strom, T.M.; Monticone, S.; Amar, L.; Meatchi, T.; et al. Genetic spectrum and clinical correlates of somatic mutations in aldosterone-producing adenoma. *Hypertension* **2014**, *64*, 354–361. [CrossRef] [PubMed]

31. Lenzini, L.; Rossitto, G.; Maiolino, G.; Letizia, C.; Funder, J.W.; Rossi, G.P. A Meta-Analysis of Somatic KCNJ5 K($^+$) Channel Mutations In 1636 Patients With an Aldosterone-Producing Adenoma. *J. Clin. Endocrinol. Metab.* **2015**, *100*, E1089–E1095. [CrossRef] [PubMed]

32. Williams, T.A.; Lenders, J.W.; Burrello, J.; Beuschlein, F.; Reincke, M. KCNJ5 Mutations: Sex, Salt and Selection. *Horm. Metab. Res.* **2015**, *47*, 953–958. [CrossRef] [PubMed]

33. Tissier, F.; Cavard, C.; Groussin, L.; Perlemoine, K.; Fumey, G.; Hagnere, A.M.; Rene-Corail, F.; Jullian, E.; Gicquel, C.; Bertagna, X.; et al. Mutations of beta-catenin in adrenocortical tumors: Activation of the Wnt signaling pathway is a frequent event in both benign and malignant adrenocortical tumors. *Cancer Res.* **2005**, *65*, 7622–7627. [CrossRef] [PubMed]

34. Lin, Y.H.; Huang, K.H.; Lee, J.K.; Wang, S.M.; Yen, R.F.; Wu, V.C.; Chung, S.D.; Liu, K.L.; Chueh, S.C.; Lin, L.Y.; et al. Factors influencing left ventricular mass regression in patients with primary aldosteronism post adrenalectomy. *J. Renin-Angiotensin-Aldosterone Syst.* **2011**, *12*, 48–53. [CrossRef] [PubMed]

35. Rossi, G.P.; Cesari, M.; Letizia, C.; Seccia, T.M.; Cicala, M.V.; Zinnamosca, L.; Kuppusamy, M.; Mareso, S.; Sciomer, S.; Iacobone, M.; et al. KCNJ5 gene somatic mutations affect cardiac remodelling but do not preclude cure of high blood pressure and regression of left ventricular hypertrophy in primary aldosteronism. *J. Hypertens.* **2014**, *32*, 1514–1522. [CrossRef] [PubMed]

36. Arnesen, T.; Glomnes, N.; Stromsoy, S.; Knappskog, S.; Heie, A.; Akslen, L.A.; Grytaas, M.; Varhaug, J.E.; Gimm, O.; Brauckhoff, M. Outcome after surgery for primary hyperaldosteronism may depend on KCNJ5 tumor mutation status: A population-based study from Western Norway. *Langenbeck's Arch. Surg.* **2013**, *398*, 869–874. [CrossRef] [PubMed]

37. Rossi, G.P.; Auchus, R.J.; Brown, M.; Lenders, J.W.; Naruse, M.; Plouin, P.F.; Satoh, F.; Young, W.F., Jr. An expert consensus statement on use of adrenal vein sampling for the subtyping of primary aldosteronism. *Hypertension* **2014**, *63*, 151–160. [CrossRef] [PubMed]

38. Monticone, S.; Viola, A.; Rossato, D.; Veglio, F.; Reincke, M.; Gomez-Sanchez, C.; Mulatero, P. Adrenal vein sampling in primary aldosteronism: Towards a standardised protocol. *Lancet Diabetes Endocrinol.* **2015**, *3*, 296–303. [CrossRef]

39. Bassett, M.H.; Mayhew, B.; Rehman, K.; White, P.C.; Mantero, F.; Arnaldi, G.; Stewart, P.M.; Bujalska, I.; Rainey, W.E. Expression profiles for steroidogenic enzymes in adrenocortical disease. *J. Clin. Endocrinol. Metab.* **2005**, *90*, 5446–5455. [CrossRef] [PubMed]

40. Saner-Amigh, K.; Mayhew, B.A.; Mantero, F.; Schiavi, F.; White, P.C.; Rao, C.V.; Rainey, W.E. Elevated expression of luteinizing hormone receptor in aldosterone-producing adenomas. *J. Clin. Endocrinol. Metab.* **2006**, *91*, 1136–1142. [CrossRef] [PubMed]

41. Williams, T.A.; Monticone, S.; Morello, F.; Liew, C.C.; Mengozzi, G.; Pilon, C.; Asioli, S.; Sapino, A.; Veglio, F.; Mulatero, P. Teratocarcinoma-derived growth factor-1 is upregulated in aldosterone-producing adenomas and increases aldosterone secretion and inhibits apoptosis in vitro. *Hypertension* **2010**, *55*, 1468–1475. [CrossRef] [PubMed]

42. Assie, G.; Auzan, C.; Gasc, J.M.; Baviera, E.; Balaton, A.; Elalouf, J.M.; Jeunemaitre, X.; Plouin, P.F.; Corvol, P.; Clauser, E. Steroidogenesis in aldosterone-producing adenoma revisited by transcriptome analysis. *J. Clin. Endocrinol. Metab.* **2005**, *90*, 6638–6649. [CrossRef] [PubMed]

43. Lenzini, L.; Seccia, T.M.; Aldighieri, E.; Belloni, A.S.; Bernante, P.; Giuliani, L.; Nussdorfer, G.G.; Pessina, A.C.; Rossi, G.P. Heterogeneity of aldosterone-producing adenomas revealed by a whole transcriptome analysis. *Hypertension* **2007**, *50*, 1106–1113. [CrossRef] [PubMed]

44. Nishimoto, K.; Tomlins, S.A.; Kuick, R.; Cani, A.K.; Giordano, T.J.; Hovelson, D.H.; Liu, C.J.; Sanjanwala, A.R.; Edwards, M.A.; Gomez-Sanchez, C.E.; et al. Aldosterone-stimulating somatic gene mutations are common in normal adrenal glands. *Proc. Natl. Acad. Sci. USA* **2015**, *112*, E4591–E4599. [CrossRef] [PubMed]

45. Omata, K.; Tomlins, S.A.; Rainey, W.E. Aldosterone-Producing Cell Clusters in Normal and Pathological States. *Horm. Metab. Res.* **2017**, *49*, 951–956. [CrossRef] [PubMed]

46. Gomez-Sanchez, C.E.; Kuppusamy, M.; Reincke, M.; Williams, T.A. Disordered CYP11B2 Expression in Primary Aldosteronism. *Horm. Metab. Res.* **2017**, *49*, 957–962. [CrossRef] [PubMed]

47. Azizan, E.A.; Lam, B.Y.; Newhouse, S.J.; Zhou, J.; Kuc, R.E.; Clarke, J.; Happerfield, L.; Marker, A.; Hoffman, G.J.; Brown, M.J. Microarray, qPCR, and KCNJ5 sequencing of aldosterone-producing adenomas reveal differences in genotype and phenotype between zona glomerulosa- and zona fasciculata-like tumors. *J. Clin. Endocrinol. Metab.* **2012**, *97*, E819–E829. [CrossRef] [PubMed]

48. Dekkers, T.; ter Meer, M.; Lenders, J.W.; Hermus, A.R.; Schultze Kool, L.; Langenhuijsen, J.F.; Nishimoto, K.; Ogishima, T.; Mukai, K.; Azizan, E.A.; et al. Adrenal nodularity and somatic mutations in primary aldosteronism: One node is the culprit? *J. Clin. Endocrinol. Metab.* **2014**, *99*, E1341–E1351. [CrossRef] [PubMed]

49. Monticone, S.; Castellano, I.; Versace, K.; Lucatello, B.; Veglio, F.; Gomez-Sanchez, C.E.; Williams, T.A.; Mulatero, P. Immunohistochemical, genetic and clinical characterization of sporadic aldosterone-producing adenomas. *Mol. Cell. Endocrinol.* **2015**, *411*, 146–154. [CrossRef] [PubMed]

50. Scholl, U.I.; Healy, J.M.; Thiel, A.; Fonseca, A.L.; Brown, T.C.; Kunstman, J.W.; Horne, M.J.; Dietrich, D.; Riemer, J.; Kucukkoylu, S.; et al. Novel somatic mutations in primary hyperaldosteronism are related to the clinical, radiological and pathological phenotype. *Clin. Endocrinol.* **2015**, *83*, 779–789. [CrossRef] [PubMed]

51. Kitamoto, T.; Suematsu, S.; Yamazaki, Y.; Nakamura, Y.; Sasano, H.; Matsuzawa, Y.; Saito, J.; Omura, M.; Nishikawa, T. Clinical and Steroidogenic Characteristics of Aldosterone-Producing Adenomas With ATPase or CACNA1D Gene Mutations. *J. Clin. Endocrinol. Metab.* **2016**, *101*, 494–503. [CrossRef] [PubMed]

52. Williams, T.A.; Peitzsch, M.; Dietz, A.S.; Dekkers, T.; Bidlingmaier, M.; Riester, A.; Treitl, M.; Rhayem, Y.; Beuschlein, F.; Lenders, J.W.; et al. Genotype-Specific Steroid Profiles Associated With Aldosterone-Producing Adenomas. *Hypertension* **2016**, *67*, 139–145. [CrossRef] [PubMed]

53. Monticone, S.; Hattangady, N.G.; Nishimoto, K.; Mantero, F.; Rubin, B.; Cicala, M.V.; Pezzani, R.; Auchus, R.J.; Ghayee, H.K.; Shibata, H.; et al. Effect of KCNJ5 mutations on gene expression in aldosterone-producing adenomas and adrenocortical cells. *J. Clin. Endocrinol. Metab.* **2012**, *97*, E1567–E1572. [CrossRef] [PubMed]

54. Akerstrom, T.; Willenberg, H.S.; Cupisti, K.; Ip, J.; Backman, S.; Moser, A.; Maharjan, R.; Robinson, B.; Iwen, K.A.; Dralle, H.; et al. Novel somatic mutations and distinct molecular signature in aldosterone-producing adenomas. *Endocr. Relat. Cancer* **2015**, *22*, 735–744. [CrossRef] [PubMed]

55. Lenders, J.W.M.; Williams, T.A.; Reincke, M.; Gomez-Sanchez, C.E. DIAGNOSIS OF ENDOCRINE DISEASE: 18-Oxocortisol and 18-hydroxycortisol: Is there clinical utility of these steroids? *Eur. J. Endocrinol.* **2018**, *178*, R1–R9. [CrossRef] [PubMed]

56. Bassett, M.H.; Suzuki, T.; Sasano, H.; De Vries, C.J.; Jimenez, P.T.; Carr, B.R.; Rainey, W.E. The orphan nuclear receptor NGFIB regulates transcription of 3beta-hydroxysteroid dehydrogenase. implications for the control of adrenal functional zonation. *J. Biol. Chem.* **2004**, *279*, 37622–37630. [CrossRef] [PubMed]

57. Reincke, M.; Beuschlein, F.; Lalli, E.; Arlt, W.; Vay, S.; Sassone-Corsi, P.; Allolio, B. DAX-1 expression in human adrenocortical neoplasms: Implications for steroidogenesis. *J. Clin. Endocrinol. Metab.* **1998**, *83*, 2597–2600. [CrossRef] [PubMed]

58. Ferraz-de-Souza, B.; Hudson-Davies, R.E.; Lin, L.; Parnaik, R.; Hubank, M.; Dattani, M.T.; Achermann, J.C. Sterol O-acyltransferase 1 (SOAT1, ACAT) is a novel target of steroidogenic factor-1 (SF-1, NR5A1, Ad4BP) in the human adrenal. *J. Clin. Endocrinol. Metab.* **2011**, *96*, E663–E668. [CrossRef] [PubMed]

59. Williams, T.A.; Monticone, S.; Crudo, V.; Warth, R.; Veglio, F.; Mulatero, P. Visinin-like 1 is upregulated in aldosterone-producing adenomas with KCNJ5 mutations and protects from calcium-induced apoptosis. *Hypertension* **2012**, *59*, 833–839. [CrossRef] [PubMed]

60. Ye, P.; Mariniello, B.; Mantero, F.; Shibata, H.; Rainey, W.E. G-protein-coupled receptors in aldosterone-producing adenomas: A potential cause of hyperaldosteronism. *J. Endocrinol.* **2007**, *195*, 39–48. [CrossRef] [PubMed]

61. Albiger, N.M.; Sartorato, P.; Mariniello, B.; Iacobone, M.; Finco, I.; Fassina, A.; Mantero, F. A case of primary aldosteronism in pregnancy: Do LH and GNRH receptors have a potential role in regulating aldosterone secretion? *Eur. J. Endocrinol.* **2011**, *164*, 405–412. [CrossRef] [PubMed]

62. Zhou, J.; Lam, B.; Neogi, S.G.; Yeo, G.S.; Azizan, E.A.; Brown, M.J. Transcriptome Pathway Analysis of Pathological and Physiological Aldosterone-Producing Human Tissues. *Hypertension* **2016**, *68*, 1424–1431. [CrossRef] [PubMed]

63. Maniero, C.; Garg, S.; Zhao, W.; Johnson, T.I.; Zhou, J.; Gurnell, M.; Brown, M.J. NEFM (Neurofilament Medium) Polypeptide, a Marker for Zona Glomerulosa Cells in Human Adrenal, Inhibits D1R (Dopamine D1 Receptor)-Mediated Secretion of Aldosterone. *Hypertension* **2017**, *70*, 357–364. [CrossRef] [PubMed]

64. Itcho, K.; Oki, K.; Kobuke, K.; Yoshii, Y.; Ohno, H.; Yoneda, M.; Hattori, N. Aberrant G protein-receptor expression is associated with DNA methylation in aldosterone-producing adenoma. *Mol. Cell. Endocrinol.* **2018**, *461*, 100–104. [CrossRef] [PubMed]

65. Kobuke, K.; Oki, K.; Gomez-Sanchez, C.E.; Ohno, H.; Itcho, K.; Yoshii, Y.; Yoneda, M.; Hattori, N. Purkinje cell protein 4 expression is associated with DNA methylation status in aldosterone-producing adenoma. *J. Clin. Endocrinol. Metab.* **2017**. [CrossRef] [PubMed]

66. Rasmussen, H.; Isales, C.M.; Calle, R.; Throckmorton, D.; Anderson, M.; Gasalla-Herraiz, J.; McCarthy, R. Diacylglycerol production, Ca^{2+} influx, and protein kinase C activation in sustained cellular responses. *Endocr. Rev.* **1995**, *16*, 649–681. [CrossRef] [PubMed]

67. Kobuke, K.; Oki, K.; Gomez-Sanchez, C.E.; Gomez-Sanchez, E.P.; Ohno, H.; Itcho, K.; Yoshii, Y.; Yoneda, M.; Hattori, N. Calneuron 1 Increased $Ca(^{2+})$ in the Endoplasmic Reticulum and Aldosterone Production in Aldosterone-Producing Adenoma. *Hypertension* **2018**, *71*, 125–133. [CrossRef] [PubMed]

68. Wang, T.; Satoh, F.; Morimoto, R.; Nakamura, Y.; Sasano, H.; Auchus, R.J.; Edwards, M.A.; Rainey, W.E. Gene expression profiles in aldosterone-producing adenomas and adjacent adrenal glands. *Eur. J. Endocrinol.* **2011**, *164*, 613–619. [CrossRef] [PubMed]

69. Li, X.; Wang, B.; Tang, L.; Zhang, Y.; Chen, L.; Gu, L.; Zhang, F.; Ouyang, J.; Zhang, X. GSTA1 expression is correlated with aldosterone level in KCNJ5-mutated adrenal aldosterone-producing adenoma. *J. Clin. Endocrinol. Metab.* **2017**. [CrossRef] [PubMed]

70. El Wakil, A.; Lalli, E. The Wnt/beta-catenin pathway in adrenocortical development and cancer. *Mol. Cell. Endocrinol.* **2011**, *332*, 32–37. [CrossRef] [PubMed]

71. Berthon, A.; Drelon, C.; Ragazzon, B.; Boulkroun, S.; Tissier, F.; Amar, L.; Samson-Couterie, B.; Zennaro, M.C.; Plouin, P.F.; Skah, S.; et al. WNT/beta-catenin signalling is activated in aldosterone-producing adenomas and controls aldosterone production. *Hum. Mol. Genet.* **2014**, *23*, 889–905. [CrossRef] [PubMed]

72. Teo, A.E.; Garg, S.; Johnson, T.I.; Zhao, W.; Zhou, J.; Gomez-Sanchez, C.E.; Gurnell, M.; Brown, M.J. Physiological and Pathological Roles in Human Adrenal of the Glomeruli-Defining Matrix Protein NPNT (Nephronectin). *Hypertension* **2017**, *69*, 1207–1216. [CrossRef] [PubMed]

73. Murthy, M.; Xu, S.; Massimo, G.; Wolley, M.; Gordon, R.D.; Stowasser, M.; O'Shaughnessy, K.M. Role for germline mutations and a rare coding single nucleotide polymorphism within the KCNJ5 potassium channel in a large cohort of sporadic cases of primary aldosteronism. *Hypertension* **2014**, *63*, 783–789. [CrossRef] [PubMed]

74. Li, N.F.; Li, H.J.; Zhang, D.L.; Zhang, J.H.; Yao, X.G.; Wang, H.M.; Abulikemu, S.; Zhou, K.M.; Zhang, X.Y. Genetic variations in the KCNJ5 gene in primary aldosteronism patients from Xinjiang, China. *PLoS ONE* **2013**, *8*, e54051. [CrossRef] [PubMed]

75. Xie, L.F.; Ouyang, J.Z.; Wang, A.P.; Wang, W.B.; Li, X.T.; Wang, B.J.; Mu, Y.M. Gene Expression Profile of Persistent Postoperative Hypertension Patients with Aldosterone-producing Adenomas. *Chin. Med. J.* **2015**, *128*, 1618–1626. [PubMed]

76. Chang, C.H.; Hu, Y.H.; Tsai, Y.C.; Wu, C.H.; Wang, S.M.; Lin, L.Y.; Lin, Y.H.; Satoh, F.; Wu, K.D.; Wu, V.C. Arterial stiffness and blood pressure improvement in aldosterone-producing adenoma harboring KCNJ5 mutations after adrenalectomy. *Oncotarget* **2017**, *8*, 29984–29995. [CrossRef] [PubMed]

77. Kitamoto, T.; Suematsu, S.; Matsuzawa, Y.; Saito, J.; Omura, M.; Nishikawa, T. Comparison of cardiovascular complications in patients with and without KCNJ5 gene mutations harboring aldosterone-producing adenomas. *J. Atheroscler. Thromb.* **2015**, *22*, 191–200. [CrossRef] [PubMed]

International Journal of
Molecular Sciences

Review

Non-Coding RNA in the Pathogenesis, Progression and Treatment of Hypertension

Christiana Leimena and Hongyu Qiu *

Department of Basic Sciences, Physiological Division, School of Medicine, Loma Linda University,
Loma Linda, CA 92324, USA; cleimena@llu.edu
* Correspondence: hqiu@llu.edu; Tel.: +1-909-651-5894; Fax: +909-558-0119

Received: 31 January 2018; Accepted: 16 March 2018; Published: 21 March 2018

Abstract: Hypertension is a complex, multifactorial disease that involves the coexistence of multiple risk factors, environmental factors and physiological systems. The complexities extend to the treatment and management of hypertension, which are still the pursuit of many researchers. In the last two decades, various genes have emerged as possible biomarkers and have become the target for investigations of specialized drug design based on its risk factors and the primary cause. Owing to the growing technology of microarrays and next-generation sequencing, the non-protein-coding RNAs (ncRNAs) have increasingly gained attention, and their status of redundancy has flipped to importance in normal cellular processes, as well as in disease progression. The ncRNA molecules make up a significant portion of the human genome, and their role in diseases continues to be uncovered. Specifically, the cellular role of these ncRNAs has played a part in the pathogenesis of hypertension and its progression to heart failure. This review explores the function of the ncRNAs, their types and biology, the current update of their association with hypertension pathology and the potential new therapeutic regime for hypertension.

Keywords: non-coding RNA; micro RNA; hypertension

---

1. Introduction

Hypertension is a major risk factor for the development of cardiovascular disease (CVD). According to a report in 2016 of population-based studies on the global disparities of hypertension, between 2000 and 2010, globally, 31.1% or 1.39 billion people were estimated to suffer from hypertension [1,2]. In 2017, the American College of Cardiology (ACC) and the American Heart Association (AHA) presented new guidelines that further lowered the definition of high blood pressure at 130/80 mmHg rather than 140/90 mmHg, which further highlights the importance of the early detection and intervention of hypertension [3]. There are two major types of systemic hypertension: essential hypertension, which accounts for 95% of all cases, and secondary hypertension [4]. Essential hypertension, also referred to as primary hypertension, is a multifactorial disease where environmental factors and genetic factors coexist. Essential hypertension is characterized by high blood pressure mainly developed at middle or elderly age, while in childhood, essential hypertension is becoming more common due to the obesity epidemic [4]. Secondary hypertension on the other hand has a younger disease onset, with an absence of family history and underlying causes such as endocrine, or renal disorder, or an iatrogenic trigger, or from different medications, including oral contraceptives, steroids, nonsteroid anti-inflammatory drug (NSAIDs) and cyclosporine [5].

The development of hypertension is complex and multifactorial, attributed to both or either of the genetic and/or environmental factors involving at least the renin-angiotensin-aldosterone system, thrombogenesis, impaired platelet function and the sympathetic nervous system [6–10]. The therapeutic drug designs have been based on genes and their encoded proteins that are involved in these signaling pathways. Although pharmacotherapies using various classes of drugs have been shown to have some efficacy in reducing cardiovascular mortality (by 33%), major adverse cardiovascular events (by 29%) and heart failure (by 37%), hypertension remains one of the world's great public health problems [11]. There is a greater need to further understand the disease mechanism of hypertension and targeted therapeutic treatment [12]. Due to the growing technology of genomics, such as microarrays and next-generation sequencing, the non-protein-coding RNA (ncRNAs) have increasingly gained attention in normal cellular processes, as well as in disease progression. ncRNA is a functional RNA molecule that is transcribed from DNA, but not translated into proteins and has been shown to be involved in regulating gene expression and inhibiting the translation and degradation of messenger RNAs [13]. Two major types of ncRNA, namely microRNAs (miRNAs) and long non-coding RNAs (lncRNAs), have been extensively studied in both hypertensive patients and animal models as outlined in a number of review papers [14,15]. This review will present an update of the most recent progress in both miRNAs and lncRNA focusing on their links to the physiological regulation and therapeutic potential in systemic hypertension.

2. Discovery and Application of Non-Coding RNAs

Since the 1950s, various types of ncRNAs have been uncovered in eukaryotic cells, including transfer RNAs (tRNAs), which make up the greatest number of RNA molecules with 10 tRNAs per ribosomal RNA (rRNA), rRNA, messenger RNA (mRNA), small nucleolar RNA (snoRNA), small nuclear RNA (snRNA), miRNA, the RNA component of the signal recognition particle (7SL RNA), other lncRNAs, circular RNA, heterogeneous nuclear RNA (hnRNA) and X-inactive-specific transcript RNA (Xist RNA) [16,17]. The ncRNAs could be classified based on size: small (around 20 base pairs (bp)), intermediate (less than 200 bp) and long (longer than 200 bp). Small ncRNAs have attracted many investigations such as: piwi-interacting RNAs, small interfering RNAs (siRNAs) and miRNAs [14]. Intermediate ncRNAs include small nuclear RNAs that are involved in splicing during protein synthesis, nucleolar RNAs that modify ribosome RNA, transcription start site (TSS)-associated RNAs and promoter-associated small RNAs [14]. The rest of the ncRNAs that are greater than 200 bp have been grouped as lncRNAs. The research on lncRNAs has gained momentum, and these partake in the epigenetic regulation of transcripts and inactivation of X-chromosomes [14].

Historically, the discovery of tRNA and rRNA began in the 1950s, and the existence of other ncRNA such as snRNAs and 7SL was uncovered from the late 1970s. However, it was in the 1980s that the transcription regulatory function of miRNA began to emerge. It began with the first discovery of micF RNA in *Escherichia coli* [18–20]. Following this, in the 1990s, the first regulatory miRNA, lin-4, in eukaryotes was discovered in *Caenorhabditis elegans* (*C. elegans*) [21]. Within the same decade, the lncRNA, Xist, became known as the regulator of the X-chromosome [22]. It was not until in 2000 when the second *C. elegans* miRNA, let-7, was discovered with sequence conservation amongst humans and animals, that the research into miRNAs in *Drosophila* and human cell lines increased exponentially [23,24]. In 2002 was the first report of human miRNAs miR-15a and miR-16-1 that were downregulated or deleted in B cell chronic lymphocytic leukemia cases [25]. Following this finding was the first human oncogenic miRNAs, miR-17-92 cluster and miR-155, which were overexpressed in other cancers and B cell lymphomas, as well as in hematological malignancies, respectively [26–28].

The various functions of ncRNAs, in particular, miRNAs and lncRNAs, have unlocked opportunities and developments in clinical trials for RNA interference (RNAi) as the next medical therapy. RNAi medicine is currently via the utilization of siRNAs and miRNA mimics. Strategies of therapeutic design using miRNAs could be in the form of repressing/inhibiting the upregulated miRNA by using antagomirs, which are synthetic antisense 21–23-base pair (bp) oligonucleotides. Alternatively, deficient miRNAs could be replaced or enhanced by the overexpression of miRNAs or utilizing synthetic miRNAs. The antisense technologies have also been trialed to repress lncRNA. The advantages that miRNAs provide over the conventional drug molecules are their potency, action on any gene of interest and accessibility to repression, which some traditional drug molecules could not access due to the encoded protein's folding conformation and/or, the protein's lack of enzymatic function [29,30].

There are promising prospects of therapeutic miRNA in the pharmaceutical industry. According to a recent review on therapeutic miRNA and siRNA, there are currently 10 existing miRNA therapeutics in pre-clinical trials [31]. Only one miRNA therapy has proceeded to the phase II clinical trials, miravirsen, utilizing LNA, an antisense oligonucleotide, against miR-122 for hepatitis C infection treatment [32,33]. A therapeutic miRNA, known as MRX34, was developed for cancerous cells, but also other disorders, such as: Alport syndrome, myocardial infarction on remodeling, cardiac fibrosis, abnormal red blood cell production, cardiometabolic disease and chronic heart failure [31]. The MRX34 miRNA therapy is a miRNA mimic, introducing miR-34, which is suppressed in tumor cells. MRX34 therapy is the only one that had entered a phase I clinical trial, but due to the severe adverse immune responses, further progress was halted. The treatment for remodeling in post-myocardial infarction, known as an anti-miR candidate drug named MGN-1374, targets miR-15 and miR-195, has entered into the preclinical stage. For more information on the ongoing or completed clinical trials, the National Institute of Health (NIH), USA, has provided an accessible database (http://www.clinicaltrials.gov).

3. Recent Progress of miRNAs in Hypertension

miRNAs (18–24 bp) are master gene regulators controlling the expressions of specific genes by their binding to the 3′ untranslated region (UTR) of a messenger RNA (mRNA), which triggers either repression or degradation of the translation mechanism, and thus gene expression. A single miRNA regulates one to several hundred genes, and a single gene could be regulated by more than one miRNA. miRNAs can be sourced from tissues, urine, serum, plasma and blood cells (which include peripheral blood mononuclear and vascular endothelial cells). The accessibility of miRNA from serum, plasma or urine stems from studies that validated the presence of circulating miRNAs packaged in exosomes, microvesicles or apoptotic bodies [34–37]. The functions of the circulating miRNAs are mainly to communicate to neighboring or remote target cells and to provide gene expression regulation. An example for this is the communication between endothelial cells and the VSMCs, via the circulating miRNAs in these mentioned extracellular vesicles [34]. The various miRNAs that will be described are also listed in Table 1.

3.1. miRNAs in the Regulation of the Renin-Angiotensin Aldosterone System

The renin-angiotensin aldosterone system (RAAS) is a hormonal system that is paramount in the regulation of blood pressure by its influence on cardiac contractility, blood volume and resistance in the vasculature. The RAAS is a collaboration of the physiological workings of various organs and systems: the renal system, the cardiovascular system, the central nervous system and adrenal glands [38]. The RAAS involves a number of molecular players: peptides (angiotensin II), substrate (angiotensinogen (AGT), enzymes (angiotensin converting enzymes 1 and 2 (ACE1 and ACE2, respectively), aldosterone and vasopressin (known as anti-diuretic hormone (ADH)) and receptors (angiotensin II receptor type 1 and type 2 (AT_1R and AT_2R encoded by AGTR1 and AGTR2 mRNAs, respectively), bradykinin receptor 2 (B2R) and thromboxane A2 receptor (TBXA2R)). Hypertension develops when this well-balanced system of RAAS is over-activated. A number of microRNAs interact with the major players of RAAS in the hypertensive human cases and animal and in vitro experiments, as shown in Table 1.

Table 1. miRNAs associated with hypertension.

miRNA	miRNA Expression	Species	Conditions/Treatment	Sample Size	Source	Ref.
RAAS						
miR-155	Down	Human	young HT, reporter silencing assay	$n = 19$–25	Blood; HEK293T cells	[39,40]
miR-526, miR-578, miR-34a, miR-34c-5p, miR-449b, miR-571, miR-765	Up	Human	SNP genotyping on miRNA binding sites in genes of RAAS that influence blood pressure	$n = 1246$	Blood, HUH7/HELA cells	[41]
miR-483-3p	Up	Human, rats, mice	MiRNA array, reporter luciferase assay	-	HASMC, RASMC, HL-1 cells	[42]
miR-143/145	Up	Mice	Shear stress on EC of Ampkα2$^{-/-}$ mice	-	EC	[43]
	-	Mice	MiR-143/145 KO mice: AngII-infusion for vascular injury		Mesenteric arteries	[44]
miR-132, miR-212	Up	Rats	AngII-infused and Endothelin	$n = 3$–5	Heart, aorta, kidney	[45]
	down	Human	AGTR1 blocker treatmt	$n = 16$	Artery	[45]
miR-21	Up	human	AngII-induced cells		Cell line	[46]
miR-4516		Human	HT iSS/SS/SR	$n = 3$–4	Exosomes in urine	[47]
miR-361-5p, miR-362-5p	Down	Human	SSH vs. SRH	$n = 6$	Whole blood	[48]
miR-638,181a,663, let-7c	Down	Human	qPCR on HT/NT	$n = 16$–22	Renal medulla	[49]
miR-21,126, 196a,451	Up					
miR-181a	Down	Mouse	Effect of RAAS on hypertension in BPH/2J mouse circadian HT	$n = 7$–13	Kidneys	[50]
Endothelial cells						
miR-122	Up	Human	HT	$n = 278$–498	Blood	[51]
miR-505	Up	Human	HT	$n = 11$–19	Plasma, HUVEC	[52,53]
miR-UL112, 296-5p, let-7e	Up		Microarray, qPCR	$n = 67$–127	Plasma	[54]
miR-155	Up	Human	-	$n = 6$	HUVEC	[55]
miR-221/222			Dicer silencing by siRNA on HUVEC, hy.926 cells			[56–58]
miR-146a/b	Up	Human, mice	miR 146a$^{-/-}$ mice exposed to inflammatory cytokines	-	HUVEC	[59]
miR-126	-	Mouse / Zebrafish	miR-126$^{-/-}$ mice	-	Stem cells, zebrafish	[60–62]
VSMC and other cells						
miR-21	Up	Human	HT patients and post antihypertensive treatment	$n = 95$	Peripheral blood mononuclear cells	[63]
miR-143,	Down	Human				
miR-145,miR-133	Down	Human	Expression analysis of miRNAs involved in VSMC plasticity	$n = 29$–60	Blood cells	[64,65]
miR-21, miR-1	Up	Human				
miR-9,126	Down	Human	HT	$n = 29$–60	Blood cells	[66]
miR-126	Up	Human	HUVEC	$n = 6$	HUVEC	[67]
miR-223	Up	Human	High density lipoprotein		HCAEC	[68]
miR-34b	Down	Rats	SHR vs. Wky	$n = 36$	VSMC	[69]
miR-29a/b/c	Up	Human	Untreated essential hypertension vs. healthy individuals	$n = 30$–54	Plasma	[70]
miR-510	Up	Human	HT vs. NT	$n = 208$–220	Blood	[71]
let-7	Up	Human	Expression of let-7 in HT vs. NT with normal/increased CMIT	$n = 60$	Plasma	[72]
miR-92a	Up	Human	Expression of miR-92a in HT vs. NT with normal/increased CMIT	$n = 60$	Plasma	[73]

Abbreviations. HT: hypertensive; NT: normotensive; ISS: inverse salt sensitive; SS: salt sensitive; SR: salt-resistant; SNP: single nucleotide polymorphism; RAAS: renin-angiotensin aldosterone system; VSMC: vascular smooth muscle cells; EC: endothelial cells; KO: knockout; SSH: salt sensitive hypertension; SSR: salt sensitive resistance; HDL: high density lipoprotein; HCAEC: human coronary artery endothelial cells; HEK293T: human embryonic kidney 293T; HUVEC: human umbilical vein endothelial cells; PWV: pulse wave velocity; Wky: Wistar-Kyoto rats; SHR: spontaneous hypertensive rats; n/iCMIT: normal/increased carotid intima-media thickness; BPH/2J: hypertensive blood pressure mice; BPN/3J: normotensive blood pressure mice.

It has been shown that many major players of RAAS are regulated by miRNAs. From single-nucleotide polymorphism (SNP) datasets, the *AGTR1* gene, which encodes for angiotensin II receptor type 1, is regulated by miR-155 by its preferential binding to the A allele at position +1166 of the 3′UTR of *AGTR1* [40]. Interestingly, in hypertension cases, there is a higher prevalence for the C allele than the A allele, which reduces the ability of miR-155 to bind to *AGTR1* [40], and individuals who were homozygous for the C allele showed lower miR-155 expression and higher *AGTR1* expression, which resulted in elevated blood pressure [39]. The role of miR-155 in repressing *AGTR1* was tested and confirmed in rat cardiomyocytes, which also reduced cardiac hypertrophy [74]. Furthermore, within a hypertensive cohort, SNPs found in the miRNA binding sites of the RAAS protein genes were associated with elevated blood pressure (in *AVPR1A*) or lower blood pressure (in *BDKRB2* and *TBXA2R*). This finding suggests the role of miRNAs in blood pressure regulation via the genes of RAAS [41]. There has been a collection of reports of SNPs identified in miRNA binding sites, as well the miRNA promoter, and these have modified the binding proficiency of miRNAs to the corresponding target gene and are associated with elevated blood pressure [75–77]. The most recent SNP in the 3′UTR gene is found in the Chinese Han population, within the miRNA miR-495 binding site. The mutant C allele has increased the hypertension susceptibility, but further tests are required to determine if this SNP has altered the miRNA binding efficiency [78]. Another report of an association study of hypertensives (156) and normotensives (187) has discovered SNP rs4705342 in the miRNA promoter with a lower frequency of minor C allele among the hypertensive group. However, no further tests have been done to verify the effect of the SNP on miR-143 expression or binding [79]. A list of more of these SNPs that alter miRNA binding sites is provided in Table 2. In addition, microarray data from human, rat and mice showed that miR-483-3p can downregulate AT2R, AGT, ACE1 and ACE2 [42]. Moreover, the cluster of miR-143/145 has been shown to increase under shear stress by activation of the AMPK-p53 pathway, which in turn downregulates the ACE expression [43]. Interestingly, an earlier study has found that expression of miR-143/145 is vital for the maintenance of the VSMC contractile phenotype [80]. A recent study showed that knockout of miR-143/145 in mice resulted in them developing the loss of myogenic tone, and under induction of AngII for increased blood pressure, these knockout mice developed severe vascular inflammation and fibrosis, compared to their wildtype littermates [44].

Table 2. SNPs associated with hypertension.

SNPs	ncRNA	Gene	SNP site	Ref.
rs3749585	miR-495	CORIN	miR-495 site	[78]
rs10757274, rs2383207, rs10757278, rs1333049	CDKN2B-AS1 (lncRNA)	-	9p21.3	[81]
rs4705342	-	-	miR-143 promoter	[79]
rs17228616	-	ACHE	miR-608	[77]
rs5068	-	NPPA	miR-425 site	[76]
rs938671	-	ATP6V0A1	miR-637 site	[75]
rs5186 (A1166C)	miR-155	AGTR1	miR-155 site	[39, 40]
rs11174811	miR-526, miR-578	AVPR1A	miR-536, miR-578 sites	
rs5225, rs2069591	miR-34a, miR-34c-5p, miR-449b	BDKRN2	miR-34a, miR-34c-5p, miR-449b sites	[41]
rs13306046	miR-571, miR-765	TBXA2R	miR-571, miR-765 sites	
ss52051869	miR-122	SLC7A1	miR-122 site	[51]

On the other side, the increase of AngII as a vasoconstrictor that induces the release of aldosterone and vasopressin has also altered miRNA expression. A study on Sprague-Dawley rats showed that 10 days of AngII intravenous infusion induces cardiac hypertrophy and fibrosis and increases miR-132 and miR-212 expression in rat hearts, aortas and kidneys [45]. When applied to the human setting, blocking AngII activity by AGTR1 blocker treatment in hypertensive patients reduces the expression of miR-132 and miR-212 in the internal mammalian artery compared to the control group [45]. The β-blocker drug was also trialed, but the AGTR1 blocker was more potent in its attenuating effect. These studies suggest that miR-132 and miR-212 assist AngII-induced hypertension. Within the in vitro setting, AngII treatment in human adrenocortical cell lines increases miR-21 expression. This releases aldosterone secretion, but not cortisol. This result suggests the possibility that miR-21 could influence the abnormal aldosterone secretion in the hypertension setting and contribute to primary aldosteronism [46].

A number of studies has reported changes in the expressions of miRNAs and its possible use as a biomarker in connection with sodium homeostasis, blood pressure and renin expression. In a small cohort study (n = 10) that evaluated, using a microarray, the correlation of salt intake and blood pressure, 45 differentially-expressed miRNAs were found with miR-4516 displaying the highest expression change across salt intake variation [47]. Interestingly, the exosomes from the urine samples showed a reduction of miR-4516 expression in the inverse salt-sensitive (mean arterial pressure (MAP) decreases ≥7 mmHg with high salt intake) vs. the salt-resistant subjects (control; <7 mmHg MAP change with high salt intake) and, conversely, an increase in the expression in the salt-sensitive (≥7 mmHg increase in MAP) vs. the salt-resistant group [47]. Similarly, a recent report showed other miRNAs, miR-361-5p and miR-362-5p, being associated with salt sensitivity. Both of these miRNAs were downregulated in the salt-sensitive hypertensive group, compared to the salt-resistant essential hypertensive group [48]. Within another cohort of white hypertensive European subjects, microarray profiling on the medulla and cortex of kidney tissues along with qPCR validation showed: downregulation of miR-638 and let-7c, in the medulla, and in the renal cortex, downregulation of miR-181a, miR-638 and miR-663 and upregulation of miR-21, miR-126, miR-196 and miR-451 [49].

The above findings of the downregulation of miR-181a and miR-663 in the hypertensive human cohort can be correlated further to some in vitro and animal studies. In the human embryonic kidney cell cline (HEK293), miR-181a and miR-663 regulate the endogenous renin expression by targeting the 3′UTR of the human renin (REN) mRNA, and specifically also apoptosis-inducing factor mitochondrion-associated 1 (AIFM1) and apolipoprotein (APO E), respectively [49]. In the BHP/2J mouse circadian hypertension model [50], they found downregulation of miR-181a and increased renin expression in the active period [50].

3.2. miRNAs in Endothelial Dysfunction

Vascular endothelial dysfunction is highly associated with hypertension. Endothelial cells function to release vasodilators into the blood stream to reduce vascular resistance. As these cells face continuous hemodynamic forces with shear stress and stretch, they play an important role in development, regulation and remodeling of the vasculature. Increased blood pressure can alter the phenotype and function of the endothelial cells [82]. With endothelial dysfunction, there is a reduced vasodilatation, activation and release of inflammatory factors, an increase of reactive oxygen species (ROS), a reduction of nitric oxide, which then develops into increased vascular tone, increased vascular stiffness and pulse pressure, and sustained elevated blood pressure [83]. The maintenance of the endothelial cell phenotype involves a number of molecular players such as protein kinases, integrins, endothelial nitric oxidase synthase (eNOS) and NO, vascular endothelial growth factor (VEGF) and miRNAs [84,85]. Evidence has emerged that miRNAs are involved in angiogenesis, the proliferation and function of endothelial cells and their dysfunction.

Several human studies have shown the association of miRNAs and hypertension in the aspect of endothelial dysfunction. In experimental models, change in expression of an L-arginine transporter gene, *SLC7A1*, alters NO production and induces endothelial dysfunction. An SNP of a novel C/T polymorphism in the 3′UTR of *SLC7A1* is found in 278 essential hypertensive subjects (T allele frequency: 13.3%), compared with the 498 normotensive subjects (T allele frequency: 7.6%) [51]. The impact of the T allele is such that it disrupts the binding of transcription factor SP1, extends the 3′UTR length to increase the binding site of miR-122, which in turn reduces *SLC7A1* expression.

A tumor-suppressive miRNA, miR-505, was found to be increased in hypertensive patients from three independent cohorts [52]. The increase of miR-505 represses fibroblast growth factor 18 (FGF18), a proangiogenic factor in the endothelial cells that promotes endothelial migration and thus disables the migration and tube formation of endothelial cells [53]. Another study shows that vascular endothelial cells release circulating miRNAs to combat pathogenic virus. From plasma collection of a hypertensive and normotensive Chinese population, three significant miRNAs were isolated: human cytomegalovirus (HCMV) miRNA (hcmv-miR-UL112), miR-296-5p and let-7e. Based on in vitro transfection of HEK293 cells with the reporter gene constructs, the interferon regulatory factor 1 has been reported to be the target of hcmv-miR-UL112 [54]. Interestingly, hypertensive patients showed increased HCMV seropositivity and quantitative titers (52.7% vs. 30%, $p = 0.0005$; 1870 vs. 54 copies per 1 mL plasma, $p < 0.0001$), which suggest that cytomegalovirus causes the release of hcmv-miR-UL112 from vascular endothelial cells. The group suggested a possible new link between HCMV infection and essential hypertension [54].

There is some evidence of the involvement of miRNA and eNOS production in hypertension and in endothelial dysfunction. eNOS is responsible for the production of NO within the endothelium. Inhibition of eNOS decreases NO availability and increases oxidative stress, endothelium dysfunction and hypertension. A study reported that inflammatory release of tumor necrosis factor-α (TNF-α) induces the transcription of miR-155, which targets the 3'UTR of eNOS to inhibit its expression. Inhibition of miR-155 reverses the effect of eNOS downregulation [55]. Furthermore, the miR-221/222 cluster, known as the sensitive regulator in the endothelium, regulates the NO release by its binding to the 3'UTR of eNOS mRNA in endothelial cells, as well as other genes, such as activator STAT5a, transcription factors Ets1, Es2 and cyclin-dependent kinase cell cycle regulators p21^{Cip1} and p27^{Kip1} [56–58]. Recently, it was found that the presence of pro-inflammatory cytokines upregulates miR-146a and miR-146b. Their upregulation not only inhibits endothelial activation and represses NF-κB and MAP kinase pathways, but also targets HuR, which is an RNA binding protein that suppresses eNOS [59].

Interestingly, Dicer, the RNaseIII enzyme, not only plays a role in processing the premiRNA sequence into a shorter double-stranded miRNA, but also contributes to angiogenesis. The absence of Dicer in both the in vitro and in vivo system causes dysregulation in angiogenesis. There is a number of miRNAs that are also involved in angiogenesis. One of them is the endothelial-specific miR-126. The role of miR-126 includes: influencing the integrity and regulation of the growth of the vasculature by controlling the endothelial response to VEGF, inhibiting the negative regulators of the VEGF pathway and controlling the level of the vascular cell adhesion molecule (VCAM-1) for inflammatory adhesion [60–62].

3.3. miRNAs Involved in VSMCs and Other Cells in Hypertension

Understanding the role of vascular smooth muscle cells (VSMC) and the factors that influence their function is important for clarity on the pathogenesis and treatment of hypertension. Within the blood vessels, there is a complex interplay of neurotransmitters, circulating hormones and endothelium-derived factors for vasodilation and vasoconstriction. VSMC influences the vascular tone and regulates blood pressure, vascular resistance and tissue perfusion. Various antihypertensive agents have been designed to target the VSMCs (such as the ACE inhibitor, calcium channel blockers) [86]. miRNAs have been found to be involved in the function of VSMC, the development of arterial stiffness and the progression to hypertension [87]. The miRNA miR-21, which has been known to regulate arterial fibrosis, was reported to have a correlation with improvement in arterial stiffness [88,89]. A study with 95 essential hypertensive patients that underwent antihypertensive treatment showed a negative correlation between miR-21 and the pulse wave velocity readings [63]. A study involving 89 individuals, of which 60 had essential hypertension and 29 were normotensive [64], showed that lower miR-143, miR-145 and miR-133, but higher miR-21 and miR-1 were found in peripheral blood mononuclear cells from the hypertensive group, compared to the normotensives. Negative correlation of the diastolic blood pressure (DBP) was found with miR-143, miR-145 and miR-21, but there was a positive correlation with miR-133. Interestingly, the miR-145 expression level was also overexpressed in 22 human atherosclerotic plaques (15 hypertensive and seven control) [90]. Furthermore, miR-145 was reported to have dual role in its binding to TGFβ receptor II (TGFBR2) [65]. The modulation of TGFβ receptor 2 signaling affects the downstream expression of the matrix genes in VSMC [65]. The same group that investigated the hypertensive cohort also compared the expression of miR-9 and miR-126 in peripheral blood mononuclear cells between the hypertensive and normotensive group [66]. There was a significant lower expression found in both miR-9 and miR-126 in the hypertensive group, and their expression level showed positive correlation with the 24-h mean pulse pressure [66]. The expression level of miR-9 showed a positive correlation with the left ventricular mass index. Another study reported that both miR-126 and miR-223 are involved in regulating vascular inflammation by repressing vascular cell adhesion molecule-1 (VCAM-1) and intercellular adhesion molecule-1 (ICAM-1), respectively [67,68]. The VSMC proliferation has also been found to be influenced by the level of miR-34b. Through qPCR, in silico analysis and the luciferase assay, miR-34b was found to target cyclin-dependent kinase 6 (CDK6), which controls cell cycle progression and proliferation [69].

Recently, a number of miRNAs has been detected to be differentially expressed in essential hypertensive individuals, compared to the healthy individuals. Increased expression of circulating miR-29a/b/c and miR-510 by qPCR was found in hypertensive individuals, compared to the normotensive individuals [70,71]. Another group has reported two sets of data of the upregulation of miRNAs: let-7 and miR-92a, in correlation with the increase in carotid intima-media thickness (CMIT), compared to the normal CMIT [72,73]. This shows that the miRNA levels could reveal the development of subclinical atherosclerosis with the thickening of the CMIT. Thus, here is another piece of evidence for the possibilities of miRNAs to be used as biomarker and in this case for the detection of end-organ damage in hypertension.

A summary of the miRNAs associated with essential hypertension is further tabulated based on species (see Table 3), and the essential miRNAs with their targets are presented in Figure 1.

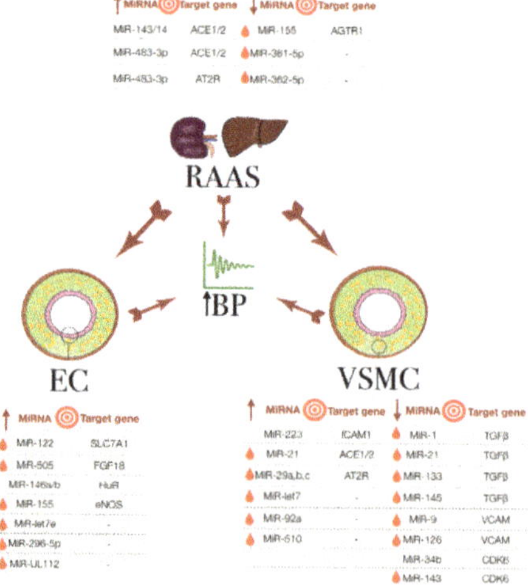

Figure 1. The miRNAs involved in essential hypertension, their association with the RAAS, EC and VSMC and their known target genes. The big central arrows indicate a system/cell type's influence on another and on the increase in blood pressure (BP). Small arrows in the tables indicate the upregulated (up arrow) or downregulated (down arrow) miRNA expression. Red droplets represent the biomarker potential from detectability in blood samples. EC: endothelial cells; RAAS: renin-angiotensin aldosterone system; VSMC: vascular smooth muscle cells.

Table 3. miRNAs associated with hypertension based on species.

Species	miRNA	SNPs/Target Gene	Subject/Model	Ref.
Human	miR-155	AGTR1: rs5186 (A1166C)	qPCR on blood mononuclear cells from 64 HT (AA: 25; AC: 20; CC: 19); HUVEC cells	[39,55]
		-	Reporter silencing assay on HEK293T	[40]
Human	miR-638, -181a, -663, let-7c	-		
Human	miR-21, -126, -196a, -451	-	Microarray. Validated by qPCR. Functional studies with HEK293 cells. qPCR HT vs. NT	[49,64,65]
Human	miR-145,133	TGF-β	qPCR HT vs. NT	[64,65]
Human	miR-122	SLC7A1: ss52051869	Genotyping, sequencing, in vitro on HT	[51]
Human	miR-505	FGF18	qPCR HT vs. NT from plasma, luciferase reporter assay	[52,53]
Human	miR-UL112,296-5p,let-7e	-	Microarray and validated by qPCR on HT vs. NT	[54]
Human	let-7	-	qPCR on let-7 in HT vs. NT with normal/increased CMIT	[72]
Human	miR-155	eNOS	qPCR and in vitro assay on HUVEC	[55]
Human	miR-143	-	qPCR HT vs. NT	[64]
Human	miR-9,126	VCAM-1, ICAM-1	qPCR HT vs. NT	[66]
Human	miR-126	VCAM	Microarray, northern blot and fucntional assay on HUVEC	[67]
Human	miR-223	ICAM-1	Whole genome and miRNA microarray on HDL treated HCAEC, qPCR, luciferase reporter assay	[68]
Human	miR-361-5p, miR-362-5p	-	qPCR on SSH vs. SRH	[48]
Human	miR-21	-	1.HT patients and post antihypertensive treatment. 2 AngII-induced H295R cells	[63]
			AngII-induced H295R cells and luciferase reporter assay	[46]
Human	miR-29a/b/c	-	untreated essential hypertension vs. healthy individuals	[70]
Human	miR-510	-	qPCR on HT vs. NT	[71]
Human	miR-92a	-	qPCR on miR-92a in HT vs. NT with normal/increased CMIT	[73]
Human	miR-4516		qPCR from exosomes of urine of HT ISS/SS/SR	[47]
Human	miR-221/222	eNOS, STAT5a, Ets1, Ets2, p21Cip1, p27Kip1	Mcroarray, Northern blotting on Dicer silenced HUVEC and and EA.hy.926 cells	[56,58]
Human, rats, mice	miR-132, 212	-	Microarray. Validated by qPCR. Humans treated: AngII blocker, β-blocker; rats treated with endothelin, mice treated with AngII	[45]
Human, rats, mice	miR-483-3p	AT2R, AGT, ACE1, ACE2	miRNA array, luciferase reporter assay on HASMC, RASMC, HL-1 cells	[42]
Human, mice	miR-146a/b	HuR	qPCR and intro assay on HUVEC and mice tissues induced by inflammatory cytokines	[59]
Rats	miR-34b	Cdk6	qPCR on SHR vs. Wky	[69]
Rats	miR-22	Chga	Luciferase reporter assay, miR-22 antagomir	[91]
Mice	miR-143/145	ACE	Shear stress on EC of Ampkα2$^{-/-}$ mice, qPCR. MiRagen Therapeutics: MGN-2677	[31,43]
Mouse	miR-181a	-	qPCR on BHP/2J mouse circadian HT	[50]
Mouse/Zebrafish	miR-126	VCAM1, SPRED-1, PIK3 regulatory subunit-2	miR-126$^{-/-}$ mice, mouse ES cells, antisense to miR-126	[60,61]

4. Recent Progress of lncRNAs in Hypertension

lncRNAs are typically greater than 200 bp in length. Though they are transcribed, 3′polyadenylated, 5′ capped and spliced, lncRNAs do not translate into proteins. There are four different types of lncRNAs based on their relative genomic location to the coding region, including: intergenic lncRNAs (or lincRNAs), intronic lncRNAs, sense lncRNAs and antisense lncRNAs [14]. Unlike miRNA, lncRNAs have numerous functions in regulating gene expression with transcription and translation (upregulating and downregulating), in splicing, imprinting and cell cycle development. lncRNAs can silence multiple genes through their interaction with chromatin, or even recruit promoters to a target gene and induce transcription. lncRNAs could also be detected in urine and blood, providing a promising future as biomarkers for disease [92].

Although the molecular mechanisms of lncRNAs are not fully understood, some studies showed that they play a role in normal physiology, as well as the development of hypertension and cardiovascular diseases. Recent studies showed that there were 68 lncRNA upregulated and 167 lncRNAs downregulated in spontaneously hypertensive rats (SHR) compared to their normotensive control (Wistar-Kyoto (WKY) rats) [93]. One particular lncRNA, XR007793, was validated to be upregulated in vitro in VSMC of hypertension. Reciprocally, knocking down of XR007793 attenuated VSMC proliferation and migration. Absence of XR007793 also inhibited interferon regulatory factor 7 (irf7), signal transducers and activators of transcription 2 (stat2) and LIM only domain 2 (limo2) [93]. Furthermore, the different expressions of 749 lncRNAs were identified between Dahl salt-sensitive vs. spontaneously hypertensive rats [94]. From these, four candidate target lncRNA-mRNA-associated genes were selected: Ankyrin repeat and SPCS box containing 3 (Asb3), cation transport regulator homolog 2 (Chac2), peroxisomal membrane 11B (Pex11b) and Sp5 transcription factor (Sp5) [94]. The lncRNAs that are specific for these genes were upregulated, while the protein of these candidate genes was downregulated. Recently, lncRNA AK098656 was detected to be upregulated in the plasma of hypertensive patients [95]. This lncRNA mediates the VSMC synthetic phenotype, which is a common characteristic in hypertension pathophysiology. A human genome-wide association study (GWAS) presented a strong association between systolic mean arterial blood pressure and lncRNA H19 locus [96]. The H19 lncRNA is expressed only during embryonic development, but it is upregulated in cardiovascular conditions [96]. In addition, a group genotyped a hypertensive vs. a normotensive cohort for SNPS that were found in a long non-coding RNA, CDKN2B-AS1 [81]. CDKN2B-AS1 is noted to contribute in some ways to regulating the cell cycle and senescence. The genotyping performed in this population showed a significant difference in the genotype frequency of the SNPs between the hypertensive and normotensive groups and strong association between rs10757274 and rs2383207 (AA) and SBP [81].

MALAT1 [97] has been reported to control vessel growth and endothelial cell function. A vascular cell-rich lncRNA, SENCR, has been shown to play a part in the smooth muscle cell phenotype [98]. Furthermore, growth arrest-specific 5 (GAS5) was found to regulate artery remodeling in caudal, renal, thoracic and carotid arteries [99]. GAS5 was also found to be downregulated, which affected endothelial proliferation and activation in hypertensive condition in the rats [99]. Interestingly, a study reported on the effects of goji berries (*Lycium barbarum* L.) on lncRNA in rats with a high salt diet. The consumption of the berries ameliorated the hypertensive condition on the borderline hypertensive rats and reduced the sONE lncRNA, which then reciprocally improved the eNOS expression [100]. The lncRNAs that have been found to be associated with essential hypertension are summarized in Table 4.

Table 4. lncRNA associated with essential hypertension.

Species	lncRNA	Cohort/Model	Function	Detection/Evaluation	Outcome	Ref.
Human	CDKN2B-AS1	HT vs. NT (Turkish)	Interacts with PRC1 & PRC2 to repress CDKN2A/B locus. Regulate VSMC stiffness	qPCR to test if published 9p21.3 SNPs are associated with BP	Significant difference in genotype freq of the 4 SNPs betw HT and NT. Association betw rs10757274 & rs2383207 (AA) and SBP.	[81]
Human	H19	87,736 indiv. + 68,368 indiv. from European ancestry	Regulator of mammlain development, inhibits cell proliferation. Methylation of H19 associated with preeclampsia and imprinting syndrome and growth retardation.	Discovery meta analysis, genome-wide SNP genotype	11 Loci with 31 genes uncovered with H19 as a lncRNA.	[96]
Human/Rat	GAS5	Transfecton of HUVEC, human VSMC, GAS5 viral knockdown in SHR vs. Wky	Regulate remodelling of arteries (caudal, carotid, renal and thoracic); regulate transcription of androgen, progesterone, mineralcorticoid receptors; involved in cellular growth arrest and apoptosis	BP measurement, tissue staining for arterial remodeling evaluation, qPCR for GAS5 expression	GAS5 expression down regulated in HT. knockdown increased SBP and DBP and mean arterial BP (in SHR) retinal neovascularization and capillary leakage, endothelial activation and proliferation	[99]
Human/Rat	AK098656	HT vs. NT (China); AK098656 transgenic rat model	Induce VSMC synthetic phenotye. Bind to myosin heavy chain-11, fibronectin-1, 26S proteasome non-ATPase regulatory subunit 11, actin, actin-binding protein	LncRNA microarray, whole-genome microarray	Upregulated in plasma of HT group vs. NT, increase VSMC proliferation & migration, upregulate extracellular matrix but downregulate contractile proteins.	[95]
Human/Mouse	MALAT1	HUVEC and MALAT1 KO model	Control cellular proliferation through histone modification	RNASeq, Microarray, qPCR	Vessel growth, endothelial cell function	[97]
Rats	XR007793	Wky/SHR and VSMC subjected to hypertensive level cyclic strain	No known predicted target	Microarray and qPCR	Kncockdown of XR007793 repress VSMC proliferation & migration. Reduced transcript expression of stat2, lmo2 and irf7.	[93]
Rats	749 lncRNAs	Dahl SS/SR and SHR	-	RNASeq, mRNA trasncrptome analysis	Asb3, Chac2, Pex11b, Sp5	[94]
Rats	sONE	Borderline hypertensive rats (BHR) fed high, medium and low salt diets	From transcription unit (NOS3AS) on opposing strand of human eNOS. Inhibiton of sONE increases eNOS and vice versa when sONE is overexpressed.	qPCR	Lycium Barbarum L. ameliorated hypertension, reduced sONE expression and improved eNOS expression compared to high salt diet rats.	[100]

Abbreviations. HT: hypertensive; NT: normotensive; SS: salt sensitive; SR: sat-resistant; SNP: single nucleotide polymorphism; VSMC: vascular smooth muscle cells; KO: knockout; SS: salt sensitive; SR: salt resistant; HUVEC: human umbilical vein endothelial cells; PWV: pulse wave velocity; Wky: Wistar-Kyoto rats; SHR: spontaneous hypertensive rats; n/iCMIT: normal/increased carotid intima-media thickness.

5. Detection of Non-Coding RNAs

miRNAs and lncRNAs have the potential to be biomarkers and targets for therapy design. Drug design on miRNA and lncRNA replacement, modulation and enhancement is a promising outlook. In the last decade, studies have reported that miRNAs and lncRNAs are packaged in microvesicles, exosomes, apoptotic bodies, high-density lipoproteins (HDL) or Ago2 as RNA-binding proteins and carried in the bloodstream [34–37,92,101,102]. This makes the miRNAs protected from degradation by RNase activity in the blood. The functions of these circulating ncRNAs are mainly to communicate to neighboring or remote target cells and to provide gene expression regulation. An example for the existence of the transfer of miRNAs in extracellular microvesicles is the communication between endothelial cells and the VSMCs of miRNAs such as miR-143 and miR-145 [34]. The facts that miRNAs and lncRNAs can be detected in circulation (serum and plasma) and urine and that they are stable in blood during transportation and storage enable sample collection to be simpler, faster and non-invasive with no requirement of tissue collection. Despite this promising advantage, more research is required to detect the origin (cell/tissue type) of these circulating miRNAs.

6. Non-Coding RNAs in the Treatment of Hypertension

As mis-expression or mutation of miRNAs has been implicated in various diseases, including hypertension, treatments utilizing ncRNAs are still in their infancy. However, there is a promising outlook with the progress in understanding their molecular mechanism. In an animal experimental model, miR-22 antagomir (LNA oligonucleotide) was administered intraperitoneally in SHR and WKY rats [91]. This administration reduced the SBP and DBP by about 18 mmHg in the SHR. Furthermore, in human microRNAome screening, miR-25 was significantly upregulated, and its upregulation delays calcium uptake kinetics specifically for SERCA 2a, in both human and mice failing cardiomyocytes [103]. When miR-25 was overexpressed by in vivo administration using adeno-associated virus 9 (AAV9), there was a confirmation of the loss of contractile function. Interestingly, using antagomirs against miR-25 in a mouse model, cardiac function improved with increased survival compared to the animals that received a control antagomir oligonucleotide [103]. According to a previously-mentioned review on miRNA therapeutics, there are two therapeutic miRNAs correlated with vasculature disorders that are currently in the pipeline of the development phase. The miRagen Therapeutics company has in the pipeline MGN-2677 for the treatment of vascular disease, which involves miR-143/145 [31]. For peripheral arterial disease treatment, mirage Therapeutics utilizes the function of miR-92 for the MGN-6114 therapy design [31].

lncRNAs have not fallen behind with respect to the interest in their use for therapeutic purposes. To overexpress lncRNA, adeno-associated viral vectors have been selected due to their low pathogenicity [104]. The inhibition or downregulation of cytosolic lncRNA has been explored using siRNA or aptamers with antisense oligonucleotide (ASO)-mediated knockdown [105–107]. Nuclear lncRNAs, on the other hand, could be downregulated using the GapmeRs system, by heteroduplex formation with the target lncRNAs for cleavage by RNase H [108,109].

Although the accessibility of these ncRNAs in circulation has promising biomarker potential, their efficiency and safe delivery for therapy still remain as challenges. The drug design utilizing these small RNA molecules has so far met the challenges of low serum stability, non-specific targeting, innate immune responses and poor pharmacological properties. Developments have emerged in the design of various delivery systems for greater bioavailability: biodegradable polymers, PEGylated liposomes and lipidoids [31]. Furthermore, the vesicles (50–500 nm) have been designed for protection from the kidney filtering system for more efficient intracellular delivery [110].

Despite the promising outlook for therapy utilizing the biology of miRNA, there are some limitations that need to be overcome and considered. As previously mentioned, a single miRNA can target several other genes, and the suppression of one gene may/may not be by more than one miRNA. A treatment using one miRNA may affect other genes that need not be dysregulated in their expression. Development of techniques for detecting microvesicles membrane markers to determine where the circulating miRNAs came from will be useful to ascertain the cell/tissue source of the miRNA. Furthermore, there are racial differences in the miRNA profiling in the disease and non-disease state. A recent study profiled the miRNA expression between the hypertensive and normotensive African American and white American cohorts [111]. There were significant mRNA/miRNA pair expression differences between the AA and white female hypertensives.

7. Conclusions

Owing to whole genome sequencing and RNA sequencing technologies, greater information of the non-coding sequences in the genome has been uncovered. Various pharmacotreatment strategies have been employed for hypertension, but due to the complexity of the disease, there still exists room for greater understanding of the disease mechanism and better therapeutic drug treatment. There are more studies that have been performed to elucidate the role of miRNAs in normal vasculature, the development of hypertension and cardiovascular disease. The current findings of miRNA have shown that there is a promising way to modulate the miRNA expression and its repressing action. The understanding of lncRNA is still in its infancy and requires greater work to understand its molecular mechanism. However, both miRNAs and lncRNAs could be detected in urine, blood and plasma, which allows them to be used as biomarkers for disease diagnosis. More research is needed to overcome the listed limitations for the miRNA mechanism for usage in future therapy.

Acknowledgments: This work is supported by NIH Grants 1R01 HL115195-01 (Hongyu Qiu) and 1R56HL137962 (Hongyu Qiu).

Author Contributions: Christiana Leimena and Hongyu Qiu wrote the manuscript and approved the final version.

Conflicts of Interest: The authors declare no conflict of interest.

References

1. Mills, K.T.; Bundy, J.D.; Kelly, T.N.; Reed, J.E.; Kearney, P.M.; Reynolds, K.; Chen, J.; He, J. Global Disparities of Hypertension Prevalence and Control: A Systematic Analysis of Population-Based Studies From 90 Countries. *Circulation* **2016**, *134*, 441–450. [CrossRef] [PubMed]
2. Rosendorff, C.; Lackland, D.T.; Allison, M.; Aronow, W.S.; Black, H.R.; Blumenthal, R.S.; Cannon, C.P.; de Lemos, J.A.; Elliott, W.J.; Findeiss, L.; et al. Treatment of Hypertension in Patients With Coronary Artery Disease. A Scientific Statement From the American Heart Association, American College of Cardiology, and American Society of Hypertension. *Circulation* **2015**, *131*, e435–e470. [CrossRef] [PubMed]
3. Whelton, P.K.; Carey, R.M.; Aronow, W.S.; Casey, D.E.; Collins, K.J.; Dennison Himmelfarb, C.; DePalma, S.M.; Gidding, S.; Jamerson, K.A.; Jones, D.W.; et al. 2017 ACC/AHA/AAPA/ABC/ACPM/AGS/ APhA/ASH/ASPC/NMA/PCNA Guideline for the Prevention, Detection, Evaluation, and Management of High Blood Pressure in Adults. A Report of the American College of Cardiology/American Heart Association Task Force on Clinical Practice Guidelines. *Hypertension* **2017**. [CrossRef]
4. Carretero, O.A.; Oparil, S. Essential Hypertension. Part I: Definition and Etiology. *Circulation* **2000**, *101*, 329–335. [CrossRef] [PubMed]
5. Poulter, N.R.; Prabhakaran, D.; Caulfield, M. Hypertension. *Lancet* **2015**, *386*, 801–812. [CrossRef]
6. Derhaschnig, U.; Testori, C.; Riedmueller, E.; Aschauer, S.; Wolzt, M.; Jilma, B. Hypertensive emergencies are associated with elevated markers of inflammation, coagulation, platelet activation and fibrinolysis. *J. Hum. Hypertens.* **2012**, *27*, 368–373. [CrossRef] [PubMed]
7. Schlaich, M.P.; Lambert, E.; Kaye, D.M.; Krozowski, Z.; Campbell, D.J.; Lambert, G.; Hastings, J.; Aggarwal, A.; Esler, M.D. Sympathetic augmentation in hypertension: Role of nerve firing, norepinephrine reuptake, and Angiotensin neuromodulation. *Hypertension* **2004**, *43*, 169–175. [CrossRef] [PubMed]
8. Cadwgan, T.M.; Benjamin, N. Evidence for altered platelet nitric oxide synthesis in essential hypertension. *J. Hypertens.* **1993**, *11*, 417–420. [CrossRef] [PubMed]

9. Brunini, T.; Moss, M.; Siqueira, M.; Meirelles, L.; Rozentul, A.; Mann, G.; Ellory, J.; Soares de Moura, R.; Mendes-Ribeiro, A. Inhibition of L-arginine transport in platelets by asymmetric dimethylarginine and N-monomethyl-L-arginine: Effects of arterial hypertension. *Clin. Exp. Pharmacol. Physiol.* **2004**, *31*, 738–740. [CrossRef] [PubMed]

10. Perticone, F.; Sciacqua, A.; Maio, R.; Perticone, M.; Maas, R.; Boger, R.H.; Tripepi, G.; Sesti, G.; Zoccali, C. Asymmetric dimethylarginine, L-arginine, and endothelial dysfunction in essential hypertension. *J. Am. Coll. Cardiol.* **2005**, *46*, 518–523. [CrossRef] [PubMed]

11. Bavishi, C.; Bangalore, S.; Messerli, F.H. Outcomes of Intensive Blood Pressure Lowering in Older Hypertensive Patients. *J. Am. Coll. Cardiol.* **2017**, *69*, 486–493. [CrossRef] [PubMed]

12. Patel, R.S.; Masi, S.; Taddei, S. Understanding the role of genetics in hypertension. *Eur. Heart J.* **2017**, *38*, 2309–2312. [CrossRef] [PubMed]

13. Kellis, M.; Wold, B.; Snyder, M.P.; Bernstein, B.E.; Kundaje, A.; Marinov, G.K.; Ward, L.D.; Birney, E.; Crawford, G.E.; Dekker, J.; et al. Defining functional DNA elements in the human genome. *Proc. Natl. Acad. Sci. USA* **2014**, *111*, 6131–6138. [CrossRef] [PubMed]

14. Murakami, K. Non-coding RNAs and hypertension-unveiling unexpected mechanisms of hypertension by the dark matter of the genome. *Curr. Hypertens. Rev.* **2015**, *11*, 80–90. [CrossRef] [PubMed]

15. Batkai, S.; Thum, T. MicroRNAs in hypertension: Mechanisms and therapeutic targets. *Curr. Hypertens. Rep.* **2012**, *14*, 79–87. [CrossRef] [PubMed]

16. Palazzo, A.F.; Lee, E.S. Non-coding RNA: What is functional and what is junk? *Front. Genet.* **2015**, *6*, 2. [CrossRef] [PubMed]

17. Cech, T.R.; Steitz, J.A. The noncoding RNA revolution-trashing old rules to forge new ones. *Cell* **2014**, *157*, 77–94. [CrossRef] [PubMed]

18. Andersen, J.; Delihas, N.; Ikenaka, K.; Green, P.J.; Pines, O.; Ilercil, O.; Inouye, M. The isolation and characterization of RNA coded by the *micF* gene in Escherichia coli. *Nucleic Acids Res.* **1987**, *15*, 2089–2101. [CrossRef] [PubMed]

19. Andersen, J.; Forst, S.A.; Zhao, K.; Inouye, M.; Delihas, N. The function of micF RNA. micF RNA is a major factor in the thermal regulation of OmpF protein in Escherichia coli. *J. Biol. Chem.* **1989**, *264*, 17961–17970. [PubMed]

20. Mizuno, T.; Chou, M.Y.; Inouye, M. A unique mechanism regulating gene expression: Translational inhibition by a complementary RNA transcript (micRNA). *Proc. Natl. Acad. Sci. USA* **1984**, *81*, 1966–1970. [CrossRef] [PubMed]

21. Lee, R.C.; Feinbaum, R.L.; Ambros, V. The *C. elegans* heterochronic gene lin-4 encodes small RNAs with antisense complementarity to lin-14. *Cell* **1993**, *75*, 843–854. [CrossRef]

22. Brockdorff, N.; Ashworth, A.; Kay, G.F.; McCabe, V.M.; Norris, D.P.; Cooper, P.J.; Swift, S.; Rastan, S. The product of the mouse Xist gene is a 15 kb inactive X-specific transcript containing no conserved ORF and located in the nucleus. *Cell* **1992**, *71*, 515–526. [CrossRef]

23. Pasquinelli, A.E.; Reinhart, B.J.; Slack, F.; Martindale, M.Q.; Kuroda, M.I.; Maller, B.; Hayward, D.C.; Ball, E.E.; Degnan, B.; Müller, P.; et al. Conservation of the sequence and temporal expression of let-7 heterochronic regulatory RNA. *Nature* **2000**, *408*, 86–89. [PubMed]

24. Reinhart, B.J.; Slack, F.J.; Basson, M.; Pasquinelli, A.E.; Bettinger, J.C.; Rougvie, A.E.; Horvitz, H.R.; Ruvkun, G. The 21-nucleotide let-7 RNA regulates developmental timing in Caenorhabditis elegans. *Nature* **2000**, *403*, 901–906. [CrossRef] [PubMed]

25. Calin, G.A.; Dumitru, C.D.; Shimizu, M.; Bichi, R.; Zupo, S.; Noch, E.; Aldler, H.; Rattan, S.; Keating, M.; Rai, K.; et al. Frequent deletions and down-regulation of micro-RNA genes miR15 and miR16 at 13q14 in chronic lymphocytic leukemia. *Proc. Natl. Acad. Sci. USA* **2002**, *99*, 15524–15529. [CrossRef] [PubMed]

26. Eis, P.S.; Tam, W.; Sun, L.; Chadburn, A.; Li, Z.; Gomez, M.F.; Lund, E.; Dahlberg, J.E. Accumulation of miR-155 and BIC RNA in human B cell lymphomas. *Proc. Natl. Acad. Sci. USA* **2005**, *102*, 3627–3632. [CrossRef] [PubMed]

27. He, L.; Thomson, J.M.; Hemann, M.T.; Hernando-Monge, E.; Mu, D.; Goodson, S.; Powers, S.; Cordon-Cardo, C.; Lowe, S.W.; Hannon, G.J.; et al. A microRNA polycistron as a potential human oncogene. *Nature* **2005**, *435*, 828–833. [CrossRef] [PubMed]

28. Costinean, S.; Zanesi, N.; Pekarsky, Y.; Tili, E.; Volinia, S.; Heerema, N.; Croce, C.M. Pre-B cell proliferation and lymphoblastic leukemia/high-grade lymphoma in Eμ-miR155 transgenic mice. *Proc. Natl. Acad. Sci. USA* **2006**, *103*, 7024–7029. [CrossRef] [PubMed]

29. Daka, A.; Peer, D. RNAi-based nanomedicines for targeted personalized therapy. *Adv. Drug Deliv. Rev.* **2012**, *64*, 1508–1521. [CrossRef] [PubMed]

30. Lam, J.K.W.; Chow, M.Y.T.; Zhang, Y.; Leung, S.W.S. siRNA Versus miRNA as Therapeutics for Gene Silencing. *Mol. Ther. Nucleic Acids* **2015**, *4*, e252. [CrossRef] [PubMed]

31. Chakraborty, C.; Sharma, A.R.; Sharma, G.; Doss, C.G.P.; Lee, S.-S. Therapeutic miRNA and siRNA: Moving from Bench to Clinic as Next Generation Medicine. *Mol. Ther. Nucleic Acids* **2017**, *8*, 132–143. [CrossRef] [PubMed]

32. Gebert, L.F.; Rebhan, M.A.; Crivelli, S.E.; Denzler, R.; Stoffel, M.; Hall, J. Miravirsen (SPC3649) can inhibit the biogenesis of miR-122. *Nucleic Acids Res.* **2014**, *42*, 609–621. [CrossRef] [PubMed]

33. Janssen, H.L.; Reesink, H.W.; Lawitz, E.J.; Zeuzem, S.; Rodriguez-Torres, M.; Patel, K.; van der Meer, A.J.; Patick, A.K.; Chen, A.; Zhou, Y.; et al. Treatment of HCV infection by targeting microRNA. *N. Engl. J. Med.* **2013**, *368*, 1685–1694. [CrossRef] [PubMed]

34. Hergenreider, E.; Heydt, S.; Tréguer, K.; Boettger, T.; Horrevoets, A.J.G.; Zeiher, A.M.; Scheffer, M.P.; Frangakis, A.S.; Yin, X.; Mayr, M.; et al. Atheroprotective communication between endothelial cells and smooth muscle cells through miRNAs. *Nat. Cell Biol.* **2012**, *14*, 249–256. [CrossRef] [PubMed]

35. Vickers, K.C.; Palmisano, B.T.; Shoucri, B.M.; Shamburek, R.D.; Remaley, A.T. MicroRNAs are transported in plasma and delivered to recipient cells by high-density lipoproteins. *Nat. Cell Biol.* **2011**, *13*, 423–433. [CrossRef] [PubMed]

36. Valadi, H.; Ekström, K.; Bossios, A.; Sjöstrand, M.; Lee, J.J.; Lötvall, J.O. Exosome-mediated transfer of mRNAs and microRNAs is a novel mechanism of genetic exchange between cells. *Nat. Cell Biol.* **2007**, *9*, 654–659. [CrossRef] [PubMed]

37. Arroyo, J.D.; Chevillet, J.R.; Kroh, E.M.; Ruf, I.K.; Pritchard, C.C.; Gibson, D.F.; Mitchell, P.S.; Bennett, C.F.; Pogosova-Agadjanyan, E.L.; Stirewalt, D.L.; et al. Argonaute2 complexes carry a population of circulating microRNAs independent of vesicles in human plasma. *Proc. Natl. Acad. Sci. USA* **2011**, *108*, 5003–5008. [CrossRef] [PubMed]

38. Coffman, T.M.; Crowley, S.D. Kidney in Hypertension: Guyton Redux. *Hypertension* **2008**, *51*, 811–816. [CrossRef] [PubMed]

39. Ceolotto, G.; Papparella, I.; Bortoluzzi, A.; Strapazzon, G.; Ragazzo, F.; Bratti, P.; Fabricio, A.S.C.; Squarcina, E.; Gion, M.; Palatini, P.; et al. Interplay Between miR-155, AT1R A1166C Polymorphism, and AT1R Expression in Young Untreated Hypertensives. *Am. J. Hypertens.* **2011**, *24*, 241–246. [CrossRef] [PubMed]

40. Sethupathy, P.; Borel, C.; Gagnebin, M.; Grant, G.R.; Deutsch, S.; Elton, T.S.; Hatzigeorgiou, A.G.; Antonarakis, S.E. Human microRNA-155 on chromosome 21 differentially interacts with its polymorphic target in the AGTR1 3′ untranslated region: A mechanism for functional single-nucleotide polymorphisms related to phenotypes. *Am. J. Hum. Genet.* **2007**, *81*, 405–413. [CrossRef] [PubMed]

41. Nossent, A.Y.; Hansen, J.L.; Doggen, C.; Quax, P.H.A.; Sheikh, S.P.; Rosendaal, F.R. SNPs in MicroRNA Binding Sites in 3′-UTRs of *RAAS* Genes Influence Arterial Blood Pressure and Risk of Myocardial Infarction. *Am. J. Hypertens.* **2011**, *24*, 999–1006. [CrossRef] [PubMed]

42. Kemp, J.R.; Unal, H.; Desnoyer, R.; Yue, H.; Bhatnagar, A.; Karnik, S.S. Angiotensin II-regulated microRNA 483-3p directly targets multiple components of the renin-angiotensin system. *J. Mol. Cell. Cardiol.* **2014**, *75*, 25–39. [CrossRef] [PubMed]

43. Kohlstedt, K.; Trouvain, C.; Boettger, T.; Shi, L.; Fisslthaler, B.; Fleming, I. AMP-activated protein kinase regulates endothelial cell angiotensin-converting enzyme expression via p53 and the post-transcriptional regulation of microRNA-143/145. *Circ. Res.* **2013**, *112*, 1150–1158. [CrossRef] [PubMed]

44. Holmberg, J.; Bhattachariya, A.; Alajbegovic, A.; Rippe, C.; Ekman, M.; Dahan, D.; Hien, T.T.; Boettger, T.; Braun, T.; Sward, K.; et al. Loss of Vascular Myogenic Tone in miR-143/145 Knockout Mice Is Associated With Hypertension-Induced Vascular Lesions in Small Mesenteric Arteries. *Arterioscler. Thromb. Vasc. Biol.* **2018**, *38*, 414–424. [CrossRef] [PubMed]

45. Eskildsen, T.; Jeppesen, P.; Schneider, M.; Nossent, A.; Sandberg, M.; Hansen, P.; Jensen, C.; Hansen, M.; Marcussen, N.; Rasmussen, L.; et al. Angiotensin II Regulates microRNA-132/-212 in Hypertensive Rats and Humans. *Int. J. Mol. Sci.* **2013**, *14*, 11190–11207. [CrossRef] [PubMed]

46. Romero, D.G.; Plonczynski, M.W.; Carvajal, C.A.; Gomez-Sanchez, E.P.; Gomez-Sanchez, C.E. Microribonucleic Acid-21 Increases Aldosterone Secretion and Proliferation in H295R Human Adrenocortical Cells. *Endocrinology* **2008**, *149*, 2477–2483. [CrossRef] [PubMed]

47. Gildea, J.J.; Carlson, J.M.; Schoeffel, C.D.; Carey, R.M.; Felder, R.A. Urinary exosome miRNome analysis and its applications to salt sensitivity of blood pressure. *Clin. Biochem.* **2013**, *46*, 1131–1134. [CrossRef] [PubMed]

48. Qi, H.; Liu, Z.; Liu, B.; Cao, H.; Sun, W.; Yan, Y.; Zhang, L. micro-RNA screening and prediction model construction for diagnosis of salt-sensitive essential hypertension. *Medicine* **2017**, *96*, e6417. [CrossRef] [PubMed]

49. Marques, F.Z.; Campain, A.E.; Tomaszewski, M.; Zukowska-Szczechowska, E.; Yang, Y.H.; Charchar, F.J.; Morris, B.J. Gene expression profiling reveals renin mRNA overexpression in human hypertensive kidneys and a role for microRNAs. *Hypertension* **2011**, *58*, 1093–1098. [CrossRef] [PubMed]

50. Jackson, K.L.; Marques, F.Z.; Watson, A.M.; Palma-Rigo, K.; Nguyen-Huu, T.P.; Morris, B.J.; Charchar, F.J.; Davern, P.J.; Head, G.A. A novel interaction between sympathetic overactivity and aberrant regulation of renin by miR-181a in BPH/2J genetically hypertensive mice. *Hypertension* **2013**, *62*, 775–781. [CrossRef] [PubMed]

51. Yang, Z.; Kaye, D.M. Mechanistic insights into the link between a polymorphism of the 3′UTR of the *SLC7A1* gene and hypertension. *Hum. Mutat.* **2009**, *30*, 328–333. [CrossRef] [PubMed]

52. Yang, Q.; Jia, C.; Wang, P.; Xiong, M.; Cui, J.; Li, L.; Wang, W.; Wu, Q.; Chen, Y.; Zhang, T. MicroRNA-505 identified from patients with essential hypertension impairs endothelial cell migration and tube formation. *Int. J. Cardiol.* **2014**, *177*, 925–934. [CrossRef] [PubMed]

53. Antoine, M.; Wirz, W.; Tag, C.G.; Gressner, A.M.; Wycislo, M.; Müller, R.; Kiefer, P. Fibroblast growth factor 16 and 18 are expressed in human cardiovascular tissues and induce on endothelial cells migration but not proliferation. *Biochem. Biophys. Res. Commun.* **2006**, *346*, 224–233. [CrossRef] [PubMed]

54. Li, S.; Zhu, J.; Zhang, W.; Chen, Y.; Zhang, K.; Popescu, L.M.; Ma, X.; Bond Lau, W.; Rong, R.; Yu, X.; et al. Signature microRNA Expression Profile of Essential Hypertension and Its Novel Link to Human Cytomegalovirus Infection. *Circulation* **2011**, *124*, 175–184. [CrossRef] [PubMed]

55. Sun, H.; Zeng, D.; Li, R.; Pang, R.; Yang, H.; Hu, Y.; Zhang, Q.; Jiang, Y.; Huang, L.; Tang, Y.; et al. Essential role of microRNA-155 in regulating endothelium-dependent vasorelaxation by targeting endothelial nitric oxide synthase. *Hypertension* **2012**, *60*, 1407–1414. [CrossRef] [PubMed]

56. Suárez, Y.; Fernández-Hernando, C.; Pober, J.S.; Sessa, W.C. Dicer Dependent MicroRNAs Regulate Gene Expression and Functions in Human Endothelial Cells. *Circ. Res.* **2007**, *100*, 1164–1173. [CrossRef] [PubMed]

57. Dentelli, P.; Rosso, A.; Orso, F.; Olgasi, C.; Taverna, D.; Brizzi, M.F. microRNA-222 controls neovascularization by regulating signal transducer and activator of transcription 5A expression. *Arterioscler. Thromb. Vasc. Biol.* **2010**, *30*, 1562–1568. [CrossRef] [PubMed]

58. Celic, T.; Metzinger-Le Meuth, V.; Six, I.; Massy, Z.A.; Metzinger, L. The mir-221/222 Cluster is a Key Player in Vascular Biology via the Fine-Tuning of Endothelial Cell Physiology. *Curr. Vasc. Pharmacol.* **2017**, *15*, 40–46. [CrossRef] [PubMed]

59. Cheng, H.S.; Sivachandran, N.; Lau, A.; Boudreau, E.; Zhao, J.L.; Baltimore, D.; Delgado-Olguin, P.; Cybulsky, M.I.; Fish, J.E. MicroRNA-146 represses endothelial activation by inhibiting pro-inflammatory pathways. *EMBO Mol. Med.* **2013**, *5*, 1017–1034. [CrossRef] [PubMed]

60. Wang, S.; Aurora, A.B.; Johnson, B.A.; Qi, X.; McAnally, J.; Hill, J.A.; Richardson, J.A.; Bassel-Duby, R.; Olson, E.N. The Endothelial-Specific MicroRNA miR-126 Governs Vascular Integrity and Angiogenesis. *Dev. Cell* **2008**, *15*, 261–271. [CrossRef] [PubMed]

61. Fish, J.E.; Santoro, M.M.; Morton, S.U.; Yu, S.; Yeh, R.-F.; Wythe, J.D.; Ivey, K.N.; Bruneau, B.G.; Stainier, D.Y.R.; Srivastava, D. miR-126 Regulates Angiogenic Signaling and Vascular Integrity. *Dev. Cell* **2008**, *15*, 272–284. [CrossRef] [PubMed]

62. Togliatto, G.; Trombetta, A.; Dentelli, P.; Gallo, S.; Rosso, A.; Cotogni, P.; Granata, R.; Falcioni, R.; Delale, T.; Ghigo, E.; et al. Unacylated ghrelin induces oxidative stress resistance in a glucose intolerance and peripheral artery disease mouse model by restoring endothelial cell miR-126 expression. *Diabetes* **2015**, *64*, 1370–1382. [CrossRef] [PubMed]

63. Parthenakis, F.; Marketou, M.; Kontaraki, J.; Patrianakos, A.; Nakou, H.; Touloupaki, M.; Vernardos, M.; Kochiadakis, G.; Chlouverakis, G.; Vardas, P. Low Levels of MicroRNA-21 Are a Marker of Reduced Arterial Stiffness in Well-Controlled Hypertension. *J. Clin. Hypertens.* **2017**, *19*, 235–240. [CrossRef] [PubMed]

64. Kontaraki, J.E.; Marketou, M.E.; Zacharis, E.A.; Parthenakis, F.I.; Vardas, P.E. Differential expression of vascular smooth muscle-modulating microRNAs in human peripheral blood mononuclear cells: Novel targets in essential hypertension. *J. Hum. Hypertens.* **2014**, *28*, 510–516. [CrossRef] [PubMed]

65. Zhao, N.; Koenig, S.N.; Trask, A.J.; Lin, C.-H.; Hans, C.P.; Garg, V.; Lilly, B. mir145 Regulates TGFBR2 Expression and Matrix Synthesis in Vascular Smooth Muscle Cells. *Circ. Res.* **2015**, *116*, 23–34. [CrossRef] [PubMed]

66. Kontaraki, J.E.; Marketou, M.E.; Zacharis, E.A.; Parthenakis, F.I.; Vardas, P.E. MicroRNA-9 and microRNA-126 expression levels in patients with essential hypertension: Potential markers of target-organ damage. *J. Am. Soc. Hypertens.* **2014**, *8*, 368–375. [CrossRef] [PubMed]

67. Harris, T.A.; Yamakuchi, M.; Ferlito, M.; Mendell, J.T.; Lowenstein, C.J. MicroRNA-126 regulates endothelial expression of vascular cell adhesion molecule 1. *Proc. Natl. Acad. Sci. USA* **2008**, *105*, 1516–1521. [CrossRef] [PubMed]

68. Tabet, F.; Vickers, K.C.; Cuesta Torres, L.F.; Wiese, C.B.; Shoucri, B.M.; Lambert, G.; Catherinet, C.; Prado-Lourenco, L.; Levin, M.G.; Thacker, S.; et al. HDL-transferred microRNA-223 regulates ICAM-1 expression in endothelial cells. *Nat. Commun.* **2014**, *5*, 3292. [CrossRef] [PubMed]

69. Yang, F.; Li, H.; Du, Y.; Shi, Q.; Zhao, L. Downregulation of microRNA34b is responsible for the elevation of blood pressure in spontaneously hypertensive rats. *Mol. Med. Rep.* **2017**, *15*, 1031–1036. [CrossRef] [PubMed]

70. Huang, Y.; Tang, S.; Huang, C.; Chen, J.; Li, J.; Cai, A.; Feng, Y. Circulating miRNA29 family expression levels in patients with essential hypertension as potential markers for left ventricular hypertrophy. *Clin. Exp. Hypertens.* **2017**, *39*, 119–125. [CrossRef] [PubMed]

71. Krishnan, R.; Mani, P.; Sivakumar, P.; Gopinath, V.; Sekar, D. Expression and methylation of circulating microRNA-510 in essential hypertension. *Hypertens. Res.* **2017**, *40*, 361–363. [CrossRef] [PubMed]

72. Huang, Y.; Huang, C.; Chen, J.; Li, J.; Feng, Y. Plasma expression level of miRNA let-7 is positively correlated with carotid intima-media thickness in patients with essential hypertension. *J. Hum. Hypertens.* **2017**, *31*, 843–847. [CrossRef] [PubMed]

73. Huang, Y.; Tang, S.; Ji-Yan, C.; Huang, C.; Li, J.; Cai, A.; Feng, Y. Circulating miR-92a expression level in patients with essential hypertension: A potential marker of atherosclerosis. *J. Hum. Hypertens.* **2017**, *31*, 200–205. [CrossRef] [PubMed]

74. Yang, Y.; Zhou, Y.; Cao, Z.; Tong, X.; Xie, H.; Luo, T.; Hua, X.; Wang, H. miR-155 functions downstream of angiotensin II receptor subtype 1 and calcineurin to regulate cardiac hypertrophy. *Exp. Ther. Med.* **2016**, *12*, 1556–1562. [CrossRef] [PubMed]

75. Wei, Z.; Biswas, N.; Wang, L.; Courel, M.; Zhang, K.; Soler-Jover, A.; Taupenot, L.; O'Connor, D.T. A common genetic variant in the 3′-UTR of vacuolar H⁺-ATPase ATP6V0A1 creates a micro-RNA motif to alter chromogranin A processing and hypertension risk. *Circ. Cardiovasc. Genet.* **2011**, *4*, 381–389. [CrossRef] [PubMed]

76. Arora, P.; Wu, C.; Khan, A.M.; Bloch, D.B.; Davis-Dusenbery, B.N.; Ghorbani, A.; Spagnolli, E.; Martinez, A.; Ryan, A.; Tainsh, L.T.; et al. Atrial natriuretic peptide is negatively regulated by microRNA-425. *J. Clin. Investig.* **2013**, *123*, 3378–3382. [CrossRef] [PubMed]

77. Hanin, G.; Shenhar-Tsarfaty, S.; Yayon, N.; Hoe, Y.Y.; Bennett, E.R.; Sklan, E.H.; Rao, D.C.; Rankinen, T.; Bouchard, C.; Geifman-Shochat, S.; et al. Competing targets of microRNA-608 affect anxiety and hypertension. *Hum. Mol. Genet.* **2014**, *23*, 4569–4580. [CrossRef] [PubMed]

78. Chen, Y.; Li, T.; Hao, Y.; Wu, B.; Li, H.; Geng, N.; Sun, Z.; Zheng, L.; Sun, Y. Association of rs2271037 and rs3749585 polymorphisms in CORIN with susceptibility to hypertension in a Chinese Han population: A case-control study. *Gene* **2018**, *651*, 79–85. [CrossRef] [PubMed]

79. Fu, X.; Guo, L.; Jiang, Z.; Zhao, L.; Xu, A. An miR-143 promoter variant associated with essential hypertension. *Int. J. Clin. Exp. Med.* **2014**, *7*, 1813–1817. [PubMed]

80. Boettger, T.; Beetz, N.; Kostin, S.; Schneider, J.; Kruger, M.; Hein, L.; Braun, T. Acquisition of the contractile phenotype by murine arterial smooth muscle cells depends on the Mir143/145 gene cluster. *J. Clin. Investig.* **2009**, *119*, 2634–2647. [CrossRef] [PubMed]

81. Bayoglu, B.; Yuksel, H.; Cakmak, H.A.; Dirican, A.; Cengiz, M. Polymorphisms in the long non-coding RNA CDKN2B-AS1 may contribute to higher systolic blood pressure levels in hypertensive patients. *Clin. Biochem.* **2016**, *49*, 821–827. [CrossRef] [PubMed]

82. Laughlin, M.H.; Newcomer, S.C.; Bender, S.B. Importance of hemodynamic forces as signals for exercise-induced changes in endothelial cell phenotype. *J. Appl. Physiol.* **2008**, *104*, 588–600. [CrossRef] [PubMed]

83. Brandes, R.P. Endothelial Dysfunction and Hypertension. *Hypertension* **2014**, *64*, 924–928. [CrossRef] [PubMed]

84. Chien, S. Mechanotransduction and endothelial cell homeostasis: The wisdom of the cell. *Am. J. Physiol. Heart Circ. Physiol.* **2007**, *292*, H1209–H1224. [CrossRef] [PubMed]

85. Nemecz, M.; Alexandru, N.; Tanko, G.; Georgescu, A. Role of MicroRNA in Endothelial Dysfunction and Hypertension. *Curr. Hypertens. Rep.* **2016**, *18*, 87. [CrossRef] [PubMed]

86. Brozovich, F.V.; Nicholson, C.J.; Degen, C.V.; Gao, Y.Z.; Aggarwal, M.; Morgan, K.G. Mechanisms of Vascular Smooth Muscle Contraction and the Basis for Pharmacologic Treatment of Smooth Muscle Disorders. *Pharmacol. Rev.* **2016**, *68*, 476–532. [CrossRef] [PubMed]

87. Nanoudis, S.; Pikilidou, M.; Yavropoulou, M.; Zebekakis, P. The Role of MicroRNAs in Arterial Stiffness and Arterial Calcification. An Update and Review of the Literature. *Front. Genet.* **2017**, *8*, 209. [CrossRef] [PubMed]

88. Lorenzen, J.M.; Schauerte, C.; Hubner, A.; Kolling, M.; Martino, F.; Scherf, K.; Batkai, S.; Zimmer, K.; Foinquinos, A.; Kaucsar, T.; et al. Osteopontin is indispensible for AP1-mediated angiotensin II-related miR-21 transcription during cardiac fibrosis. *Eur. Heart J.* **2015**, *36*, 2184–2196. [CrossRef] [PubMed]

89. Marquez, R.T.; Bandyopadhyay, S.; Wendlandt, E.B.; Keck, K.; Hoffer, B.A.; Icardi, M.S.; Christensen, R.N.; Schmidt, W.N.; McCaffrey, A.P. Correlation between microRNA expression levels and clinical parameters associated with chronic hepatitis C viral infection in humans. *Lab. Investig.* **2010**, *90*, 1727–1736. [CrossRef] [PubMed]

90. Santovito, D.; Mandolini, C.; Marcantonio, P.; de Nardis, V.; Bucci, M.; Paganelli, C.; Magnacca, F.; Ucchino, S.; Mastroiacovo, D.; Desideri, G.; et al. Overexpression of microRNA-145 in atherosclerotic plaques from hypertensive patients. *Expert Opin. Ther. Targets* **2013**, *17*, 217–223. [CrossRef] [PubMed]

91. Friese, R.S.; Altshuler, A.E.; Zhang, K.; Miramontes-Gonzalez, J.P.; Hightower, C.M.; Jirout, M.L.; Salem, R.M.; Gayen, J.R.; Mahapatra, N.R.; Biswas, N.; et al. MicroRNA-22 and promoter motif polymorphisms at the Chga locus in genetic hypertension: Functional and therapeutic implications for gene expression and the pathogenesis of hypertension. *Hum. Mol. Genet.* **2013**, *22*, 3624–3640. [CrossRef] [PubMed]

92. Bolha, L.; Ravnik-Glavač, M.; Glavač, D. Long Noncoding RNAs as Biomarkers in Cancer. *Dis. Markers* **2017**, *2017*, 7243968. [CrossRef] [PubMed]

93. Yao, Q.P.; Xie, Z.W.; Wang, K.X.; Zhang, P.; Han, Y.; Qi, Y.X.; Jiang, Z.L. Profiles of long noncoding RNAs in hypertensive rats: Long noncoding RNA XR007793 regulates cyclic strain-induced proliferation and migration of vascular smooth muscle cells. *J. Hypertens.* **2017**, *35*, 1195–1203. [CrossRef] [PubMed]

94. Gopalakrishnan, K.; Kumarasamy, S.; Mell, B.; Joe, B. Genome-Wide Identification of Long Noncoding RNAs in Rat Models of Cardiovascular and Renal Disease. *Hypertension* **2015**, *65*, 200–210. [CrossRef] [PubMed]

95. Jin, L.; Lin, X.; Yang, L.; Fan, X.; Wang, W.; Li, S.; Li, J.; Liu, X.; Bao, M.; Cui, X.; et al. AK098656, a Novel Vascular Smooth Muscle Cell–Dominant Long Noncoding RNA, Promotes Hypertension. *Hypertension* **2018**, *71*, 262–272. [CrossRef] [PubMed]

96. Tragante, V.; Barnes, M.R.; Ganesh, S.K.; Lanktree, M.B.; Guo, W.; Franceschini, N.; Smith, E.N.; Johnson, T.; Holmes, M.V.; Padmanabhan, S.; et al. Gene-centric Meta-analysis in 87,736 Individuals of European Ancestry Identifies Multiple Blood-Pressure-Related Loci. *Am. J. Hum. Genet.* **2014**, *94*, 349–360. [CrossRef] [PubMed]

97. Michalik, K.M.; You, X.; Manavski, Y.; Doddaballapur, A.; Zornig, M.; Braun, T.; John, D.; Ponomareva, Y.; Chen, W.; Uchida, S.; et al. Long noncoding RNA MALAT1 regulates endothelial cell function and vessel growth. *Circ. Res.* **2014**, *114*, 1389–1397. [CrossRef] [PubMed]

98. Bell, R.D.; Long, X.; Lin, M.; Bergmann, J.H.; Nanda, V.; Cowan, S.L.; Zhou, Q.; Han, Y.; Spector, D.L.; Zheng, D.; et al. Identification and initial functional characterization of a human vascular cell-enriched long noncoding RNA. *Arterioscler. Thromb. Vasc. Biol.* **2014**, *34*, 1249–1259. [CrossRef] [PubMed]

99. Wang, Y.; Shan, K.; Yao, M.; Yao, J.; Wang, J.; Li, X.; Liu, B.; Zhang, Y.; Ji, Y.; Jiang, Q.; et al. Long Noncoding RNA-GAS5: A Novel Regulator of Hypertension-Induced Vascular Remodeling. *Hypertension* **2016**, *68*, 736–748. [CrossRef] [PubMed]

100. Zhang, X.; Yang, X.; Lin, Y.; Suo, M.; Gong, L.; Chen, J.; Hui, R. Anti-hypertensive effect of *Lycium barbarum* L. with down-regulated expression of renal endothelial lncRNA sONE in a rat model of salt-sensitive hypertension. *Int. J. Clin. Exp. Pathol.* **2015**, *8*, 6981–6987. [PubMed]

101. Gezer, U.; Ozgur, E.; Cetinkaya, M.; Isin, M.; Dalay, N. Long non-coding RNAs with low expression levels in cells are enriched in secreted exosomes. *Cell Biol. Int.* **2014**, *38*, 1076–1079. [CrossRef] [PubMed]

102. Hewson, C.; Capraro, D.; Burdach, J.; Whitaker, N.; Morris, K.V. Extracellular vesicle associated long non-coding RNAs functionally enhance cell viability. *Non-Coding RNA Res.* **2016**, *1*, 3–11. [CrossRef] [PubMed]

103. Wahlquist, C.; Jeong, D.; Rojas-Muñoz, A.; Kho, C.; Lee, A.; Mitsuyama, S.; van Mil, A.; Jin Park, W.; Sluijter, J.P.G.; Doevendans, P.A.F.; et al. Inhibition of miR-25 improves cardiac contractility in the failing heart. *Nature* **2014**, *508*, 531–535. [CrossRef] [PubMed]

104. Gomes, C.P.C.; Spencer, H.; Ford, K.L.; Michel, L.Y.M.; Baker, A.H.; Emanueli, C.; Balligand, J.-L.; Devaux, Y. The Function and Therapeutic Potential of Long Non-coding RNAs in Cardiovascular Development and Disease. *Mol. Ther. Nucleic Acids* **2017**, *8*, 494–507. [CrossRef] [PubMed]

105. Robb, G.B.; Brown, K.M.; Khurana, J.; Rana, T.M. Specific and potent RNAi in the nucleus of human cells. *Nat. Struct. Mol. Biol.* **2005**, *12*, 133–137. [CrossRef] [PubMed]

106. Bennett, C.F.; Swayze, E.E. RNA targeting therapeutics: Molecular mechanisms of antisense oligonucleotides as a therapeutic platform. *Annu. Rev. Pharmacol. Toxicol.* **2010**, *50*, 259–293. [CrossRef] [PubMed]

107. Arun, G.; Diermeier, S.; Akerman, M.; Chang, K.-C.; Wilkinson, J.E.; Hearn, S.; Kim, Y.; MacLeod, A.R.; Krainer, A.R.; Norton, L.; et al. Differentiation of mammary tumors and reduction in metastasis upon Malat1 lncRNA loss. *Genes Dev.* **2016**, *30*, 34–51. [CrossRef] [PubMed]

108. Fluiter, K.; Mook, O.R.; Vreijling, J.; Langkjaer, N.; Hojland, T.; Wengel, J.; Baas, F. Filling the gap in LNA antisense oligo gapmers: The effects of unlocked nucleic acid (UNA) and 4′-C-hydroxymethyl-DNA modifications on RNase H recruitment and efficacy of an LNA gapmer. *Mol. Biosyst.* **2009**, *5*, 838–843. [CrossRef] [PubMed]

109. Viereck, J.; Kumarswamy, R.; Foinquinos, A.; Xiao, K.; Avramopoulos, P.; Kunz, M.; Dittrich, M.; Maetzig, T.; Zimmer, K.; Remke, J.; et al. Long noncoding RNA Chast promotes cardiac remodeling. *Sci. Transl. Med.* **2016**, *8*. [CrossRef] [PubMed]

110. Scherberich, J.E. Immunological and ultrastructural analysis of loss of tubular membrane-bound enzymes in patients with renal damage. *Clin. Chim. Acta* **1989**, *185*, 271–282. [CrossRef]

111. Dluzen, D.F.; Noren Hooten, N.; Zhang, Y.; Kim, Y.; Glover, F.E.; Tajuddin, S.M.; Jacob, K.D.; Zonderman, A.B.; Evans, M.K. Racial differences in microRNA and gene expression in hypertensive women. *Sci. Rep.* **2016**, *6*, 35815. [CrossRef] [PubMed]

International Journal of
Molecular Sciences

Review

Liddle Syndrome: Review of the Literature and Description of a New Case

Martina Tetti [1], **Silvia Monticone** [1], **Jacopo Burrello** [1], **Patrizia Matarazzo** [2], **Franco Veglio** [1], **Barbara Pasini** [3], **Xavier Jeunemaitre** [4,5,6] **and Paolo Mulatero** [1,*]

1   Division of Internal Medicine and Hypertension Unit, Department of Medical Sciences, University of Torino, Via Genova 3, 10126 Torino, Italy; tetti.martina@gmail.com (M.T.); silvia.monticone@unito.it (S.M.); jacopo.burrello@gmail.com (J.B.); franco.veglio@unito.it (F.V.)
2   Department of Pediatric Endocrinology, Regina Margherita Children Hospital, Città della Salute e della Scienza, 10126 Torino, Italy; pmatarazzo@cittadellasalute.to.it
3   Unit of Medical Genetics, Department of Medical Sciences, University of Torino, Via Genova 3, 10126 Torino, Italy; barbara.pasini@unito.it
4   INSERM U970, Paris Cardiovascular Research Center, 75015 Paris, France; xavier.jeunemaitre@aphp.fr
5   Faculty of Medicine, University Paris Descartes, Paris Sorbonne Cité, 75006 Paris, France
6   AP-HP, Department of Genetics, Hôpital Européen Georges Pompidou, 75015 Paris, France
*   Correspondence: paolo.mulatero@unito.it; Tel.: +39-0116-336-959; Fax: +39-0116-336-931

Received: 31 January 2018; Accepted: 7 March 2018; Published: 11 March 2018

Abstract: Liddle syndrome is an inherited form of low-renin hypertension, transmitted with an autosomal dominant pattern. The molecular basis of Liddle syndrome resides in germline mutations of the *SCNN1A*, *SCNN1B* and *SCNN1G* genes, encoding the α, β, and γ-subunits of the epithelial Na^+ channel (ENaC), respectively. To date, 31 different causative mutations have been reported in 72 families from four continents. The majority of the substitutions cause an increased expression of the channel at the distal nephron apical membrane, with subsequent enhanced renal sodium reabsorption. The most common clinical presentation of the disease is early onset hypertension, hypokalemia, metabolic alkalosis, suppressed plasma renin activity and low plasma aldosterone. Consequently, treatment of Liddle syndrome is based on the administration of ENaC blockers, amiloride and triamterene. Herein, we discuss the genetic basis, clinical presentation, diagnosis and treatment of Liddle syndrome. Finally, we report a new case in an Italian family, caused by a *SCNN1B* p.Pro618Leu substitution.

Keywords: hypertension; hypokalemia; low renin hypertension; monogenic hypertension; Liddle syndrome; *SCNN1A*; *SCNN1B*; *SCNN1G*

1. Introduction

Arterial hypertension, affecting about one billion people worldwide, is the most prevalent modifiable risk factor for cardiovascular diseases and related disability [1]. Essential hypertension is a multifactorial condition, resulting from a complex interaction between lifestyle and genetic factors. A positive family history increases the overall risk of developing high blood pressure and genetic factors account for 30–50% of the individual risk [2]. A minority of the hypertensive patients are affected by an inherited disease, resulting from single gene germline mutations affecting mineralocorticoid, glucocorticoid or sympathetic pathways [2,3]. Among these diseases, Liddle syndrome (LS) is caused by point mutations of the epithelial sodium channel (ENaC), that cause renal aldosterone-independent sodium reabsorption. The aim of this review is to provide an update on the current knowledge of LS, including the genetic and pathophysiological basis, the clinical features, the diagnostic and medical management and a new case is reported.

2. Liddle Syndrome

2.1. Historical Description

The first family affected by a new clinical syndrome that mimicked primary aldosteronism was reported by Liddle et al. [4,5] in 1963 (pseudoaldosteronism, subsequently named Liddle syndrome, OMIM #177200). The index case was a 16-year-old Caucasian girl, who presented with low renin resistant hypertension (180/110 mmHg), severe hypokalemia (2.8 mmol/L) and metabolic alkalosis. These features could resemble those of primary aldosteronism, but this disease was ruled out on the basis of suppressed plasma aldosterone. At the time, Liddle et al. conducted many clinical and biochemical analyses in order to further characterize this peculiar disorder. Under the condition of low sodium intake, aldosterone secretion did not increase and urinary sodium excretion rate decreased, but not to the level that would have been expected in a normal subject. Instead, the urinary sodium levels fell maximally after the administration of exogenous aldosterone. These pieces of evidence suggested an inadequate level of plasma aldosterone rather than an intrinsic renal defect in sodium reabsorption as a cause of the incapability to maximally retain sodium. Compared with patients affected by Addison's disease, subjects with LS presented a lower urinary Na$^+$ excretion, indicating a greater renal reabsorption due to a mechanism different from mineralocorticoid activity. Urinary mineralocorticoid and glucocorticoid metabolites resulted within the physiologic range. Notably, spironolactone administration neither modified urinary electrolytes excretion nor corrected hypokalemia, hence an intrinsic renal defect was hypothesized. Instead, the index case responded to the administration of triamterene, an inhibitor of the epithelial Na$^+$ channel, that induced an increase in urinary sodium and a reduction in urinary potassium excretion [4,5]. Moreover, the association of triamterene (100 mg every 8 h) and a low sodium diet normalized blood pressure (diastolic blood pressure dropped to 80 mmHg) and hypokalemia (serum K$^+$ rose to 5 mmol/L) [4,6]. Considering the biochemical profile and the response to triamterene, Liddle et al. hypothesized that the distal nephron could be the site of sodium retention. The index case developed chronic renal failure due to hypertensive nephrosclerosis and she underwent renal transplantation in 1989. This intervention corrected the disorder, normalizing blood pressure (140/79 mmHg) and the kalemia (4.2 mmol/L). After the transplantation, a regimen of low salt intake resulted in a normal increase in plasma renin activity and in plasma aldosterone concentration [6]. In 1994, Botero-Velez et al. described the extended pedigree of the family reported by Liddle et al., thus, demonstrating the autosomal dominant inheritance of the disorder [6]. Indeed, in the original manuscript, Liddle et al. described the index case and her two siblings while Botero-Velez et al. studied the index case again at the age of 49 (20 months after kidney transplantation), in addition to 43 family members. They considered as affected 18 relatives presenting with arterial hypertension, as unaffected 15 normotensive subjects with an affected parent and ten not-at-risk subjects (partners and offspring of unaffected parents) [6]. A great variability in clinical features (hypertension severity, age at onset, plasma potassium concentration, urinary aldosterone excretion levels) suggested a variable penetrance of the disease [7]. Considering the clinical response to epithelial sodium channel inhibitor triamterene and the lack of improvement using mineralocorticoid receptor antagonist and a low sodium diet, it was hypothesized that the candidate gene could be involved in the pathway of sodium handling in the distal nephron. Thus, in 1994, a complete linkage of LS to the *SCNN1B* gene (encoding the β subunit of epithelial sodium channel, ENaC) was demonstrated and the first causative mutation was identified in Liddle's original kindred as a premature stop codon, p.Arg566* (originally referred as p.Arg564* according to the homologous rat sequence) [7]. In the following years, several different germinal mutations in the *SCNN1A*, *SCNN1B* and *SCNN1G* genes, encoding, respectively, for the α, β and γ subunits of ENaC were identified, as described below.

2.2. Pathophysiology and Genetics

ENaC is a amiloride-sensitive epithelial sodium channel, localized in the apical portion of epithelial cells of distal nephron, distal colon, lung and ducts of exocrine glands [8]. Under physiological conditions, its expression and activity in the distal nephron are positively regulated by aldosterone and antidiuretic hormone and they are influenced by numerous extracellular factors, such as sodium, chloride, protons and proteases [9,10]. This channel is crucial, together with ROMK (renal outer medullary K^+) channels and Na^+/K^+ ATPase, for Na^+ reabsorption and, thus, for electrolytes homeostasis [9] (Figure 1A). The channel is a heteromeric complex constituted of three homologous subunits, α, β and γ [8,11,12], encoded by the *SCNN1A*, *SCNN1B* and *SCNN1G* genes, respectively. *SCNN1A* is located on chromosome 12p13.31, while *SCNN1B* and *SCNN1G* are located on chromosome 16p12.2 [9]. Although the α subunit alone is sufficient to induce a Na^+ current, the expression of the three subunits induces a maximal amiloride-sensitive Na^+ current [8]. The amino acid sequences of the three homologous subunits share 30–40% identity [8,9] and the protein structures are very similar, composed of two short intracellular N-terminus and C-terminus, two transmembrane domains (identified as TM1 and TM2) and a big extracellular loop [9,13]. Within the C-terminus of all three ENaC subunits, there is a highly conserved sequence, named the PY (Proline Tyrosine) motif [14]. This proline-rich sequence, PPxY, is a binding site for a member of the ubiquitin ligase family, Nedd4 (Neural precursor cell expressed, developmentally down-regulated 4), that mediates the internalization and the proteasomal degradation of the channel [9,14–16].

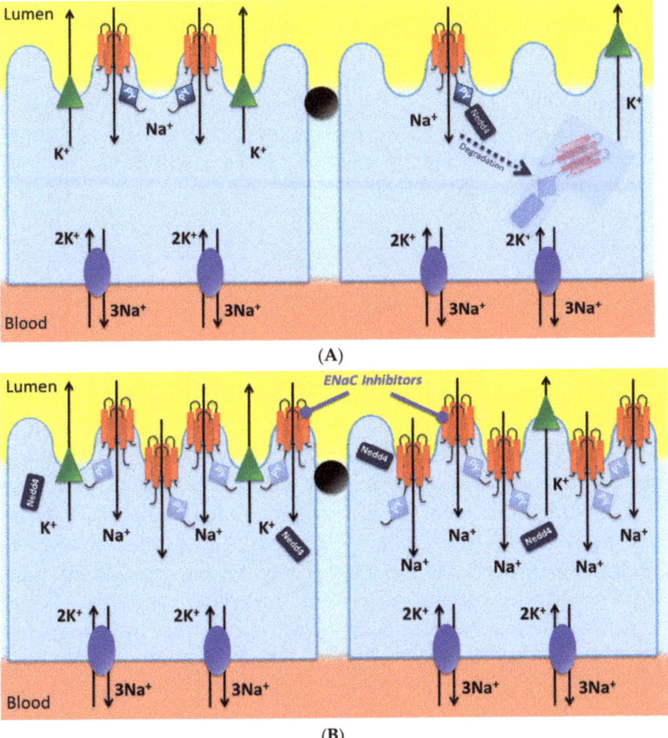

Figure 1. *Cont.*

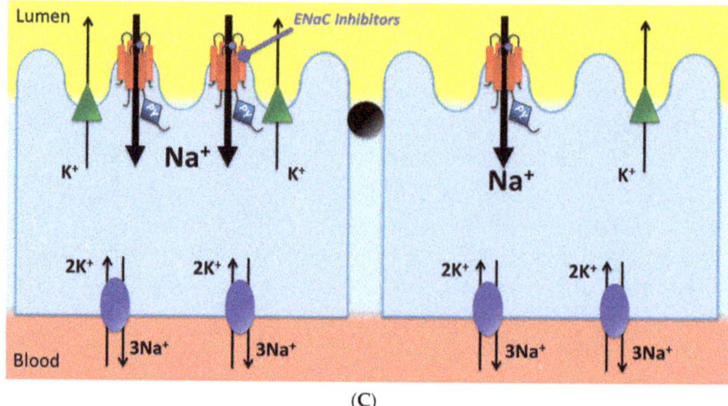

(C)

Figure 1. (**A**) Physiopathology of ENaC. Under physiological conditions, the epithelial Na⁺ channel (ENaC) is expressed on the luminal side of distal nephron epithelium. ENaC is positively regulated by aldosterone and antidiuretic hormone and allows the passage of Na⁺ ions from lumen toward cytoplasm. The proline-rich sequence (indicated as PY), located at the C-terminus of each subunit, regulates channel internalization and degradation, through Nedd4 binding and ubiquitination. ENaC function is combined with K⁺ channel ROMK (green triangles) and Na⁺/K⁺ ATPase (blue ovals) and it is crucial for hydroelectrolytic homeostasis, consisting in sodium renal reabsorption and potassium excretion; (**B**) β and γ subunits mutations. The germline mutations of the *SCNN1B* and *SCNN1G* genes causes the loss or disruption of proline-rich sequence that has a pivotal role in negative regulation of the channel. These mutations are gain-of-function and determine an increased membrane density of ENaC and a consequent increase in renal Na⁺ reabsorption; (**C**) α subunit mutation. The germline mutation of the *SCNN1A* gene affects the extracellular domain, causing the disruption of a disulphide bridge. It is a gain-of-function mutation that leads to an increase of the open probability of the channel and a consequent increase in Na⁺ current, without affecting the PY motif.

Liddle syndrome results from germline mutations in *SCNN1A*, *SCNN1B* or *SCNN1G* genes. The first mutation to be identified was the nonsense p.Arg566* substitution of the β subunit, in the large kindred described by Liddle et al. and subsequently by Botero-Velez et al. [4,6,7,17]. This mutation causes a truncation of the C-terminus of the β subunit with loss of the PY motif.

The first germinal mutation in the *SCNN1G* gene, resulting in the nonsense substitution p.Trp573*, was identified by Hansson et al. in 1995 [18]. Again, this mutation erases the γ subunit's C-terminus, causing the loss of the PY motif. In the following years, 24 different mutations of the β subunit and six different mutations of the γ subunit were identified in 72 families from different countries (Table 1). The vast majority of the reported cases are determined by missense (ten different in 30 families), nonsense (eight in 21 families) or frameshift mutations (12 in 20 families) in *SCNN1B* or *SCNN1G* genes, that cause loss or disruption of the PY motif [9,19]. The loss of the proline-rich sequence prevents the internalization and degradation of the channel via the ubiquitination-proteasomal pathway and allows the accumulation of ENaC in the distal nephron apical membrane leading to an increase in sodium reabsorption [9,20,21]. The mutations are in fact responsible for an augmented apical membrane channel density and a subsequent increase in amiloride-sensitive inward sodium current, as demonstrated by in vitro studies in *Xenopus laevis* oocytes (Figure 1B) [21]. In 1996, Firsov et al. developed a quantitative method, based on the binding of a monoclonal antibody against a FLAG epitope inserted in the extracellular domain of α, β and γ subunits, and demonstrated a significant correlation between the entity of Na⁺ inward current and the number of ENaC on the cellular membrane [22].

Interestingly, additional mechanisms have been implicated in the augmented Na⁺ reabsorption, including an increase in channel open probability [23], an increase in the fraction of proteolitically cleaved channel (active) [24], together with a reduced feedback inhibition of ENaC by intracellular Na⁺ [25].

Table 1. Clinical and biochemical phenotype of patients affected by Liddle's syndrome. *—in the original manuscript the mutation is reported according to the homologous rat sequence, HT—hypertension, SD—sudden death, LVH—left ventricular hypertrophy, TOD—target organ damage, CV—cardio vascular, n.a.—not available. Hypokalemia defined as serum K⁺ <3.5 mmol/L, hypoaldosteronemia defined as serum aldosterone <5 ng/dL or urinary aldosterone <5 µg/24 h.

Study	Country	Families (n)	Patients (Genetic/Clinical Diagnosis)	Sex (M/F, not Available)	Hypertension (n/tot Available)	Spontaneous Hypokalemia (n/tot Available)	Low Aldosterone (n/tot Available)	Reported Symptoms/TOD/CV Events/Other
			SCNN1A mutations (NM_001038.5→NP_001029.1 isoform 1)					
			p.Cys479Arg					
Salih M. (2017) [26]	The Netherlads	1	2/0	1/1	1/2	1/1	2/2	None
			SCNN1B mutations (NM_000336.2→NP_000327.2)					
			p.Gln564*					
Liu K. (2017) [27]	China	1	1/0	1/0	1/1	0/1	0/1	Stroke
			p.Arg566*					
Shimkets R.A. (1994) [7] *	USA	2	19/4	10/8, 1	19/19	3/3	3/3	Renal failure, history of juvenile CV accidents
Melander O. (1998) [28] *	Sweden	1	6/0	2/4	4/6	2/6	3/5	None
Kyuma M. (2001) [29] *	Japan	1	3/0	2/1	3/3	2/3	0/1	Muscular weakness, retinopathy
Shi J.Y. (2010) [30] *	China	1	1/0	1/0	1/1	1/1	1/1	History of stroke
Gong L. (2014) [31]	China	1	3/0	2/1	3/3	3/3	0/1	LVH
Wang L.P. (2015) [32]	China	1	1/0	1/0	1/1	1/1	0/1	None
Polfus L.M. (2016) [33]	USA	1	2/0	1/1	2/2	1/1	1/1	Asthenia, palpitation, LVH, proteinuria
Cui Y. (2017) [34] *	China	3	3/6	½	3/3	3/3	0/3	Dizziness, headache, history of SD and stroke
Liu K. (2017) [27]	China	1	3/0	2/1	3/3	3/3	0/3	LVH
			p.Gln568*					
Cui Y. (2017) [34]	China	1	1/0	1/0	1/1	1/1	0/1	Dizziness, headache
			p. Ser570Tyrfs*589					
Freerks R. (2017) [35]	South Africa (Black origin)	1	1/0	1/0	1/1	1/1	1/1	Headache, muscle fatigue, exertional dyspnoea, retinopathy, LVH
			p.Pro575Argfs*591					
Phoojaroenchanachai M. (2015) [36]	Thailand	1	2/1	1/1	2/2	2/2	1/1	Lightheadedness, proximal muscle weakness
			p. Ala579Leufs*582					
Jeunemaitre X. (1997) [37]	France	1	4/0	3/1	4/4	4/4	4/4	History of SD
			p.Gln591*					
Shimkets R.A. (1994) [7] *	USA	1	1/0	0/0, 1	1/1	0/0	0/0	n.a.
			p.Thr594Hisfs*607					
Shimkets R.A. (1994) [7] *	USA	1	1/0	0/0, 1	1/1	0/0	0/0	n.a.
			p.Ala595Argfs*607					
Findling J.W. (1997) [38] *	USA	1	8/2	1/7	5/8	2/7	7/7	Myocardial infarction

Table 1. *Cont.*

Study	Country	Families (n)	Patients (Genetic/Clinical Diagnosis)	Sex (M/F, not Available)	Hypertension (n/tot Available)	Spontaneous Hypokalemia (n/tot Available)	Low Aldosterone (n/tot Available)	Reported Symptoms/TOD/CV Events/Other
p.Arg597Profs*607								
Inoue T. (1998) [39]	Japan	1	2/4	0/2	2/2	1/2	2/2	None
Jackson S.N. (1998) [40] *	UK	1	4/1	3/1	3/4	1/2	4/4	None
Nakano Y. (2002) [41] *	Japan	1	1/0	0/1	1/1	1/1	1/1	None
Gong L. (2014) [31]	China	1	1/1	0/1	1/1	1/1	0/1	Dizziness, chronic kidney disease, SD, history of stroke
Awadalla M. (2017) [42]	USA (Black origin)	1	1/6	0/1	1/1	1/1	1/1	Proteinuria
p.Arg597Alafs*675								
Shimkets R.A. (1994) [7] *	USA	1	2/0	0/2	2/2	0/0	0/0	n.a.
p.Thr601Aspfs*607								
Ma X. (2001) [43]	China	1	8/0	5/3	8/8	4/8	8/8	Fatigue, headache, nycturia, history of cerebral hemorrhage
Hiltunen T.P. (2002) [44]	Finland	1	4/0	1/3	3/4	1/1	0/0	None
p.Pro603Alafs*607								
Cui Y. (2017) [34]	China	2	3/0	3/0	3/3	3/3	0/3	Headache, dizziness, history of stroke and SD
p.Pro616Leu								
Gao L. (2013) [45] *	China	1	7/3	5/2	7/7	7/7	0/7	Tachycardia, LVH, history of stroke
Liu K. (2017) [27]	China	3	7/0	2/5	7/7	5/7	1/7	LVH
Kuang Z.M. (2017) [46]	China	1	2/1	1/1	2/2	1/2	0/1	Muscular weakness, history of cerebral hemorrhage
p.Pro617His								
Sawathiparnich P. (2009) [47] *	Thailand	1	4/2	0/4	4/4	1/2	3/3	Headache, LVH
p.Pro617Leu								
Rossi E. (2008) [48]	Italy	1	1/2	1/0	1/1	1/1	1/1	LVH
Rossi E. (2011) [49]	Italy	1	4/4	2/2	4/4	0/4	4/4	LVH
Caretto A. (2014) [50]	Italy	1	4/1	1/3	3/3	1/3	2/2	Headache, visual scotoma, fetal growth retardation, history of stroke and cerebral hemorrhage
p.Pro617Ser								
Inoue J. (1998) [51] *	Japan	1	4/2	4/0	1/4	3/4	0/0	None
Cui Y. (2017) [34]	China	1	1/0	0/1	1/1	1/1	0/1	Headache, dizziness
p.Pro617Serfs*621								
Cui Y. (2017) [34]	China	1	1/0	1/0	1/1	0/1	0/1	Headache, dizziness
p.Pro618His								
Freundlich M. (2005) [52]	USA (Black)	1	2/0	1/1	2/2	0/1	1/1	None
Wang W. (2006) [53] *	China	1	5/0	2/3	4/5	5/5	1/1	Muscular weakness
Yang K.Q. (2018) [54]	China	1	6/2	3/3	4/6	4/6	6/6	Syncope, microalbuminuria, LVH, headache, history of stroke

Table 1. *Cont.*

Study	Country	Families (*n*)	Patients (Genetic/Clinical Diagnosis)	Sex (M/F, not Available)	Hypertension (n/tot Available)	Spontaneous Hypokalemia (n/tot Available)	Low Aldosterone (n/tot Available)	Reported Symptoms/TOD/CV Events/Other
				p.Pro618Leu				
Hansson J.H. (1995) [55] *	USA (Black origin)	1	3/0	1/2	3/3	3/3	2/2	Stroke
Uehara Y. (1998) [56] *	Japan	1	1/0	0/1	1/1	1/1	1/1	None
Takeda I. (1999) [57] *	Japan	1	1/0	0/0, 1	1/1	1/1	0/0	n.a.
Gao P.J.(2001) [58] *	China	1	5/2	4/1	4/5	1/5	1/5	History of stroke and SD
Yamashita Y. (2001) [59] *	Japan	1	1/0	1/0	1/1	0/0	0/0	n.a.
Cui Y. (2017) [34]	China	2	2/0	2/0	2/2	2/2	0/2	None
Büyükkaragöz B. (2016) [60]	Turkey	1	5/0	1/4	5/5	3/5	4/4	Headache, dizziness, LVH, retinopathy, history of SD
This manuscript	Italy	1	1/1	1/0	1/1	1/1	0/1	History of SD
				p.Pro618Arg				
Ciechanowicz A. (2005) [61] *	Czech Republic	1	2/0	2/0	2/2	2/2	2/2	None
Furuhashi M. (2005) [62] *	Japan	1	2/0	0/2	1/2	2/2	2/2	Asthenia
				p.Pro618Ser				
Uehara Y. (1998) [56] *	Japan	1	2/2	1/1	2/2	2/2	1/2	None
Bogdanović R. (2012) [63]	Serbia	1	3/3	2/1	3/3	1/3	3/3	LVH, history of SD
Wang L.P. (2012) [64] *	China	1	3/0	0/3	3/3	3/3	0/3	None
				p. Asn619Glnfs*621				
Yang K.Q. (2015) [65]	China	1	1/0	1/0	1/1	1/1	1/1	None
Cui Y. (2017) [34]	China	1	1/0	1/0	1/1	1/1	0/1	Headache, dizziness, history of stroke
Liu K. (2017) [27]	China	1	1/0	1/0	0/1	1/1	1/1	Mild impairment of renal function
				p.Tyr620His				
Tamura H. (1996) [66] *	Japan	1	5/0	3/2	5/5	2/4	4/4	Chronic kidney disease
			SCNN1G mutations (NM_001039.3→NP_001030.2)					
				p.Asn530Ser				
Hiltunen T.P. (2002) [44]	Finland	1	2/0	1/1	2/2	1/2	0/1	None
				p.Gln567*				
Shi J.Y. (2010) [30]	China	1	3/0	2/1	1/1	1/1	0/1	n.a.
Zhang P. (2017) [67]	China	1	1/0	1/0	1/1	1/1	1/1	Muscular weakness, polyuria, polydipsia, LVH
				p.Glu571*				
Wang L.P. (2015) [32]	China	1	6/1	2/4	5/6	6/6	0/6	History of stroke
Liu K. (2017) [27]	China	1	5/0	1/4	5/5	5/5	0/5	Stroke, aortic stenosis
				p.Trp573*				
Hansson J.H. (1995) [18] *	Japan	1	6/0	2/4	6/6	5/6	4/6	Leg numbness
				p.Trp575*				
Yamashita Y. (2001) [59] *	Japan	1	1/0	0/0, 1	1/1	0/0	0/0	n.a.
				p.Glu583Aspfs*585				
Wang Y. (2007) [68]	China	1	1/0	1/0	1/1	1/1	0/1	None
				TOTAL				
	-	72	200/51	97/98, 5	182/197 (92,4%)	117/163 (71,8%)	85/146 (58,2%)	-

As an example, the reported mutation p.Asn530Ser in the γ subunit [44] which is located in the TM2 segment and does not affect the PY motif, causes a two-fold increase in amiloride-sensitive Na$^+$ current, that was not associated to an increase in cell surface expression of the channel [44].

Recently, a germline mutation in the α subunit (p.Cys479Arg) was identified in a Caucasian family affected by Liddle syndrome (Table 1) [26]. This missense mutation is localized in the highly conserved extracellular domain of the subunit and leads to the disruption of a disulphide bridge. The p.Cys479Arg substitution increases the open conformation of the channel, resulting in a two-fold increase in Na$^+$ current, without affecting channel density at the plasma membrane [26].

In vivo studies conducted on mice homozygous for the *SCNN1B* p.Arg566* mutation, indicate that the transition zone between the late distal convoluted tubule and the connecting tubule, is the main nephron site of ENaC hyperactivity in LS [69], where its activity is largely aldosterone independent [70]. However, ENaC is also expressed in several brain structures, including the supraoptic nucleus, magnocellular paraventricular nucleus, hippocampus, choroid plexus, ependyma, and brain blood vessels [71]. Mice lacking Nedd4-2 (Nedd4$^{-/-}$) develop a phenotype of LS and display an increased ENaC expression in the central nervous system together with an increased blood pressure response after the infusion of Na$^+$-rich cerebrospinal fluid compared to wild-type animals [72]. Similarly, Nedd4-2$^{-/-}$ mice display a marked increase in cerebrospinal fluid Na$^+$ concentration, following a high sodium diet. Both effects were largely prevented by the intra-cerebro-ventricular infusion of the ENaC blocker benzamil, raising the question as to whether a similar mechanism could be implicated in the pathogenesis of arterial hypertension in patients affected by LS as well [72].

Interestingly, specific β ENaC single nucleotide polymorphisms (SNPs) have been associated with arterial hypertension. In particular, the SNP rs3743966 in intron 12 (c.1543-112A>T) was significantly associated with essential hypertension in Chinese hypertensive families [73] and the intronic variants rs7205273 (c.-9+11091C>T) and rs8044970 (c.311+1599T>G) were associated with blood pressure in a large Korean population [74]. The missense SNPs (rs1799979, rs149868979 and rs1799980 leading to the substitutions p.Thr594Met, p.Arg563Gln and p.Gly442Val), have been found to be associated with arterial hypertension and with increased markers of Na$^+$ channel activity [75–78]. In particular, the p.Thr594Met substitution was highly prevalent in a large population of black African origin, its frequency increased with the severity of hypertension [77] and was significantly associated with low plasma renin activity [76].The association of α ENaC polymorphisms (rs2228576, rs11542844, rs3741913) (resulting in the substitutions p.Thr663Ala, p.Ala334Thr and p.Cys618Phe) have been associated with high blood pressure in some studies, but not in others [79]. Functional studies in *Xenopus laevis* oocytes showed that the p.Cys618Phe and p.Ala663Thr polymorphisms (but not the p.Thr633Ala) increased channel activity by 3.3 and 1.6-fold, respectively [79]. Similarly, after different studies showed an association between *SCNN1G* locus and blood pressure variation [80,81], four *SCNN1G* intronic SNPs, rs13331086 (c.914-468T>G), rs11074553 (c.1077+2571G>A), rs4299163 (c.1077+3271C>G) and rs5740 (c.1176+14A>G) resulted to be associated to systolic blood pressure in the general Australian white population, after adjustment for age, sex and body mass index [82]. In particular, the association of rs13331086 was confirmed in a much larger cohort including more than 8000 individuals and the minor allele of this SNP was associated with a 1 mmHg increase in systolic blood pressure and 0.52 mmHg increase in diastolic blood pressure [83].

2.3. Diagnosis Prevalence and Phenotypes

The prevalence of Liddle syndrome across the general hypertensive population is unknown. In two recent studies, including 330 and 766 Chinese patients affected by arterial hypertension, after the exclusion of the most common secondary forms, the prevalence of Liddle syndrome resulted to be 1.52% (5/330) [32] and 0.91% (7/766) [27], respectively. Through genome-wide analysis, Pagani et al. demonstrated the presence of a common ancestor for three apparently unrelated Italian families carrying the p.Pro617Leu β mutation. Estimating the number of generations intervening between LS patients reported as unrelated, the authors suggested a much higher prevalence of LS than currently estimated [84].

The diagnosis of Liddle syndrome is based on *SCNN1A*, *SCNN1B* and *SCNN1G* gene sequencing. The genetic test is appropriate in the presence of early onset hypertension, hypokalemia, low renin and low aldosterone, with or without a positive family history. Genetic screening has to be performed also in first-degree relatives of a mutation carrier given the autosomal dominant inheritance (50% risk of transmission) and the variable phenotype reported in some families.

The typical clinical feature is resistant, early onset salt-sensitive arterial hypertension, often associated with a family history for early onset hypertension and sudden death. Biochemically, the characteristic findings are hypokalemia, metabolic alkalosis, suppressed PRA (plasma renin activity) and low serum aldosterone levels (Table 1). Hypertension results from increased Na^+ reabsorption at the distal nephron level, leading to volume expansion, which is also responsible for the observed biochemical phenotype of low renin and low serum aldosterone. At the cellular level, following ENaC opening, 3 Na^+ ions are actively exchanged for 2 K^+ ions across the basolateral membrane by the Na^+/K^+-ATPase (Figure 1A), which exit the apical membrane through different K^+ channels and are lost in the urine (resulting in hypokalemia and metabolic alkalosis) [85]. Other signs and symptoms frequently reported arise as a consequence of hypokalemia and include muscular weakness, polyuria (as low K^+ concentrations in the tubular fluid prevent the $Na^+/2Cl^-/K^+$ pump of the thick ascending limb of the loop of Henle and the Na^+/K^+ pump of the collecting duct from working properly [86] and downregulate aquaporin-2 channels [87]), polydipsia (secondary to polyuria), and as a consequence of hypertension, including headache, dizziness, retinopathy, chronic kidney disease, left ventricular hypertrophy and sudden death (supposed to be caused by malignant arrhythmias elicited by severe hypokalemia).

However, extremely severe phenotypes and mild forms can coexist, with some patients carrying a causative mutation who are normotensive (Table 1) or in whom a clinical diagnosis of LS was made in old age [88]. Systematic review of the reported cases revealed that hypertension is present in 92.4% of the patients, hypokalemia (defined as serum K^+ <3.5 mmol/L) in 71.8% and hypoaldosteronemia (defined as serum aldosterone <5 ng/dL) in 58.2% of the cases. As reported for other forms of monogenic hypertension [89,90], this variability is not only observed between kindreds carrying different mutations, but also between affected members of the same family (Table 1). It is likely that both environmental and genetic factors, including Na^+ intake and polymorphisms in genes involved in Na^+ handling could influence the phenotypic manifestation of the disease [63].

The variable expression of the clinical phenotype can hamper the diagnosis of Liddle syndrome, that might be overlooked in patients with a mild clinical manifestation.

The specific treatment of LS is represented by K^+-sparing diuretics amiloride and triamterene, that are ENaC blockers. According to the pathophysiology, the efficacy of the ENaC blockers is enhanced by dietary low salt intake (2 g NaCl/day) [54]; indeed, the competition between these molecules and sodium at the level of the ENaC ionic pore is well known [17]. ENaC blockers are effective in normalizing both blood pressure and the typical biochemical alterations (hypokalemia, suppressed PRA and low aldosterone level). Monitoring serum electrolytes during therapy is worthwhile, although the incidence of hyperkalemia is rare if renal function is normal and potassium intake is not excessive [17]. In most countries these drugs are commercialized only in association with thiazide or loop diuretic and the fixed doses could be a disadvantage in titrating therapy. Amiloride appears to be a safe and effective medication in pregnancy in reaching optimal blood pressure values and normal kalemia [42,50]. Neither hypertension nor hypokalemia improve under treatment with spironolactone (since activation of the mineralocorticoid receptor is not implicated in Na^+ reabsorption) and this might represent an additional clinical criterion to suspect Liddle syndrome [91].

3. Description of a New Case of Liddle Syndrome

The index case is a 13-year-old Caucasian boy referred to our Hypertension Unit by the Pediatric Endocrinology Department for arterial hypertension that had been diagnosed six months before. His mother was normotensive and in general good health. His father, affected by arterial hypertension and hypokalemia, died at the age of 38 of sudden cardiac death. The patient was born at term from an uneventful pregnancy to non-consanguineous parents and his past medical history was unremarkable, with no clinical signs of abnormal sexual development. On physical examination, his height and weight were 165 cm (86th percentile for age and sex) and 55 kg (80th percentile for age and sex), respectively. At diagnosis, blood pressure was 184/109 mmHg (>99th percentile for age, gender and height (SBP, 90th percentile: 125 mmHg; 95th percentile: 129 mmHg; 99th percentile: 136 mmHg. DBP, 90th percentile: 79 mmHG; 95th percentile: 83 mmHg; 99th percentile: 91 mmHg) [92] and serum potassium was 3.2 mmol/L (normal range 3.5–5.0 mmol/L). Assessment of target organ damage revealed neither left ventricular hypertrophy nor microalbuminuria.

The patient was investigated to exclude secondary forms of hypertension. Low PRA (<0.1 ng/mL/h) and low serum aldosterone were detected on different occasions (1.0–5.9 ng/dL). 17-OH-progesterone, dehydroepiandrosterone, 4-δ-androstenedione, urinary cortisol and urinary androgen catabolites resulted in normal range for sex and age. The urinary steroid profile for apparent mineralocorticoid excess syndrome resulted negative (tetra-hydrocortisol + allo-tetra-hydrocortisol/tetra-hydrocortisone = 1.45). After three months of therapy with spironolactone (50 mg daily) without clinical and biochemical response, Liddle syndrome was hypothesized. The patient was treated with amiloride (5 mg daily) that successfully controlled blood pressure (120/65 mmHg) and normalized plasma K^+ (4.8 mmol/L). The diagnosis of Liddle syndrome was confirmed by genetic analysis that identified the β ENaC germline mutation p.Pro618Leu. Considering the clinical presentation of the index case's father, the inheritance of the mutation by paternal lineage is highly probable. However, a DNA sample from his father was not available.

4. Conclusions

Liddle syndrome is genetic autosomal dominant form of low renin arterial hypertension caused by germline mutations in the *SCNN1A*, *SCNN1B* and *SCNN1G* genes, encoding, respectively, the α, β and γ subunits of the epithelial sodium channel ENaC. Despite the typical phenotype presenting with severe hypertension and hypokalemia, the disease can be clinically heterogeneous, even with mild phenotypes. Herein, we report a new case caused by the germline p.Pro618Leu mutation of the gene *SCNN1B*. The index case presented with high blood pressure and hypokalemia at the age of 13 and a family history of sudden death. Hypertension and hypokalemia were well controlled by amiloride.

Considering the frequency of early-onset hypertension and severity of correlated complications, a well-timed diagnosis of LS is very important in order to administer the proper therapy.

In conclusion, further studies are needed to better define the clinical manifestations and the real prevalence of LS, an example of actionable genetic disease that warrant a proper therapy in order to prevent target organ damage and associated cardiovascular complications.

Acknowledgments: Paolo Mulatero and Barbara Pasini are in receipt of a grant from MIUR (ex-60% 2017).

Author Contributions: All the authors contributed extensively to the work presented in this manuscript. Martina Tetti, Jacopo Burrello and Silvia Monticone performed the literature search and wrote the manuscript. Jacopo Burrello and Martina Tetti prepared the art work. Xavier Jeunemaitre performed the genetic analysis. Franco Veglio and Paolo Mulatero conceived the work and critically revised it. Barbara Pasini revised the genetic aspects of the manuscript and the literature review.

Conflicts of Interest: The authors declare no conflict of interest.

References

1. Olsen, M.H.; Angell, S.Y.; Asma, S.; Boutouyrie, P.; Burger, D.; Chirinos, J.A.; Damasceno, A.; Delles, C.; Gimenez-Roqueplo, A.P.; Hering, D.; et al. A call to action and a lifecourse strategy to address the global burden of raised blood pressure on current and future generations: The Lancet Commission on hypertension. *Lancet* **2016**, *388*, 2665–2712. [CrossRef]

2. Burrello, J.; Monticone, S.; Buffolo, F.; Tetti, M.; Veglio, F.; Williams, T.A.; Mulatero, P. Is there a role for genomics in the management of hypertension? *Int. J. Mol. Sci.* **2017**, *18*, E1131. [CrossRef] [PubMed]

3. Zennaro, M.C.; Boulkroun, S.; Fernandes-Rosa, F. Inherited forms of mineralocorticoid hypertension. *Best Pract. Res. Clin. Endocrinol. Metab.* **2015**, *29*, 633–645. [CrossRef] [PubMed]

4. Liddle, G.W.; Bledsoe, T.; Coppage, W.S.J. A familial renal disorder simulating primary aldosteronism but with negligible aldosterone secretion. *Trans. Assoc. Am. Phys.* **1963**, *76*, 199–213.

5. Palmer, B.F.; Alpern, R.J. Liddle's syndrome. *Am. J. Med.* **1998**, *104*, 301–309. [CrossRef]

6. Botero-Velez, M.; Curtis, J.J.; Warnock, D.G. Brief report: Liddle's syndrome revisited—A disorder of sodium reabsorption in the distal tubule. *N. Engl. J. Med.* **1994**, *330*, 178–181. [CrossRef] [PubMed]

7. Shimkets, R.A.; Warnock, D.G.; Bositis, C.M.; Nelson-Williams, C.; Hansson, J.H.; Schambelan, M.; Gill, J.R.; Ulick, S.; Milora, R.V.; Findling, J.W.; et al. Liddle's syndrome: Heritable human hypertension caused by mutations in the β subunit of the epithelial sodium channel. *Cell* **1994**, *79*, 407–414. [CrossRef]

8. Canessa, C.M.; Schild, L.; Buell, G.; Thorens, B.; Gautschi, I.; Horisberger, J.D.; Rossier, B.C. Amiloride-sensitive epithelial Na^+ channel is made of three homologous subunits. *Nature* **1994**, *367*, 463–467. [CrossRef] [PubMed]

9. Hanukoglu, I.; Hanukoglu, A. Epithelial sodium channel (ENaC) family: Phylogeny, structure-function, tissue distribution, and associated inherited diseases. *Gene* **2016**, *579*, 95–132. [CrossRef] [PubMed]

10. Rotin, D.; Schild, L. ENaC and its regulatory proteins as drug targets for blood pressure control. *Curr. Drug Targets* **2008**, *9*, 709–716. [CrossRef] [PubMed]

11. Staruschenko, A.; Medina, J.L.; Patel, P.; Shapiro, M.S.; Booth, R.E.; Stockand, J.D. Fluorescence resonance energy transfer analysis of subunit stoichiometry of the epithelial Na^+ channel. *J. Biol. Chem.* **2004**, *279*, 27729–27734. [CrossRef] [PubMed]

12. Jasti, J.; Furukawa, H.; Gonzales, E.B.; Gouaux, E. Structure of acid-sensing ion channel 1 at 1.9 A resolution and low pH. *Nature* **2007**, *449*, 316–323. [CrossRef] [PubMed]

13. Kashlan, O.B.; Adelman, J.L.; Okumura, S.; Blobner, B.M.; Zuzek, Z.; Hughey, R.P.; Kleyman, T.R.; Grabe, M. Constraint-based, homology model of the extracellular domain of the epithelial Na^+ channel α subunit reveals a mechanism of channel activation by proteases. *J. Biol. Chem.* **2011**, *286*, 649–660. [CrossRef] [PubMed]

14. Schild, L.; Lu, Y.; Gautschi, I.; Schneeberger, E.; Lifton, R.P.; Rossier, B.C. Identification of a PY motif in the epithelial Na channel subunits as a target sequence for mutations causing channel activation found in Liddle syndrome. *EMBO J.* **1996**, *15*, 2381–2387. [PubMed]

15. Rotin, D.; Staub, O. Role of the ubiquitin system in regulating ion transport. *Pflugers Arch.* **2011**, *461*, 1–21. [CrossRef] [PubMed]

16. Butterworth, M.B. Regulation of the epithelial sodium channel (ENaC) by membrane trafficking. *Biohim. Biophys. Acta* **2010**, *1802*, 1166–1177. [CrossRef] [PubMed]

17. Warnock, D.G. Liddle syndrome: An autosomal dominant form of human hypertension. *Kidney Int.* **1998**, *53*, 18–24. [CrossRef] [PubMed]

18. Hansson, J.H.; Nelson-Williams, C.; Suzuki, H.; Schild, L.; Shimkets, R.; Lu, Y.; Canessa, C.; Iwasaki, T.; Rossier, B.C.; Lifton, R.P. Hypertension caused by a truncated epithelial sodium channel γ subunit: Genetic heterogeneity of Liddle syndrome. *Nat. Genet.* **1995**, *11*, 76–82. [CrossRef] [PubMed]

19. Yang, K.Q.; Xiao, Y.; Tian, T.; Gao, L.G.; Zhou, X.L. Molecular genetics of Liddle's syndrome. *Clin. Chim. Acta* **2014**, *436*, 202–206. [CrossRef] [PubMed]

20. Rotin, D. Role of the UPS in Liddle syndrome. *BMC Biochem.* **2008**, *9*, S5. [CrossRef] [PubMed]

21. Schild, L.; Canessa, C.M.; Shimkets, R.A.; Gautschi, I.; Lifton, R.P.; Rossier, B.C. A mutation in the epithelial sodiumm channel causing Liddle disease increases channel activity in the *Xenopus laevis* oocyte expression system. *Proc. Natl. Acad. Sci. USA* **1995**, *92*, 5699–5703. [CrossRef] [PubMed]

22. Firsov, D.; Schild, L.; Gautschi, I.; Mérillat, A.M.; Schneeberger, E.; Rossier, B.C. Cell surface expression of the epithelial Na channel and a mutant causing Liddle syndrome: A quantitative approach. *Proc. Natl. Acad. Sci. USA* **1996**, *93*, 15370–15375. [CrossRef] [PubMed]

23. Anantharam, A.; Tian, Y.; Palmer, L.G. Open probability of the epithelial sodium channel is regulated by intracellular sodium. *J. Physiol.* **2006**, *574*, 333–347. [CrossRef] [PubMed]

24. Knight, K.K.; Olson, D.R.; Zhou, R.; Snyder, P.M. Liddle's syndrome mutations increase Na$^+$ transport through dual effects on epithelial Na$^+$ channel surface expression and proteolytic cleavage. *Proc. Natl. Acad. Sci. USA* **2006**, *103*, 2805–2808. [CrossRef] [PubMed]

25. Kellenberger, S.; Gautschi, I.; Rossier, B.C.; Schild, L. Mutations causing Liddle syndrome reduce sodium-dependent downregulation of the epithelial sodium channel in the *Xenopus* oocyte expression system. *J. Clin. Investig.* **1998**, *101*, 2741–2750. [CrossRef] [PubMed]

26. Salih, M.; Gautschi, I.; van Bemmelen, M.X.; di Benedetto, M.; Brooks, A.S.; Lugtenberg, D.; Schild, L.; Hoorn, E.J. A missense mutation in the extracellular domain of αENaC causes Liddle syndrome. *J. Am. Soc. Nephrol.* **2017**, *28*, 3291–3299. [CrossRef] [PubMed]

27. Liu, K.; Qin, F.; Sun, X.; Zhang, Y.; Wang, J.; Wu, Y.; Ma, W.; Wang, W.; Wu, X.; Qin, Y.; et al. Analysis of the genes involved in Mendelian forms of low-renin hypertension in Chinese early-onset hypertensive patients. *J. Hypertens.* **2017**. [CrossRef] [PubMed]

28. Melander, O.; Orho, M.; Fagerudd, J.; Bengtsson, K.; Groop, P.H.; Mattiasson, I.; Groop, L.; Hulthén, U.L. Mutations and variants of the epithelial sodium channel gene in Liddle's syndrome and primary hypertension. *Hypertension* **1998**, *31*, 1118–1124. [CrossRef] [PubMed]

29. Kyuma, M.; Ura, N.; Torii, T.; Takeuchi, H.; Takizawa, H.; Kitamura, K.; Tomita, K.; Sasaki, S.; Shimamoto, K. A family with liddle's syndrome caused by a mutation in the β subunit of the epithelial sodium channel. *Clin. Exp. Hypertens.* **2001**, *23*, 471–478. [CrossRef] [PubMed]

30. Shi, J.Y.; Chen, X.; Ren, Y.; Long, Y.; Tian, H.M. Liddle's syndrome caused by a novel mutation of the γ-subunit of epithelial sodium channel gene *SCNN1G* in Chinese. *Zhonghua Yi Xue Yi Chuan Xue Za Zhi* **2010**, *27*, 132–135. (In Chinese) [CrossRef] [PubMed]

31. Gong, L.; Chen, J.; Shao, L.; Song, W.; Hui, R.; Wang, Y. Phenotype-genotype analysis in two Chinese families with Liddle syndrome. *Mol. Biol. Rep.* **2014**, *41*, 1569–1575. [CrossRef] [PubMed]

32. Wang, L.P.; Yang, K.Q.; Jiang, X.J.; Wu, H.Y.; Zhang, H.M.; Zou, Y.B.; Song, L.; Bian, J.; Hui, R.T.; Liu, Y.X.; et al. Prevalence of Liddle syndrome among young hypertension patients of undetermined cause in a Chinese population. *J. Clin. Hypertens.* **2015**, *17*, 902–907. [CrossRef] [PubMed]

33. Polfus, L.M.; Boerwinkle, E.; Gibbs, R.A.; Metcalf, G.; Muzny, D.; Veeraraghavan, N.; Grove, M.; Shete, S.; Wallace, S.; Milewicz, D.; et al. Whole-exome sequencing reveals an inherited R566X mutation of the epithelial sodium channel β-subunit in a case of early-onset phenotype of Liddle syndrome. *Cold Spring Harb. Mol. Case Stud.* **2016**, *2*, a001255. [CrossRef] [PubMed]

34. Cui, Y.; Tong, A.; Jiang, J.; Wang, F.; Li, C. Liddle syndrome: Clinical and genetic profiles. *J. Clin. Hypertens.* **2017**, *19*, 524–529. [CrossRef] [PubMed]

35. Freercks, R.; Meldau, S.; Jones, E.; Ensor, J.; Weimers-Willard, C.; Rayner, B. Liddle's syndrome in an African male due to a novel frameshift mutation in the β-subunit of the epithelial sodium channel gene. *Cardiovasc. J. Afr.* **2017**, *28*, E4–E6. [CrossRef] [PubMed]

36. Phoojaroenchanachai, M.; Buranakitjaroen, P.; Limwongse, C. Liddle's syndrome: A case report. *J. Med. Assoc. Thail.* **2015**, *98*, 1035–1040. [PubMed]

37. Jeunemaitre, X.; Bassilana, F.; Persu, A.; Dumont, C.; Champigny, G.; Lazdunski, M.; Corvol, P.; Barbry, P. Genotype-phenotype analysis of a newly discovered family with Liddle's syndrome. *J. Hypertens.* **1997**, *15*, 1091–1100. [CrossRef] [PubMed]

38. Findling, J.W.; Raff, H.; Hansson, J.H.; Lifton, R.P. Liddle's syndrome: Prospective genetic screening and suppressed aldosterone secretion in an extended kindred. *J. Clin. Endocrinol. Metab.* **1997**, *82*, 1071–1074. [CrossRef] [PubMed]

39. Inoue, T.; Okauchi, Y.; Matsuzaki, Y.; Kuwajima, K.; Kondo, H.; Horiuchi, N.; Nakao, K.; Iwata, M.; Yokogoshi, Y.; Shintani, Y.; et al. Identification of a single cytosine base insertion mutation at Arg-597 of the β subunit of the human epithelial sodium channel in a family with Liddle's disease. *Eur. J. Endocrinol.* **1998**, *138*, 691–697. [CrossRef] [PubMed]

40. Jackson, S.N.J.; Williams, B.; Houtman, P.; Trembath, R.C. The diagnosis of Liddle Syndrome by identification of a mutation in the β subunit of the epithelial sodium channel. *J. Med. Genet.* **1998**, *35*, 510–512. [CrossRef] [PubMed]

41. Nakano, Y.; Ishida, T.; Ozono, R.; Matsuura, H.; Yamamoto, Y.; Kambe, M.; Chayama, K.; Oshima, T. A frameshift mutation of β subunit of epithelial sodium channel in a case of isolated Liddle syndrome. *J. Hypertens.* **2002**, *20*, 2379–2382. [CrossRef] [PubMed]

42. Awadalla, M.; Patwardhan, M.; Alsamsam, A.; Imran, N. Management of Liddle syndrome in pregnancy: A case report and literature review. *Case Rep. Obstet. Gynecol.* **2017**, *2017*, 6279460. [CrossRef] [PubMed]

43. Ma, X.; Tian, Y.; Gao, Y.; Guo, X. A study of mutation(s) of the epithelial sodium channel gene in a Liddle's syndrome family. *Zhonghua Nei Ke Za Zhi* **2001**, *40*, 390–393. (In Chinese) [PubMed]

44. Hiltunen, T.P.; Hannila-Handelberg, T.; Petäjäniemi, N.; Kantola, I.; Tikkanen, I.; Virtamo, J.; Gautschi, I.; Schild, L.; Kontula, K. Liddle's syndrome associated with a point mutation in the extracellular domain of the epithelial sodium channel γ subunit. *J. Hypertens.* **2002**, *20*, 2383–2390. [CrossRef] [PubMed]

45. Gao, L.; Wang, L.; Liu, Y.; Zhou, X.; Hui, R.; Hu, A. A family with Liddle syndrome caused by a novel missense mutation in the PY motif of the β-subunit of the epithelial sodium channel. *J. Pediatr.* **2013**, *162*, 166–170. [CrossRef] [PubMed]

46. Kuang, Z.M.; Wang, Y.; Wang, J.J.; Liu, J.H.; Zeng, R.; Zhou, Q.; Yu, Z.Q.; Jiang, L. The importance of genetic counseling and genetic screening: A case report of a 16-year-old boy with resistant hypertension and severe hypokalemia. *J. Am. Soc. Hypertens.* **2017**, *11*, 136–139. [CrossRef] [PubMed]

47. Sawathiparnich, P.; Sumboonnanonda, A.; Weerakulwattana, P.; Limwongse, C. A novel mutation in the β-subunit of the epithelial sodium channel gene (*SCNN1B*) in a Thai family with Liddle's syndrome. *J. Pediatr. Endocrinol. Metab.* **2009**, *22*, 85–89. [CrossRef] [PubMed]

48. Rossi, E.; Farnetti, E.; Debonneville, A.; Nicoli, D.; Grasselli, C.; Regolisti, G.; Negro, A.; Perazzoli, F.; Casali, B.; Mantero, F.; et al. Liddle's syndrome caused by a novel missense mutation (P617L) of the epithelial sodium channel β subunit. *J. Hypertens.* **2008**, *26*, 921–927. [CrossRef] [PubMed]

49. Rossi, E.; Farnetti, E.; Nicoli, D.; Sazzini, M.; Perazzoli, F.; Regolisti, G.; Grasselli, C.; Santi, R.; Negro, A.; Mazzeo, V.; et al. A clinical phenotype mimicking essential hypertension in a newly discovered family with Liddle's syndrome. *Am. J. Hypertens.* **2011**, *24*, 930–935. [CrossRef] [PubMed]

50. Caretto, A.; Primerano, L.; Novara, F.; Zuffardi, O.; Genovese, S.; Rondinelli, M. A therapeutic challenge: Liddle's syndrome managed with amiloride during pregnancy. *Case Rep. Obstet. Gynecol.* **2014**, *2014*, 156250. [CrossRef] [PubMed]

51. Inoue, J.; Iwaoka, T.; Tokunaga, H.; Takamune, K.; Naomi, S.; Araki, M.; Takahama, K.; Yamaguchi, K.; Tomita, K. A family with Liddle's syndrome caused by a new missense mutation in the β subunit of the epithelial sodium channel. *J. Clin. Endocrinol. Metab.* **1998**, *83*, 2210–2213. [CrossRef] [PubMed]

52. Freundlich, M.; Ludwig, M. A novel epithelial sodium channel β-subunit mutation associated with hypertensive Liddle syndrome. *Pediatr. Nephrol.* **2005**, *20*, 512–515. [CrossRef] [PubMed]

53. Wang, W.; Zhou, W.; Jiang, L.; Cui, B.; Ye, L.; Su, T.; Wang, J.; Li, X.; Ning, G. Mutation analysis of *SCNN1B* in a family with Liddle's syndrome. *Endocrine* **2006**, *29*, 385–390. [CrossRef]

54. Yang, K.Q.; Lu, C.X.; Fan, P.; Zhang, Y.; Meng, X.; Dong, X.Q.; Luo, F.; Liu, Y.X.; Zhang, H.M.; Wu, H.Y.; et al. Genetic screening of *SCNN1B* and *SCNN1G* genes in early-onset hypertensive patients helps to identify Liddle syndrome. *Clin. Exp. Hypertens.* **2018**, *40*, 107–111. [CrossRef] [PubMed]

55. Hansson, J.H.; Schild, L.; Lu, Y.; Wilson, T.A.; Gautschi, I.; Shimkets, R.; Nelson-Williams, C.; Rossier, B.C.; Lifton, R.P. A de novo missense mutation of the β subunit of the epithelial sodium channel causes hypertension and Liddle syndrome, identifying a proline-rich segment critical for regulation of channel activity. *Proc. Natl. Acad. Sci. USA* **1995**, *92*, 11495–11499. [CrossRef] [PubMed]

56. Uehara, Y.; Sasaguri, M.; Kinoshita, A.; Tsuji, E.; Kiyose, H.; Taniguchi, H.; Noda, K.; Ideishi, M.; Inoue, J.; Tomita, K.; et al. Genetic analysis of the epithelial sodium channel in Liddle's syndrome. *J. Hypertens.* **1998**, *16*, 1131–1135. [CrossRef] [PubMed]

57. Takeda, I.; Horii, M.; Yamano, S.; Kawamoto, A.; Shiiki, H.; Fujimoto, T.; Hashimoto, T.; Doi, K. Case of nonfamilial idiopathic Liddle syndrome. *Nihon Naika Gakkai Zasshi* **1999**, *88*, 339–341. (In Japanese) [CrossRef] [PubMed]

58. Gao, P.J.; Zhang, K.X.; Zhu, D.L.; He, X.; Han, Z.Y.; Zhan, Y.M.; Yang, L.W. Diagnosis of Liddle syndrome by genetic analysis of β and γ subunits of epithelial sodium channel—A report of five affected family members. *J. Hypertens.* **2001**, *19*, 885–889. [CrossRef] [PubMed]

59. Yamashita, Y.; Koga, M.; Takeda, Y.; Enomoto, N.; Uchida, S.; Hashimoto, K.; Yamano, S.; Dohi, K.; Marumo, F.; Sasaki, S. Two sporadic cases of Liddle's syndrome caused by de novo ENaC mutations. *Am. J. Kidney Dis.* **2001**, *37*, 499–504. [CrossRef] [PubMed]

60. Büyükkaragöz, B.; Yilmaz, A.C.; Karcaaltincaba, D.; Ozdemir, O.; Ludwig, M. Liddle syndrome in a Turkish family with heterogeneous phenotypes. *Pediatr. Int.* **2016**, *58*, 801–804. [CrossRef] [PubMed]

61. Ciechanowicz, A.; Dolezel, Z.; Placha, G.; Starha, J.; Góra, J.; Gaciong, Z.; Brodkiewicz, A.; Adler, G. Liddle syndrome caused by P616R mutation of the epithelial sodium channel β subunit. *Pediatr. Nephrol.* **2005**, *20*, 837–838. [CrossRef] [PubMed]

62. Furuhashi, M.; Kitamura, K.; Adachi, M.; Miyoshi, T.; Wakida, N.; Ura, N.; Shikano, Y.; Shinshi, Y.; Sakamoto, K.; Hayashi, M.; et al. Liddle's syndrome caused by a novel mutation in the proline-rich PY motif of the epithelial sodium channel β-subunit. *J. Clin. Endocrinol. Metab.* **2005**, *90*, 340–344. [CrossRef] [PubMed]

63. Bogdanović, R.; Kuburović, V.; Stajić, N.; Mughal, S.S.; Hilger, A.; Ninić, S.; Prijić, S.; Ludwig, M. Liddle syndrome in a Serbian family and literature review of underlying mutations. *Eur. J. Pediatr.* **2012**, *171*, 471–478. [CrossRef] [PubMed]

64. Wang, L.P.; Gao, L.G.; Zhou, X.L.; Wu, H.Y.; Zhang, L.; Wen, D.; Li, Y.H.; Liu, Y.X.; Tian, T.; Fan, X.H.; et al. Genetic diagnosis of Liddle's syndrome by mutation analysis of *SCNN1B* and *SCNN1G* in a Chinese family. *Chin. Med. J.* **2012**, *125*, 1401–1404. [PubMed]

65. Yang, K.Q.; Lu, C.X.; Xiao, Y.; Liu, Y.X.; Jiang, X.J.; Zhang, X.; Zhou, X.L. A novel frameshift mutation of epithelial sodium channel β-subunit leads to Liddle syndrome in an isolated case. *Clin. Endocrinol.* **2015**, *82*, 611–614. [CrossRef] [PubMed]

66. Tamura, H.; Schild, L.; Enomoto, N.; Matsui, N.; Marumo, F.; Rossier, B.C. Liddle disease caused by a missense mutation of β subunit of the epithelial sodium channel gene. *J. Clin. Investig.* **1996**, *97*, 1780–1784. [CrossRef] [PubMed]

67. Zhang, P.; Kong, L.Q.; Huang, D.J. Liddle syndrome complicating with nonfunctional adrenal cortical adenoma: A case report. *Zhonghua Xin Xue Guan Bing Za Zhi* **2017**, *45*, 331–332. (In Chinese) [CrossRef] [PubMed]

68. Wang, Y.; Zheng, Y.; Chen, J.; Wu, H.; Zheng, D.; Hui, R. A novel epithelial sodium channel γ-subunit de novo frameshift mutation leads to Liddle syndrome. *Clin. Endocrinol.* **2007**, *67*, 801–804. [CrossRef] [PubMed]

69. Nesterov, V.; Krueger, B.; Bertog, M.; Dahlmann, A.; Palmisano, R.; Korbmacher, C. In Liddle Syndrome, Epithelial sodium channel is hyperactive mainly in the early part of the aldosterone-sensitive distal nephron. *Hypertension* **2016**, *67*, 1256–1262. [CrossRef] [PubMed]

70. Nesterov, V.; Dahlmann, A.; Krueger, B.; Bertog, M.; Loffing, J.; Korbmacher, C. Aldosterone-dependent and -independent regulation of the epithelial sodium channel (ENaC) in mouse distal nephron. *Am. J. Physiol. Renal Physiol.* **2012**, *303*, F1289–F1299. [CrossRef] [PubMed]

71. Amin, M.S.; Wang, H.W.; Reza, E.; Whitman, S.C.; Tuana, B.S.; Leenen, F.H. Distribution of epithelial sodium channels and mineralocorticoid receptors in cardiovascular regulatory centers in rat brain. *Am. J. Physiol. Regul. Integr. Comp. Physiol.* **2005**, *289*, R1787–R1797. [CrossRef] [PubMed]

72. Van Huysse, J.W.; Amin, M.S.; Yang, B.; Leenen, F.H. Salt-induced hypertension in a mouse model of Liddle syndrome is mediated by epithelial sodium channels in the brain. *Hypertension* **2012**, *60*, 691–696. [CrossRef] [PubMed]

73. Wang, Y.; Liu, Z.; Hua, Q.; Chen, Y.; Cai, Y.; Liu, R. Association of epithelial sodium channel β-subunit common polymorphism with essential hypertension families in a Chinese population. *Cell Biochem. Biophys.* **2014**, *70*, 1277–1282. [CrossRef] [PubMed]

74. Jin, H.S.; Hong, K.W.; Lim, J.E.; Hwang, S.Y.; Lee, S.H.; Shin, C.; Park, H.K.; Oh, B. Genetic variations in the sodium balance-regulating genes *ENaC*, *NEDD4L*, *NDFIP2* and *USP2* influence blood pressure and hypertension. *Kidney Blood Press. Res.* **2010**, *33*, 15–23. [CrossRef] [PubMed]

75. Persu, A.; Barbry, P.; Bassilana, F.; Houot, A.M.; Mengual, R.; Lazdunski, M.; Corvol, P.; Jeunemaitre, X. Genetic analysis of the β subunit of the epithelial Na$^+$ channel in essential hypertension. *Hypertension* **1998**, *32*, 129–137. [CrossRef] [PubMed]

76. Baker, E.H.; Dong, Y.B.; Sagnella, G.A.; Rothwell, M.; Onipinla, A.K.; Markandu, N.D.; Cappuccio, F.P.; Cook, D.G.; Persu, A.; Corvol, P.; et al. Association of hypertension with T594M mutation in β subunit of epithelial sodium channels in black people resident in London. *Lancet* **1998**, *351*, 1388–1392. [CrossRef]

77. Dong, Y.B.; Zhu, H.D.; Baker, E.H.; Sagnella, G.A.; MacGregor, G.A.; Carter, N.D.; Wicks, P.D.; Cook, D.G.; Cappuccio, F.P. T594M and G442V polymorphisms of the sodium channel β subunit and hypertension in a black population. *J. Hum. Hypertens.* **2001**, *15*, 425–430. [CrossRef] [PubMed]

78. Rayner, B.L.; Owen, E.P.; King, J.A.; Soule, S.G.; Vreede, H.; Marais, D.; Davidson, J.S. A new mutation, R563Q, of the β subunit of the epithelial sodium channel associated with low-renin. Low-aldosterone hypertension. *J. Hypertens.* **2003**, *21*, 921–926. [CrossRef] [PubMed]

79. Tong, Q.; Menon, A.G.; Stockand, J.D. Functional polymorphisms in the α-subunit of the human epithelial Na+ channel increase activity. *Am. J. Physiol. Renal Physiol.* **2006**, *290*, F821–F827. [CrossRef] [PubMed]

80. Wong, Z.Y.; Stebbing, M.; Ellis, J.A.; Lamantia, A.; Harrap, S.B. Genetic linkage of β and γ subunits of epithelial sodium channel to systolic blood pressure. *Lancet* **1999**, *353*, 1222–1225. [CrossRef]

81. De Lange, M.; Spector, T.D.; Andrew, T. Genome-wide scan for blood pressure suggests linkage to chromosome 11, and replication of loci on 16, 17, and 22. *Hypertension* **2004**, *44*, 872–877. [CrossRef] [PubMed]

82. Büsst, C.J.; Scurrah, K.J.; Ellis, J.A.; Harrap, S.B. Selective genotyping reveals association between the epithelial sodium channel γ-subunit and systolic blood pressure. *Hypertension* **2007**, *50*, 672–678. [CrossRef] [PubMed]

83. Büsst, C.J.; Bloomer, L.D.; Scurrah, K.J.; Ellis, J.A.; Barnes, T.A.; Charchar, F.J.; Braund, P.; Hopkins, P.N.; Samani, N.J.; Hunt, S.C.; et al. The epithelial sodium channel γ-subunit gene and blood pressure: Family based association, renal gene expression, and physiological analyses. *Hypertension* **2011**, *58*, 1073–1078. [CrossRef] [PubMed]

84. Pagani, L.; Diekmann, Y.; Sazzini, M.; De Fanti, S.; Rondinelli, M.; Farnetti, E.; Casali, B.; Caretto, A.; Novara, F.; Zuffardi, O.; et al. Three reportedly unrelated families with liddle syndrome inherited from a common ancestor. *Hypertension* **2018**, *71*, 273–279. [CrossRef] [PubMed]

85. Bubien, J.K. Epithelial Na$^+$ channel (ENaC), hormones, and hypertension. *J. Biol. Chem.* **2010**, *285*, 23527–23531. [CrossRef] [PubMed]

86. Unwin, R.J.; Luft, F.C.; Shirley, D.G. Pathophysiology and management of hypokalemia: A clinical perspective. *Nat. Rev. Nephrol.* **2011**, *7*, 75–84. [CrossRef] [PubMed]

87. Amlal, H.; Krane, C.M.; Chen, Q.; Soleimani, M. Early polyuria and urinary concentrating defect in potassium deprivation. *Am. J. Physiol. Renal Physiol.* **2000**, *279*, F655–F663. [CrossRef] [PubMed]

88. Pepersack, T.; Allegre, S.; Jeunemaître, X.; Leeman, M.; Praet, J.P. Liddle syndrome phenotype in an octogenarian. *J. Clin. Hypertens.* **2015**, *17*, 59–60. [CrossRef] [PubMed]

89. Monticone, S.; Buffolo, F.; Tetti, M.; Veglio, F.; Pasini, B.; Mulatero, P. Genetics in endocrinology: The expanding genetic horizon of primary aldosteronism. *Eur. J. Endocrinol.* **2018**, *178*, R101–R111. [CrossRef] [PubMed]

90. Mulatero, P.; di Cella, S.M.; Williams, T.A.; Milan, A.; Mengozzi, G.; Chiandussi, L.; Gomez-Sanchez, C.E.; Veglio, F. Glucocorticoid remediable aldosteronism: Low morbidity and mortality in a four-generation italian pedigree. *J. Clin. Endocrinol. Metab.* **2002**, *87*, 3187–3191. [CrossRef] [PubMed]

91. Mulatero, P.; Verhovez, A.; Morello, F.; Veglio, F. Diagnosis and treatment of low-renin hypertension. *Clin. Endocrinol.* **2007**, *67*, 324–334. [CrossRef] [PubMed]

92. Lurbe, E.; Agabiti-Rosei, E.; Cruickshank, J.K.; Dominiczak, A.; Erdine, S.; Hirth, A.; Invitti, C.; Litwin, M.; Mancia, G.; Pall, D.; et al. 2016 European Society of Hypertension guidelines for the management of high blood pressure in children and adolescents. *J. Hypertens.* **2016**, *34*, 1887–1920. [CrossRef] [PubMed]

 International Journal of
Molecular Sciences

Review

Advances in the Genetics of Hypertension: The Effect of Rare Variants

Alessia Russo [1,2], Cornelia Di Gaetano [1,2], Giovanni Cugliari [1,2] and Giuseppe Matullo [1,2,*]

[1]   Department of Medical Sciences, University of Turin, 10126 Turin, Italy; alessia.russo@hugef.org (A.R.);
       cornelia.digaetano@unito.it (C.D.G.); giovanni.cugliari@hugef.org (G.C.)
[2]   Italian Institute for Genomic Medicine (IIGM, Formerly HuGeF), 10126 Turin, Italy
*   Correspondence: giuseppe.matullo@unito.it; Tel.: +39-011-6705601

Received: 31 January 2018; Accepted: 26 February 2018; Published: 28 February 2018

Abstract: Worldwide, hypertension still represents a serious health burden with nine million people dying as a consequence of hypertension-related complications. Essential hypertension is a complex trait supported by multifactorial genetic inheritance together with environmental factors. The heritability of blood pressure (BP) is estimated to be 30–50%. A great effort was made to find genetic variants affecting BP levels through Genome-Wide Association Studies (GWAS). This approach relies on the "common disease–common variant" hypothesis and led to the identification of multiple genetic variants which explain, in aggregate, only 2–3% of the genetic variance of hypertension. Part of the missing genetic information could be caused by variants too rare to be detected by GWAS. The use of exome chips and Next-Generation Sequencing facilitated the discovery of causative variants. Here, we report the advances in the detection of novel rare variants, genes, and/or pathways through the most promising approaches, and the recent statistical tests that have emerged to handle rare variants. We also discuss the need to further support rare novel variants with replication studies within larger consortia and with deeper functional studies to better understand how new genes might improve patient care and the stratification of the response to antihypertensive treatments.

Keywords: essential hypertension; blood pressure; genome-wide association studies; exome microarray; next-generation sequencing; rare variants; rare-variants association testing; burden test; sequence kernel association test

---

1. Introduction

Systemic hypertension is a consistently elevated systolic or diastolic blood pressure in the systemic arteries. Systolic blood pressure (SBP) is generated by the contraction of the ventricles and represents the highest blood pressure (BP) level. Diastolic blood pressure (DBP) is the BP remaining during the relaxation of the ventricles and represents the lowest BP level. The term Pulse Pressure (PP) refers to the difference (in mmHg) between the systolic and diastolic pressures, while the Mean Arterial Pressure (MAP) is the average BP during a single cardiac cycle [1–3]. Clinicians consider 140 mmHg as the maximum normal adult SBP value, and 90 mmHg as the upper limit for normal DBP value, as suggested by the World Health Organization (WHO) [4]. Usually, high SBP is caused by the narrowing of the arterioles. This narrowing raises the peripheral resistance to blood flow, which requires a greater workload for the heart and raises arterial pressure [1]. Elevated BP levels still represent a huge public health issue worldwide, being the major risk factor for cardiovascular disease, including coronary heart disease, stroke, and heart failure. Each year, 17 million people prematurely die because of cardiovascular disease, and, among these, nine million deaths occur as a consequence of hypertension-related complications [5].

Ninety-five percent of hypertensive patients presents a type lacking an obvious identifiable cause (Essential or Primary Hypertension). Investigations of twin and family studies revealed a moderate heritability ranging between 30% and 50% [6,7]. Hypertension is a heterogeneous disease; besides genetic variation, several factors such as age, sex, and ethnicity influence this trait, in addition to other environmental factors (e.g., lipid levels and obesity).

So far, the study of hypertension has mostly been based on Genome-Wide Association Studies (GWAS). GWAS represent a valuable approach to type hundreds of thousands of Single Nucleotide Polymorphisms (SNPs) in very large cohorts. During the last ten years, many studies have been published thanks to the setting of very large consortia, including the International Consortium for Blood Pressure Genome-Wide Association Studies, Cohorts for Heart and Aging Research in Genomic Epidemiology (CHARGE) and Global BPgen, the Wellcome Trust Case Control Consortium Studies, the UK Biobank, the PBCHARGE-EchoGen consortium, the CHARGE-HF consortium [8–12], leading to the identification of more than 100 SNPs implicated in BP levels, as recently reviewed by Seidel and Scholl [13].

The cause of a complex trait, like essential hypertension, remains elusive if examined in the light of the GWAS results. There has been a step forward compared to the classic GWAS analyses thanks to system genetics approaches and related statistical methods [14]. These approaches use intermediate phenotypes, such as transcript, protein, or metabolite levels, and quantify and integrate them with several traits of interest. Several genes pathways and networks underlying common human diseases have been discovered using systems genetics studies.

For example, data derived from GWAS were integrated with expression data to provide a measure of functional variation, i.e., the expression Quantitative Trait Loci (eQTL). When one of these loci is located within ≤1 Mb from the gene encoding the transcript, it is termed a *cis*-eQTL. When an eQTL affects the expression level of a distal gene, it is called *trans*-eQTL. Disease susceptibility can be regulated by a plethora of genes controlled by *trans*-eQTLs which, for this reason, are very informative [15].

Thanks to studies based on rat, mouse, or human cells and tissues, it has been calculated that about 30% of mammalian genes are under the control of eQTLs and they heavily contribute to complex disease susceptibility [16]. Moreover, using comparative genomics between established rat models of hypertension and humans, several studies have shown that human genes found to be associated with hypertension through GWAS, when conserved in the rat, are likely to form both *cis*- and *trans*-acting eQTLs in multiple tissues [17]. These studies have also taken advantage of a statistical methods known as Weighted Gene Co-Expression Network Analysis (WGCNA) that studies biological networks based on pairwise correlations between variables and is often used to highlight clusters (modules) of highly correlated genes [18].

At the heart of the GWAS-based approaches lies the "common variant–common disease" hypothesis. However, when considering all the detected high-frequency variants in aggregate, the percentage of BP variability explained by genetic variants accounts for only 2–3%. Moreover, blood pressure changes related to different genotypes at these loci are estimated to be modest, approximately 1.0 and 0.5 mmHg for SBP and DBP, respectively [19]. Considering the moderate effects and the scarce genetic control ascribable to high-frequency variants, two possible scenarios came forward: a wrong heritability was estimated, or alleles are more likely to be heterogeneous and uncommon. Furthermore, array-based technologies were infrequently conducted for the detection of causal polymorphisms. These observations implicated strong limits in exploiting GWAS to identify druggable targets with high confidence and supported the idea that rare (frequency < 1%) and uncommon (frequency between 1% and 5%) functional variants may explain a greater fraction of hypertensive individuals. The arrival of Next-Generation Sequencing (NGS) technologies facilitated a shift in focus from common to rare variants and provided the opportunity to unravel the genomic architecture underlying hypertension risk. Along with the development of even more advanced laboratory methodologies, statistical genetic models must also evolve to meet the challenge of using rare variants to link previously unidentified genome loci to BP changes [20,21].

In this review, we first present an overview of the most recent findings regarding the role of rare and uncommon variants in BP alteration identified through the currently available technologies, moving from the candidate-gene approach to the high-throughput exome chips, and then to NGS solutions; next, we report the statistical methods proposed so far for rare variants analysis. Finally, we draw conclusions on the contribution ascribable to rare and low-frequency variants in the improvement of cardiovascular risk assessment.

2. Results

2.1. Results from Studies on Selected Single Nucleotide Variants and Genes

Conducting studies based on a candidate-gene approach is the easiest and cheapest way to investigate genetic variation. *FBN1*, a gene that is thought to be causative of vascular damage and whose mutations have been previously detected only in relation to Marfan syndrome, was selected by Jeppesen and colleagues [22] as their research focus. A sample of 4839 Danish subjects was genotyped for the rs11856553 rare variant (Minor Allele Frequency, MAF, of A allele = 0.2%, 1000 Genomes) using a PCR-based method. In the Health 2006 study, an unadjusted risk of hypertension of 2.67 (95% Confidence Interval, CI, 1.14–6.18) for the G/A genotype was reported. The adjusted risk of moderate to severe hypertension (grade 3) for the A/A–G/A genotypes (homozygous and heterozygous carriers were grouped) was 8.01 (95% CI, 3.27–19.58), $p < 0.0001$. No significant differences in BP between G/A and G/G variant carriers were described within the MONICA10 study, however, the adjusted risk of moderate to severe hypertension (grade 2) for A/A-G/A variants was 6.54 (2.12–20.2); $p < 0.01$. It is still undefined how this intronic mutation could functionally affect hypertension [22].

The cytokine Interleukin-6 (IL-6) is a fundamental mediator of the acute-phase response to endothelial injury and regulates the production of C Reactive Protein (CRP) in hepatocytes [23]; therefore, both *IL-6* and *CRP* genetic variants have been evaluated in relation to hypertension [2,20]. In the paper from Karaman et al. [24], *IL-6* rs1800795 and rs1800796 SNPs (MAF = 14.12% and 31.39%, respectively, 1000 Genomes) were genotyped in a Turkish sample of 108 controls and 111 hypertension patients. Both SNPs genotypes were not significantly related to hypertension or to IL-6 and CRP plasma levels. The CC genotype of the rs1800796 SNP is very rare in the examined population and large frequency differences among different populations and geographic regions have been reported [25].

Endothelial nitric oxide synthase (eNOS) produces Nitric Oxide, a vasodilator of vascular smooth muscles, and thus plays a crucial role in regulating BP. A four-SNP haplotype, comprising the uncommon variant rs11699009 in the *BPIFB4* gene, has been associated with notable longevity [26]. In the study of Vecchione et al. [27], 416 individuals were genotyped to determine their haplotypes. The rare variant-haplotype carriers showed a significantly increased DBP ($p = 0.013$) and a borderline increased SBP ($p = 0.067$). The authors demonstrated that the overexpression of the *BPIFB4* uncommon variant in mice impaired eNOS signaling and increased BP, opening the way for the development of new therapeutic strategies.

2.2. Results from Exome Chips-Based Studies

When the 1000 Genomes Project became publicly available, data from NGS technology allowed the development, from Affymetrix (Santa Clara, CA, USA) and Illumina (San Diego, CA, USA) companies, of array-based genotyping platforms which offer the possibility to capture a greater range of single nucleotide variability compared to GWAS. In Table 1, studies investigating common and rare variants in association with hypertension andBP phenotypes and through exome array approaches are listed. Most publications [28–38]; (Table 1) took advantage of the Illumina HumanExome BeadChip (Exome Chip; Illumina, Inc., San Diego, CA, USA).

Table 1. Results from Exome Chips-Based Studies.

N	Technology	Design	Population	BP Trait	Statistical Analysis	Main Results	References
Discovery: 517 Replication: 57,234	Infinium OmniExpress Exome-Illumina	Linkage analysis in 130 families from CFS to identify rare, coding variants	Whites	SBP, DBP, PP	Family-based burden; SKAT	Linkage peak observed on Chr. 16p13 (MLOD = 2.81) for SBPMultiple rare, coding variants in *RBFOX1* associated with reduced SBP	He et al. [35]
14,028	Illumina ExomeChip	Pleiotropic effects of lipid-associated loci on 10 cardiometabolic traits	Korean	SBP, DBP	GLM	3 SNPs associated with SBP and DBP. Effect sizes (se): -1.5 ± 0.3--0.78 ± 0.20; $p < 1.09 \times 10^{-4}$	Kim et al. [38]
~475,000	Affymetrix UK Biobank Axiom Array and Affymetrix UK BiLEVE Axiom Array	Meta-analysis of CHARGE, European-led, and UK Biobank Exome Consortia to identify BP-associated SNVs	~423,000 European	SBP, DBP, PP	EPACTS; EMMA eXpedited; GEMMA	21 SNVs associated with at least 1 BP trait with $p < 5 \times 10^{-8}$; βs(se): $-1.14(0.19)$--$0.42(0.06)$	Kraja et al. [36]
Discovery: 146,562. Follow-up: 180,726. Meta-analysis: 327,288	Illumina ExomeChip	Meta-analysis of CHARGE+, CHD Exome+, ExomeBP, T2D-Genes, GoT2DGenes consortia to identify functional coding variants	All ancestries	SBP, DBP, PP, MAP, HTN	SKAT; Burden test	31 new loci associated with BP ($p < 3 \times 10^{-7}$; 28 common and 3 low-frequency variants) explaining 0.7% and 1.3% of interindividual variation in SBP and DBP, respectively*PTPMT1*, *DBH*, *NPR1* genes had aggregated rare and low-frequency variants associated with BP ($p < 9 \times 10^{-7}$)	Liu et al. [33]
15,914	Illumina ExomeChip	Meta-analysis of AADM, ARIC, CARDIA, GenNet, GENOA, HUFS, HyperGEN, LUC cohorts to identify new genes and SNVs across the full frequency spectrum	African ancestry	SBP, DBP	SKAT; T1 burden test; burden-T1-del	9 rare SNVs (mostly missense) within 8 genes (*SLC28A3*, *KRBA1*, *SEL1L3*, *YOD1*, *COL6A1*, *CRYBA2*, *GAPDHS*, and *AFF1*) associated with SBP or DBP (Betas$_{st}$:1.16-3.97; $p < 5 \times 10^{-7}$)2 significant genes (*CCDC13*, *QSOX1*) for SBP and DBP were also described (Betas$_{sm}$: 54.38 (10.68) and 32.93 (7.13), respectively; $p < 9.95 \times 10^{-6}$)	Nandakumar et al. [34]
5453	Illumina ExomeChip	Identifying stop-coding variants	Swedish	SBP, DBP	GLM	19 SNVs associated with SBP*PDE11A* R307X mutation: 7 mmHg higher SBP-4.6 mmHg higher DBP	Ohlsson et al. [31]
2045	Illumina ExomeChip	79,578 low-frequency variants analysis within the HyperGEN cohort	African Americans	SBP	CAST; CMC; w-SUM; SST; VT; C-alpha; SKAT; SKAT-O; Minimum P; Fisher's statistic; RBS; FPCA; Higher criticism	No genome-wide significant results	Sung et al. [28]
Discovery: 192,763 Replication: 155,063	Illumina ExomeChip	Meta-analysis of CHD Exome+, ExomeBP, and GoT2D/T2D-GENES consortia with independent replication within CHARGE + consortium to identify novel coding variants	European: 290,989, South Asian: 27,487, African American, Hispanics and SAS ancestries: 29,350	SBP, DBP, PP, HTN	SKAT; Burden test	Discovery: 51 loci associated with at least one BP traits with $p < 5 \times 10^{-8}$ Replication: 30 novel SNVs ($p < 6 \times 10^{-4}$; βs: -1.43--2.70) Rare putative functional variants were identified within *A2ML1*, *COL21A1*, *RRAS*, *RBM47*, and *ENPEP* genes	Surendran et al. [32]

Table 1. *Cont.*

N	Technology	Design	Population	BP Trait	Statistical Analysis	Main Results	References
3165	AffymetrixGenome-WideHumanSNP6.0 Array and Illumina ExomeChip	Analysis of common and rare variants in *PCSK9* gene within HyperGEN study and REGARDS population	African-American	SBP, DBP, HTN	SKAT; Joint effect	GWAS: rs12048828 and rs9730100 marginally associated with DBP (βs = 1.8 and 1.0; p = 0.05). No significant associations with SBPRare variants: higher median SBP and DBP for carriers of non-synonymous SNPs compared to non-carriers (median SBP and DBP for non-carriers: 127 mmHg and 73 mmHg) Significant cumulative effect of rare variants with DBP (p = 0.04) but not with SBP (p = 0.14) in HyperGEN Significant cumulative effect of non-synonymous or stop-gain SNPs (p = 0.02) but not synonymous SNPs (p = 0.73) The joint effect of rs12048828, rs9730100, and 19 rare variants was not statistically significantly associated with DBP (p = 0.07) or SBP (p = 0.53) The joint effect of the 2 GWAS SNPs and 16 non-synonymous SNPs was significant for DBP (p = 0.03) but not for SBP (p = 0.41) *PCSK9* rare variants had a cumulative significant association with SBP (p = 0.04) but not with DBP (p = 0.36) in REGARDS data. Same results when restricted to 15 non-synonymous SNPs (p = 0.04 for SBP, p = 0.40 for DBP)	Tran et al. [29]
Discovery: 140,886 Replication: >330,000	Customized array with genome-wide imputation based on 1000 Genomes and UK10K sequence data	Analysis of SNVs with MAF $\geq$ 1% and MAF $\geq$ 0.01% within UK Biobak	European	SBP, DBP, PP	Linear regression	107 loci validated with $p < 5 \times 10^{-8}$ The impact of the combination of all loci accounts for 9.3 mmHg higher SBP and over 2-fold higher risk of HTN	Warren et al. [37]
6026	Infinium HumanExome-12 ver. 1.2 BeadChip and Infinium Exome-24 ver. 1.0-Illumina	Longitudinal EWAS for HTN	Japanese	SBP, DBP	GEE model	7 HTN-related SNVs detected, 6 of these variants were located at 12q24.1, creating an East Asian-specific haplotype comprising five derived alleles People carrying the East Asian-specific haplotype displayed a HTN prevalence significantly lower than those individuals carrying a common haplotype. A SNV in *COL6A5* gene was significantly associated with SBP	Yasukochi et al. [30]

Sample number (N), Systolic Blood Pressure (SBP), Diastolic Blood Pressure (DBP), Pulse Pressure (PP), Mean Artery Pressure (MAP), Hypertension (HTN), Single Nucleotide Polymorphism (SNP), Genome-Wide Association Studies (GWAS), Single Nucleotide Variant (SNV), Minor Allele Frequency (MAF), Exome-Wide Association Studies (EWAS), Cleveland Family Study (CFS), Cohorts for Heart and Aging Research in Genomic Epidemiology (CHARGE), Atherosclerosis Risk in Communities (ARIC), Coronary Artery Risk Development in Young Adults (CARDIA), Africa America Diabetes Mellitus (AADM), The Genetic Epidemiology Network of Arteriopathy (GENOA), Howard University Family Study (HUFS), Hypertension Genetic Epidemiology Network (HyperGEN), Loyola University Chicago (LUC), Congenital heart disease (CHD) Exome+, The Genetics of Type 2 Diabetes Consortium (GoT2D)/Type 2 Diabetes Genetic Exploration by Next-generation sequencing in multi-Ethnic Samples (T2D-GENES), REasons for Geographic And Racial Differences in Stroke (REGARDS), Framingham Heart Study (FHS), Standard Error (se), Beta as standardized mean difference (Betas$_{st}$), SeqMeta Beta (Betas$_{sm}$), Efficient and Parallelizable Association Container Toolbox (EPACTS), Efficient Mixed-Model Association eXpedited (EMMA eXpedited), Genome-Wide Efficient Mixed Model Association (GEMMA), Sequence Kernel Association Test (SKAT), Optimal Unified Test (SKAT-O), SKAT-Combined (SKAT-C), Cohort Allelic Sums Test (CAST), Combined Multivariate and Collapsing (CMC), Weighted-Sum (w-SUM), Simple Sum Test (SST), Variable-Threshold (VT), Replication-Based Weighted-Sum Statistic (RBS), Functional Principal Components Analysis (FPCA), Generalized Estimating Equation model (GEE model), burden test on deleterious variants (burden-T1-del).

This chip was produced in order to meet the need of moving from relatively frequent variants derived from GWAS to functional variants located in coding regions. The array constitutes an intermediate choice between GWAS and NGS of large number of samples in terms of both cost and practical issues. The Exome Chip was designed on genome and exome sequencing data of 16 contributing studies, reaching a total of 12,031 subjects. In the array, 247,039 markers were assayed including 84% rare variants, 9.2% low-frequency variants, and only 5.8% common variants which were identified more than three times in at least two different datasets. Most variants (>90%) are non-synonymous or splicing variants that were absent in previously available chips. Genotyped individuals were mostly of European American ancestry which led to some concerns about the evolutionary young age of variants and population-specific results. The Exome Chip consortia provided information on several common diseases, including cardiovascular disease (Available online: http://genome.sph.umich.edu/wiki/Exome_Chip_Design) [39].

In 2015, Sung and colleagues [28] reported the results of rare and low-frequency single variants and four sets of gene-based analyses using Exome Chip data on 2045 African-American subjects from the HyperGEN cohort. Neither Single Nucleotide Variants (SNVs) nor gene level analyses reached genome-wide Bonferroni-corrected thresholds ($p < 6.4 \times 10^{-7}$ for SNVs; $p < 2 \times 10^{-6}$ with MAF < 1% and $p < 3.9 \times 10^{-6}$ with MAF < 5%) [28].

The same cohort was used for analyses focused on both rare (through the Exome Chip) and common (using the AffymetrixGenome-WideHumanSNP6.0 Array) variants within the *PCSK9* gene in relation to BP traits. PCSK9 is a protease able to interact with the three subunits of the renal epithelial sodium channel (ENaC). This interaction consequently increases proteasomal degradation of the ENaC which regulates sodium reabsorption [40]. Among the 31 SNPs identified, none of the associations were statistically significant ($p > 0.05$). The cumulative effect of rare variants (mostly non-synonymous or stop-gain SNVs) detected in *PCSK9* was significantly associated with DBP in HyperGEN ($p = 0.04$) and to SBP in REGARDS data ($p = 0.04$). The disparity in the associated phenotypes was probably due to differences in the age of populations [29].

Alteration in lipid levels is strongly related to hypertension [41], and a pleiotropic effect of lipid-associated loci on hypertension could be speculated. To investigate this, the group of Kim et al. [38] interrogated 135 Exome Chip SNVs for associations with ten cardiometabolic traits in 14,028 Korean individuals. Three new common variants in the *BRAP*, *ACAD10*, and *ALDH2* genes within the 12q24.12 locus were significantly associated with both SBP and DBP ($p < 1.09 \times 10^{-4}$; effect sizes between -1.53 ± 0.32 and -0.78 ± 0.20). The locus was also associated with High-Density Lipoprotein (HDL), Low-Density Lipoprotein (LDL), triglycerides, fasting plasma glucose, body mass index, and waist–hip ratio ($p < 1.06 \times 10^{-2}$; effect sizes between -7.60 ± 1.72 and 2.55 ± 0.53) [38]. Successively, a longitudinal Exome-Wide Association Study (EWAS), which is a genotyping method restricted to exonic SNVs using Illumina exome chips, allowed the detection of six hypertension-related SNVs at the 12q24.1 locus, creating an East Asian-specific haplotype comprising five derived alleles. The study was conducted in 6026 Japanese individuals whose disease progression and physiological changes were traced for several years during annual health check-ups. The rationale of this study was the observation that SBP, DBP, and the prevalence of hypertension are significantly correlated with age, while conventional GWAS have commonly been conducted in a cross-sectional manner measuring traits at a single point in time. People carrying the East Asian-specific haplotype displayed a hypertension prevalence significantly lower than those individuals carrying a common haplotype (mean Odds Ratio (OR) = 0.78, $p < 1.0 \times 10^{-8}$). Furthermore, using a recessive model, an SNV located within the *COL6A5* gene, was significantly associated with SBP (Estimate: -2.93; $p = 2.3 \times 10^{-8}$) [30].

Stop codons are highly likely to alter protein function affecting BP-related traits, and, for this reason, Ohlsson et al. [31] focused the aim of their work on the relationship between BP and protein truncating variants in the genotypes of 5453 Swedish people. They reported 19 SNVs associated with SBP with a p value < 0.05. The *PDE11A* R307X mutation conferred a 7 mmHg higher SBP and a 4.6 mmHg higher DBP (β coefficients = 7.0 (1.8–12) for SBP corrected, $p = 0.009$ and 4.6 (1.8–7.4) for DBP corrected, $p = 0.001$) and was previously described as a loss-of-function mutation linked to familial hypertension and Cushing's syndrome [42]. The stop codon mutation caused a three-fold increased risk of hypertension in female carriers (OR = 3.1 (95% CI, 1.3–7.4), $p = 0.009$).

Two very large meta-analyses were then published at the same time to identify novel coding variants and loci influencing BP traits and hypertension. In the first meta-analysis, Surendran et al. [32] genotyped 192,763 subjects, mostly of European descent, in the discovery phase. Fifty-one genomic regions were found to be significantly associated with at least one of the following BP traits: SBP, DBP, PP, and hypertension in the discovery analysis ($p < 5 \times 10^{-8}$). Thirty novel SNVs replicated in 155,063 multiethnic populations ($p < 6.2 \times 10^{-4}$; βs: $-1.43-2.70$). Among these, rare putative functional variants were identified within *A2ML1*, *COL21A1*, *RRAS*, *RBM47*, and *ENPEP* genes. Interestingly, intersecting previous GWAS data with Exome Chip data revealed five out 35 known loci which likely had rare coding functional variants. The second large meta-analysis was conducted on all ancestry subjects from the same five consortia described in Surendran et al. [32], reaching a total sample number of 327,288. Here, the authors identified 31 additional new loci with statistically significant associations with one of the BP traits ($p < 3.4 \times 10^{-7}$). Three variants had frequencies between 1% and 5% and were non-synonymous substitutions in *NPR1* (already established), *SVEP1*, and *PTPMT1* (novel genes) with a p value less than 3.4×10^{-7} when corrected for multiple testing. To note, the BP increment attributable to any of these low-frequency variants (>1.5 mmHg) was higher than any of the novel common SNPs described here. Low-frequency and frequent SNVs with non-synonymous, stop-coding, and splicing effects were aggregated using burden tests to identify new gene-based associations. These analyses showed significant results for *NPR1* ($p = 4.4 \times 10^{-5}$) and marginally for *PTPMT1* and *DBH* genes ($p = 0.019$ and 0.053, respectively). Considering that an overlap between cardiovascular-specific pathways and metabolic disease-related factors was observed, the authors suggested a shared origin between the phenotypes that could be exploited for new drugs discovery [33].

In the most recent meta-analysis on Exome Chip data, Nandakumar et al. [34] screened 15,914 individuals of African ancestry to detect novel genes and BP-related SNVs considering the full spectrum frequency. Nine rare SNVs (mostly missense) within eight genes (*SLC28A3*, *KRBA1*, *SEL1L3*, *YOD1*, *COL6A1*, *CRYBA2*, *GAPDHS*, and *AFF1*) exhibited Bonferroni-corrected associations with SBP or DBP (SeqMeta βs ($βs_{sm}$): 21.10 (4.12)–73.65 (13.19); $p < 4.6 \times 10^{-7}$) and the *CCDC13*, *QSOX1* genes were also described through burden test including only predicted damaging variants (Betas$_{sm}$: 54.38 (10.68) and 32.93 (7.13), respectively; $p < 3.86 \times 10^{-6}$). By contrast, no significant results were obtained considering common and low-frequency variations.

Linkage analysis can have good power to detect multiple rare or lower frequency BP variants in a gene or region with relatively larger effect sizes [43]. However, the identified linkage regions from well-designed linkage family studies [44–46] did not overlap with many BP loci identified by large BP GWAS of mostly unrelated individuals. Therefore, He and colleagues [35] applied variance-component linkage analysis to the Cleveland Family Study (CFS) to identify candidate genomic regions related to SBP, DBP, and PP. Since the region identified (16p13) showed no overlapping with any SNPs derived from previous GWAS, 517 individuals from the CFS who had been genotyped using the Illumina OmniExpress Exome array [39], were screened for variants within the 16p13 locus. At a gene-based level, the association between the aggregation of five rare variants within the *RBFOX1* gene and SBP as well as PP traits replicated in the meta-analysis of a large sample of 57,234 participants ($p < 1.71 \times 10^{-2}$). This gene encodes for the Ataxin-2 Binding Protein 1 whose genetic variations were suggested to have a protective effect on BP levels, although the underlying mechanisms remain to be clarified [35].

The UK Biobank is a huge prospective cohort including 500,000 individuals of European ancestry recruited to investigate genetic and non-genetic factors underlying diseases that takes advantage of many phenotypes and biological samples [47]. Genotypes obtained through a customized array in addition to genome-wide imputation based on 1000 Genomes and UK10K sequence data, and information related to BP traits, were retrieved for 140,886 participants included in the discovery phase of the study conducted by Warren and colleagues [37]. Both GWAS and exome analyses were performed to identify SNVs with MAF $\geq$ 1% and MAF $\geq$ 0.01%. Among the 240 loci derived from the discovery phase, 102 GWAS and five exome variants with $p < 5 \times 10^{-8}$ were reported. Noteworthy, a 9.3 mmHg higher SBP was observed after comparing subjects with the highest genetic risk score (estimated on the basis of all the loci identified) and above 50 years old with those with the lowest genetic risk score (95% CI: 6.9–11.7, $p = 1 \times 10^{-13}$) [37]. In the recently published paper from Pazoki and coauthors [48], the 267 SNPs identified by Warren et al. [37] were combined with the 47 BP-associated loci reported by Hoffman et al. [11] to calculate a genetic risk score for high BP in 277,005 subjects belonging to the prospective UK Biobank cohort. A healthy lifestyle score was also constructed for all the individuals in order to investigate whether the adherence to a favorable lifestyle could counteract the high genetic susceptibility to develop hypertension and cardiovascular diseases. The authors reported an association between healthy lifestyle and lower SBP and DBP within each tertile of genetic risk. In particular, at low genetic risk, the estimated mean SBP was 140 mmHg (95% CI, 102–177) among subjects with an unhealthy lifestyle and 134 mmHg (95% CI, 95–172) among those with a healthy lifestyle.

To date, the largest meta-analysis on exome chips data was conducted on 475,000 individuals (mostly European) genotyped using the UK BiLEVE array and the UK Biobank Axiom Array. These arrays are closely related new next-generation microarrays (95% identical content) designed from the Affymetrix Company. More than 800,000 markers were included to comprehensively cover beyond common SNPs, rare and low-frequency coding variants, copy number variants, pharmacogenomics markers, Human Leukocyte Antigen (HLA), inflammation, and eQTL variants. Among rare variants, in addition to primarily missense mutations, protein truncating variants resulting in premature stop codons, frameshifts, and loss of start mutations were included as loss-of-function variants. The genomic coverage was optimized for European and British populations. The array provided the opportunity to test the association between a wide range of genetic variations and many frequent human diseases, including cardiovascular disease and cardiometabolic traits such as BP (Available online: http://www.ukbiobank.ac.uk); [49]. In the paper from Kraja and colleagues [36], 21 SNVs showed significant associations with at least one BP trait, after correcting for multiple testing ($p < 5 \times 10^{-8}$; βs(se): −1.14 (0.19)–0.42 (0.06)). Moreover, all variants had concordant directions across all the datasets. Only one SNV (in the *DBH* gene) had a MAF less than 1% and exhibited the lowest effect estimate (βs(se): −1.14 (0.19); $p = 1.23 \times 10^{-9}$). Four novel associations of common SNPs within *SLC4A1AP*, *AFAP1*, *STAB1*, and *SYNPO2L* genes were reported [36].

2.3. Results from DNA Sequencing Studies

Large-scale genotyping through high-throughput platforms opened the way to great efforts aimed at discovering the causative variants explaining the associations described.

2.3.1. Pre-Next-Generation Sequencing Era

Direct sequencing represented an easy way to characterize hypertension-related genes embedding SNPs found through GWAS. Okuda et al. [50] validated 143 SNPs identified in a small Japanese population. Among these SNPs, most had frequencies higher than 5% and caused amino acid substitutions, whereas almost all novel variants were rare (13 out of 16).

Genetic Epidemiology Network of Salt Sensitivity (GenSalt) study participants were recruited to evaluate SBP, DBP, and MAP responses to a dietary sodium intervention. The renin-angiotensin -aldosterone system (RAAS) is a hormonal cascade essential for the control of homeostasis, BP, and vascular tone [51,52]. In the first re-sequencing study focused on the RAAS pathway, Kelly and coauthors [53] analysed seven genes for putative associations with BP salt-sensitivity among participants of the GenSalt study. Carriers of 124 rare variants had 1.55-fold increased odds (95% CI: 1.15, 2.10) of salt sensitivity compared to non-carriers ($p = 0.004$). No genes showed significant associations with salt sensitivity after Bonferroni correction. No significant common and low-frequency single markers were detected when the analyses were corrected for multiple comparisons [53]. The reabsorption of sodium in epithelial cells located in the renal tubule is carried out by the renal epithelial sodium channel (ENaC) whose activity is fundamental for BP control [54]. *SCNN1A*, *SCNN1B*, and *SCNN1G* genes encode the three ENaC subunits [55]. These genes were targeted by Gu et al. [56] to identify novel common, low-frequency, and rare variants in 300 GenSalt participants with the highest MAP response to the high-sodium intervention and 300 GenSalt participants with the lowest MAP response to the high-sodium intervention. No significant associations with salt sensitivity were observed. In gene-based analyses, *SCNN1A* gene showed a significant association with salt sensitivity ($p = 0.009$). Individuals carrying rare variants in *SCNN1A* gene had an odds ratio of 0.52 (95% CI: 0.32–0.85). Neither *SCNN1G* nor *SCNN1B* associated with salt sensitivity in rare variant analyses. Three common variants in *SCNN1A* associated with salt sensitivity of BP ($p < 1.3 \times 10^{-3}$; 1.23-fold increased odds and 0.68–0.69-fold decreased odds of salt sensitivity) [56].

Another suggested candidate gene for hypertension is represented by the Cadherin-13 gene (*CDH13*). *CDH13* encodes a cell adhesion molecule involved in the protection of vascular endothelial cells from apoptosis following oxidative stress, survival, proliferation, and endothelial cells migration [57–59]. The promoter region was re-sequenced and subjected to methylation QTL (meQTL) analysis within the HYPertension in ESTonia (HYPEST) and Coronary Artery Disease in Czech (CADCZ) studies. The meQTL rs8060301 (a frequent variant) showed a pleiotropic effect on HDL and DBP (nominal $p < 0.005$), which was unconfirmed after multiple testing correction [60].

2.3.2. Results from Next-Generation Sequencing Studies

GWAS identified more than 100 genetic variants influencing BP [13]. However, the causal variants underlying the majority of genetic associations remained unknown. In recent years, three different NGS approaches have been proposed to study rare variants in hypertension and BP (Table 3).

The first approach is to check GWAS signals and describe novel associations by performing a re-sequencing of only a few genes previously indicated by GWAS. This approach, commonly called target re-sequencing, is cheaper and allows one to highlight the variations within the whole frequency spectrum in a precise genomic locus. The strategy was adopted by the CHARGE Consortium. In the frame of this consortium, the signals identified by precedent GWAS were re-sequenced with the aim of describing novel variations with large effects on several common diseases [74]. Concerning BP, within the CHARGE Targeted Sequencing Study, target re-sequencing of 4178 Europeans was performed on six BP genes identified by GWAS (*ATP2B1*, *CACNB2*, *CYP17A1*, *JAG1*, *PLEKHA7*, and *SH2B3*), however, neither common nor rare variants were consistently associated with the trait with large effect sizes, independently of the original GWAS signals [63].

Regarding hypertension, an association with rs3918226 in the *eNOS* gene promoter was described in the GWAS from Salvi et al. [75] (OR for minor allele T = 1.34 (95% CI, 1.25–1.44); $p = 1.03 \times 10^{-14}$). In 2013, a 140 kb genomic area encompassing the *eNOS* gene was re-sequenced from the same group. The study identified 338 variants, including 61 novel variants, and rs3918226 still appeared as the SNP most closely associated with hypertension. Moreover, if compared with the C major allele, the T risk allele was associated with lower *eNOS* transcriptional activity when tested in HeLa cells [64].

A second approach is whole exome sequencing (WES) in which only the coding portions of the genome, (about 2%), estimated to harbor 85% of disease-causing mutations, are sequenced [76]. A WES study was performed on DNA samples from 17,956 individuals of European and African ancestries, included in the CHARGE, National Heart, Lung, and Blood Institute GO Exome Sequencing Project, Rotterdam Study, and the Erasmus Rucphen Family cohorts. These findings implicated the effect of the aggregation of 95 rare coding variants in *CLCN6* on decreasing BP levels of 3–4 mmHg, independently of the tagging SNP rs17367504 previously reported. The effect size described here was about four- to six-fold larger than previous common BP variants from GWAS [66].

Table 2. Results from Next-Generation Sequencing Studies.

N	Technology	Design	Population	BP Trait	Statistical Analysis	Main Results	References
1851	WES	Haplotype association analysis for *ULK4* and *MAP4* genes within the GAW19 data set	Mexican American	SBP, DBP, HTN	SKAT; SKAT-O; SKAT-C	• 36 rare haplotype blocks associated with BP in *ULK4* gene and 10 in *MAP4* gene	Datta et al. [61]
1985 unrelated subjects and 1140 relatives	WES	Screening of *SLC12A3*, *SLC12A1* and *KCNJ1* genes exons to identify rare variants within FHS offspring cohort	Largely whites of European descent	SBP, DBP	Two-tailed paired t-test	• 30 different mutations observed • Mean long-term SBP among mutation carriers was 6.3 mmHg lower than the mean of the cohort ($p = 0.0009$). For DBP, mean effect was -3.4 mmHg ($p = 0.003$)	Ji et al. [62]
4178	Target-re-sequencing	Case-cohort study design within the CHARGE Targeted Sequencing Study on 6 BP loci	European	SBP, DBP, PP, MAP	Kernel association test	• None of the common variants reached statistical significance threshold of $p = 0.0001$ • Rare variation was not significantly associated with any of the BP measures	Morrison et al. [63]
92 (HYPERGENES study)2722 (BP cohort)2013 (HTN cohort)	Target-re-sequencing	Target re-sequencing of a 140-Kb DNA region of Chr. 7 to identify causal or functional variants tagged by the rs3918226 SNP	Flemish	SBP, DBP, HTN	Multivariable-adjusted models	• 61 novel variants detected by DNA sequencing and confirmed by array-based genotyping • rs3918226 remained the SNP most closely associated with HTN • The risk allele was associated with lower transcriptional activity of the *eNOS* gene	Salvi et al. [64]
103	WGS	Case-control study on rare variants in unrelated subjects within GAW18 data set	Mexican American	SBP, DBP, HTN	qMSAT; C-alpha; CMC	• Rare variants in *SETX* gene intronic region were significantly associated, as aggregate, with hypertension (OR = 9.5, 95% CI (3.43, 28.70); $p = 8.8 \times 10^{-7}$)	Wang and Wei. [65]
Discovery: 14,497 in first stage and 3459 in second stage	WES	To examine the impact of rare variants in CHARGE and ESP studies with meta-analysis of two-stage discovery cohorts	European and African ancestry	SBP, DBP, PP, MAP	T1; SKAT	• 95 rare coding variants identified in *CLCN6* associated, in aggregate, with decreased BP (3–4 mmHg), independent of the tagging SNP rs17367504 previously identified • The effect size was about four- to six-fold larger than previous common BP variants from GWAS	Yu et al. [66]
142	WGS	Test for the effects of both rare and common variants across the whole genome of unrelated individuals within the GAW18 study	Mexican American	SBP, DBP, HTN	FBAT; GCTA; SKAT	• Significant windows within Chr. 3 were reported for associations with SBP and DBP. The most represented gene was *MAP4*	Zhao et al. [67]
1509 unrelated subjects; 256 individuals in 47 families	WGS and WES	To apply CAPL-burden and CAPL-SKAT tests to the GAW19 data set using the combined family and case–control data for HTN (GAW19)	Mexican American	SBP, DBP, HTN	CAPL-burden; CAPL-SKAT	• None of the tests for the top 10 genes passed the multiple testing correction threshold ($p = 3.4 \times 10^{-6}$)	Lin et al. [68]
142	WGS	WGS and gene expression joint analysis in relation to SBP, DBP, and HTN (GAW19)	Mexican American	SBP, DBP, HTN	Weighted U approach	• No gene reached statistical significance after adjusting for multiple testing	Tong et al. [69]

Table 3. Results from Next-Generation Sequencing Studies.

N	Technology	Design	Population	BP Trait	Statistical Analysis	Main Results	References
1509 unrelated subjects; 256 individuals in 47 families	WGS and WES	To apply CAPL-burden and CAPL-SKAT tests to the GAW19 data set using the combined family and case–control data for HTN (GAW19)	Mexican American	SBP, DBP, HTN	CAPL-burden; CAPL-SKAT	• None of the tests for the top 10 genes passed the multiple testing correction threshold ($p = 3.4 \times 10^{-6}$)	Lin et al. [68]
142	WGS	WGS and gene expression joint analysis in relation to SBP, DBP, and HTN (GAW19)	Mexican American	SBP, DBP, HTN	Weighted U approach	• No gene reached statistical significance after adjusting for multiple testing	Tong et al. [69]
1851	WES	To apply W-test on real NGS data set of hypertensive disorder (GAW19)	Mexican American	SBP, DBP, HTN	W-test	• *MACROD1/LRP16* locus was associated with HTN after Bonferroni correction (OR = 3.8; $p = 6.1 \times 10^{-7}$)	Sun et al. [70]
275 trios	WGS	To analyse rare variants within *ADCY5* and *UBE2E2* genes in parent-child trios (GAW18)	Mexican American	SBP, DBP, HTN	Trio-SVM	• *ADCY5* and *UBE2E2* genes showed marginal association with HTN with $p = 3.2 \times 10^{-4}$ for *ADCY5* and $p = 0.035$ for *UBE2E2*	Lu and Cantor. [71]
103 unrelated individuals	WGS	To analyse rare variants from Chr. 3 (GAW18)	Mexican American	SBP, DBP, HTN	SKAT-O	• No significant results in the analysis of real phenotype data ($p = 5.6 \times 10^{-5}$ for coding variants; $p = 6.9 \times 10^{-5}$ for changing variants; $p = 1.1 \times 10^{-4}$ for damaging variants)	Derkach et al. [72]
783 (GWAS); 506 (WGS)	WGS	To apply USR algorithm to data from GAW18	Mexican American	SBP, DBP, HTN	USR algorithm	• 23 promising genes and 3 significant pathways relevant to HTN identified ($p < 5.28 \times 10^{-3}$)	Cao et al. [73]

Sample number (N), Systolic Blood Pressure (SBP), Diastolic Blood Pressure (DBP), Pulse Pressure (PP), Mean Artery Pressure (MAP), Hypertension (HTN), Single Nucleotide Polymorphism (SNP), Genome-Wide Association Studies (GWAS), Whole Genome Sequencing (WGS), Whole Exome Sequencing (WES), Genetic Analysis Workshop (GAW), Cohorts for Heart and Aging Research in Genomic Epidemiology (CHARGE), Exome Sequencing Project (ESP), Sequence Kernel Association Test (SKAT), Optimal Unified Test (SKAT-O), SKAT-Combined (SKAT-C), Quality-based Multivariate Score Association Test (qMSAT), Combined Multivariate and Collapsing (CMC), Family-based Association Test (FBAT), Genome-wide Complex Trait Analysis (GCTA), Combined Association in the Presence of Linkage (CAPL), support vector machine (SVM), Unified Sparse Regression (USR), Odds Ratio (OR).

Two additional studies exploited WES data to focus on selected genes. Loss-of-function mutations in *SLC12A3*, *SLC12A1*, and *KCNJ1* genes, essential for normal renal NaCl reabsorption, cause Bartter's and Gitelman's syndromes. Their exons were screened to search for rare heterozygous variants within the Framingham Heart Study offspring cohort. Thirty different mutations were observed. The mean long-term SBP among mutation carriers was 6.3 mmHg lower than the mean of the cohort ($p = 0.0009$). For DBP, the mean effect was -3.4 mmHg ($p = 0.003$) [62]. Findings from previous GWAS indicated *ULK4* and *MAP4* genes, encoding, respectively, a Serine/Threonine-Protein Kinase and a non-neuronal microtubule-associated protein, as related to BP and hypertension [8,77]. Thirty-six rare haplotype blocks were found to be significantly associated with BP in *ULK4* gene, and ten in *MAP4* gene [61]. The study described above was conducted in the frame of the Genetic Analysis Workshops (GAWs). Since 1982, GAWs were held by a group of multidisciplinary scientists to deal with the role of genetics in complex diseases. For GAW18, GT2D-GENES Consortium and the San Antonio Family Heart Study provided data on the whole genome, systolic and diastolic BP, and related covariates in two Mexican American samples. In the GAW19, new data were included reaching a collection of WGS, WES, and gene expression data from 20 large families in addition to a set of 1943 unrelated subjects whose exome sequences were available. Simulated phenotypes were also included for each sample on the basis of the real sequence data [78]. Several papers have been published so far, mostly on methodological approaches (see the following paragraph "Statistical analysis of rare variants") to handle rare variations in relation to hypertension.

The third and most comprehensive NGS approach to examine the effect of rare variants is represented by WGS. Until now, to the best of our knowledge, only studies published within the GAWs analysed WGS data (Table 3) to search for genetic variations associated to hypertension, likely because a very large sample is needed to highlight rare variants, and this feature heavily affects the costs of the study. Three studies failed to identify significant associations after correction for multiple testing [68,69,72]. In the frame of the GAW18, Zhao et al. used novel sliding window approaches and a simulated dataset to analyse 142 unrelated individuals focusing on chromosome 3. The most significant windows fell into the known *MAP4* gene, considering both SBP and DBP. Other windows were reported within *SUMF* and *ARHGF3* genes in relation to DBP, and in *FLNB* and *BTD* for SBP [67]. Wang and Wei performed a gene-based genome-wide scan of 103 unrelated individuals to search for hypertension-associated genes. After using three different methods, only the *SETX* gene exhibited significant association. This gene consists of large intronic regions; indeed, most of the rare variants detected fall in intronic regions. The risk of hypertension, estimated after collapsing all the intronic variants, was 9.5 (OR = 9.5, 95% CI (3.43, 28.70); $p = 8.8 \times 10^{-7}$) [65]. Other significant findings were reported within the *MACROD1/LRP16* locus [70], *ADCY5*, and *UBE2E2* genes [71], and in an additional 23 genes [73] using different statistical approaches.

3. Statistical Analysis of Rare Variants

Gene-based association tests evaluate the relationship of rare variants enrichment in genes and phenotype or Mendelian and common diseases [79]. Region-based analysis has become the standard approach for analyzing rare variants, since standard individual variant tests are underpowered to detect rare variant effects because of the low allele frequencies. Statistical methods to test for rare variants can be categorized as burden approach [80–82] and SKAT (Sequence Kernel Association Test) approach [74,83]. Burden tests assume all rare variants in the target region have effects on the phenotype in the same direction and of similar magnitude [84,85], but they undergo a considerable loss of power in the presence of a large number of non-causal variants or in the presence of protective, deleterious, and null variants [86,87]. SKAT aggregates genetic information across the region using a kernel function and uses a computationally efficient variance component test to test for association. CMC (Combined Multivariate and Collapsing Method) collapses variants in subgroups according to allele frequencies and combines these subgroups using a T1 test [66,88]. Compared with population-based methods, family-based methods have more power and can prevent bias induced by population substructure [89]. The optimal weight was first proposed by Sha and coauthors in 2012 [90] in a population-based test called TOW (Test for the effect of an Optimally Weighted combination of variants) by assuming the independence among rare variants. FamSKAT [91], which accounts for familial correlation based on kinship coefficients in a linear mixed model, may be able to use both family and unrelated samples (developed for quantitative traits). Wang and coauthors in 2016 [92] proposed four weighting schemes for the family-based rare variants test (FBAT-v) [93]. Lee and coauthors in 2012 [94] derived the optimal test SKAT-O by estimating the correlation parameter in the kernel matrix to maximize the power, which corresponds to the estimated weight in the linear combination of the burden test and SKAT test statistics that maximizes power. Lin and coauthors in 2016 [68] extended the CAPL (Combined Association in the Presence of Linkage) [95] test, using both case-control and family data for testing from common variants to rare variant associations. A similarity-based weighted U approach is used to model the joint association analysis of sequencing variants and gene expression [69]. Sun and coauthors in 2016 [70] introduced a W-test collapsing method to evaluate rare variant associations by measuring the distributional differences between cases and controls through combined log of odds ratio within a genomic region. Wang and coauthors in 2016 [96] developed SKAT+, an estimation method that uses only control subjects; it has superior power over SKAT, while maintaining control over the type I error rate. Lu and coauthors in 2014 [71] reported the development and application of Trio-SVM (Support Vector Machine) approach that aggregates and evaluates the transmission of rare variants. The focus of Derkach and coauthors in 2014 [72] confirmed that Fisher's method is not only robust but can also improve power over individual pooled linear and quadratic tests and is often better than other robust tests such as SKAT-O. Cao and coauthors in 2014 [73] developed a USR (Unified Sparse Regression) to incorporate prior information and jointly adjust for cryptic relatedness, population structure, and other environmental covariates; qMSAT (Quality-based Multivariate Score Association Test) [97] and SSU (Sum of Squared U) statistic tests [98] were equivalent to the SKAT.

4. Conclusions and Perspectives

Thanks to the introduction of exome arrays technologies, great efforts have been conducted to extend association analyses to rare and coding variants. Recently, the joint work of large consortia allowed the interrogation of hundreds of thousands of SNVs in up to 475,000 individuals [28–38]; (Table 1). Some new low-frequency and rare variants have been reported that are consistently associated with BP traits, with size effects higher than 1.5 mmHg, and that should undergo deep functional testing. Considering the single variant analyses described here, the largest effect, to date, was observed for a rare missense SNV in the *KLH3* gene in relation to SBP (8.2 mmHg with se = 4.1) [37]. Despite the large sample size (up to 422,604 subjects for the exome analysis), the study from Warren and coauthors was still underpowered to identify rare variants with statistical significance. When considering the joint impact of 107 mostly common variants, a 9.3 mmHg higher SBP was reported for subjects >50 years and carrying the highest genetic risk score [37]. This finding has potential implications concerning early lifestyle interventions in high-risk individuals. In summary, although several complex networks of interacting pathways controlling BP have been established (e.g., RAAS and ENaC-related pathways), the current efforts on rare variants analysis have not yet provided a clear answer on where the missing heritability lies.

The advent of NGS provided the opportunity to detect, in a high-throughput way, the entire spectrum of genomic variation ranging from rare to common variants and from SNVs to insertions, deletions, and copy number variants. Despite the undeniable advantages, few studies have been conducted so far using NGS technologies in relation to hypertension and/or BP [61–67]; (Table 3). WES and, more so, WGS costs are still too high to analyse the large sample size required to identify rare variants. Target re-sequencing allows the cutting of laboratory costs and increases the statistical power by reducing multiple signals testing, therefore, this approach could be useful to detect causative variants underlying the trait by deeply analysing BP-associated loci described by GWAS. However, the studies reported here failed to identify new rare variants, likely because of the reduced sample size compared to GWAS [63,64]. The joint effort of large consortia with available sequencing data would be helpful to meet the need of a larger sample size.

Novel statistical approaches have been developed to overcome the limit imposed by the extremely low frequencies. Also, these tests attempt to take into account the high heterogeneity of the genetic regions in which both common and rare as well as causative and non-causative variants are more likely to occur [99]. However, detecting the few true causative variants among the large number of non-coding variants arising from NGS still represents a big challenge, and additional improvements to better annotate and filter the variants are required.

Another main limitation of rare variants analysis is the study of gene-gene and gene-environment interactions at a population level, which can be investigated only in terms of burden and collapsing tests, with environmental factors playing, anyway, an important role in systemic hypertension. Functional in vitro and in vivo models should further support the statistical interactions.

Rodent models represent an attractive genetic resource to functionally evaluate previously identified rare variants overlapped with human loci. Several rat and mouse strains have been developed for complex phenotypes, including hypertension, and exploited to perform QTL analysis and genome sequencing [100–104]. Here, we reported the study of Vecchione et al. [27], in which, thanks to experimental models, the authors clarified how a rare variant within the *BPIFB4* gene, a possible genetic risk factor for high BP, was implicated in the BP homeostasis by altering eNOS signaling.

It should also be considered that, as hypertension is an age-related condition, additional longitudinal studies incorporating repeated measures of BP would be advantageous. Lastly, most findings should be treated as trait-specific (SBP, DBP, PP, MAP, or hypertension) and population-specific. The majority of studies reported findings deriving from European populations. Allele frequencies and hypertension risk may differ among different geographic regions because of a selective pressure that occurred during the Out-of-Africa Expansion [105].

In conclusion, as sequencing costs will sufficiently decrease to ensure the proper sample size, and novel bioinformatic and biostatistical tools will be available for appropriate analyses, the identification of functional rare and low-frequency variants could really contribute to solving the high complexity of the genetics of hypertension and to elucidate whether new genes might improve patients care and the stratification of patients to distinguish those who will respond best to antihypertensive treatments.

Acknowledgments: This work was supported by Compagnia di San Paolo-IIGM (to Giuseppe Matullo) and by Ministero dell'Istruzione, dell'Università e della Ricerca—MIUR project "Dipartimenti di Eccellenza 2018–2022" to Department of Medical Sciences (Giuseppe Matullo), University of Torino.

Author Contributions: Alessia Russo: defined the structure of the review, performed literature paper selection, contributed to writing and critical reading; Cornelia Di Gaetano: contributed to literature paper selection, writing, and critical reading; Giovanni Cugliari: contributed to literature paper selection, writing, and critical reading; Giuseppe Matullo: defined the structure of the review, supervised the paper selection, contributed to writing and critical reading.

Conflicts of Interest: The authors declare no conflicts of interest.

Abbreviations

SBP	Systolic Blood Pressure
DBP	Diastolic Blood Pressure
PP	Pulse Pressure
MAP	Mean Artery Pressure
HTN	Hypertension
SNV	Single Nucleotide Variant
SNP	Single Nucleotide Polymorphism
MAF	Minor Allele Frequency
EWAS	Exome Wide Association Studies
GWAS	Genome-Wide Association Studies

NGS	Next Generation Sequencing
WGS	Whole Genome Sequencing
WES	Whole Exome Sequencing
CFS	Cleveland Family Study
CHARGE	Cohorts for Heart and Aging Research in Genomic Epidemiology
ARIC	Atherosclerosis Risk in Communities
CARDIA	Coronary Artery Risk Development in Young Adults
AADM	Africa America Diabetes Mellitus
GENOA	The Genetic Epidemiology Network of Arteriopathy
HUFS	Howard University Family Study
HyperGEN	Hypertension Genetic Epidemiology Network
CHD	Congenital heart disease Exome+
T2D-GENES	The Genetics of Type 2 Diabetes Consortium (GoT2D)/Type 2 Diabetes Genetic Exploration by Next-generation sequencing in multi-Ethnic Samples
REGARDS	REasons for Geographic And Racial Differences in Stroke
FHS	Framingham Heart Study
GAW	Genetic Analysis Workshop
ESP	Exome Sequencing Project
Beta$_{sst}$	Beta as standardized mean difference
Beta$_{ssm}$	SeqMeta Beta
EPACTS	Efficient and Parallelizable Association Container Toolbox
EMMA eXpedited	Efficient Mixed-Model Association eXpedited
GEMMA	Genome-Wide Efficient Mixed Model Association
SKAT	Sequence Kernel Association Test
SKAT-O	Optimal Unified Test
SKAT-C	SKAT-Combined
CAST	Cohort Allelic Sums Test
CMC	Combined Multivariate and Collapsing
w-SUM	Weighted-Sum
SST	Simple Sum Test
VT	Variable-Threshold
RBS	Replication-Based Weighted-Sum Statistic
FPCA	Functional Principal Components Analysis
FBAT	Family-based Association Test
GEE model	Generalized Estimating Equation model
burden-T1-del	burden test on deleterious variants
qMSAT	Quality-based Multivariate Score Association Test
GCTA	Genome-wide Complex Trait Analysis
CAPL	Combined Association in the Presence of Linkage
SVM	support vector machine
USR	Unified Sparse Regression
GLM	Generalized Linear Model
OR	Odds Ratio
N	Sample number
se	Standard Error
CI	Confidence Interval

References

1. Oparil, S.; Zaman, M.A.; Calhoun, D.A. Pathogenesis of hypertension. *Ann. Intern. Med.* **2003**, *139*, 761–776. [CrossRef] [PubMed]
2. Dart, A.M.; Kingwell, B.A. Pulse pressure—A review of mechanisms and clinical relevance. *J. Am. Coll. Cardiol.* **2001**, *37*, 975–984. [CrossRef]

3. Henry, J.B.; Miller, M.C.; Kelly, K.C.; Champney, D. Mean arterial pressure (MAP): An alternative and preferable measurement to systolic blood pressure (SBP) in patients for hypotension detection during hemapheresis. *J. Clin. Apher.* **2002**, *17*, 55–64. [CrossRef] [PubMed]

4. Shrout, T.; Rudy, D.W.; Piascik, M.T. Hypertension update, JNC8 and beyond. *Curr. Opin. Pharmacol.* **2017**, *33*, 41–46. [CrossRef] [PubMed]

5. Lim, S.S.; Vos, T.; Flaxman, A.D.; Danaei, G.; Shibuya, K.; Adair-Rohani, H.; Amann, M.; Anderson, H.R.; Andrews, K.G.; Aryee, M.; et al. A comparative risk assessment of burden of disease and injury attributable to 67 risk factors and risk factor clusters in 21 regions, 1990–2010: A systematic analysis for the Global Burden of Disease Study 2010. *Lancet* **2012**, *380*, 2224–2260. [CrossRef]

6. Luft, F.C. Twins in cardiovascular genetic research. *Hypertension* **2001**, *37*, 350–356. [CrossRef] [PubMed]

7. Padmanabhan, S.; Caulfield, M.; Dominiczak, A.F. Genetic and molecular aspects of hypertension. *Circ. Res.* **2015**, *116*, 937–959. [CrossRef] [PubMed]

8. Levy, D.; Ehret, G.B.; Rice, K.; Verwoert, G.C.; Launer, L.J.; Dehghan, A.; Glazer, N.L.; Morrison, A.C.; Johnson, A.D.; Aspelund, T.; et al. Genome-wide association study of blood pressure and hypertension. *Nat. Genet.* **2009**, *41*, 677–687. [CrossRef] [PubMed]

9. Ehret, G.B.; Munroe, P.B.; Rice, K.M.; Bochud, M.; Johnson, A.D.; Chasman, D.I.; Smith, A.V.; Tobin, M.D.; Verwoert, G.C.; Hwang, S.J.; et al. Genetic variants in novel pathways influence blood pressure and cardiovascular disease risk. *Nature* **2011**, *478*, 103–109. [CrossRef] [PubMed]

10. Ehret, G.B.; Ferreira, T.; Chasman, D.I.; Jackson, A.U.; Schmidt, E.M.; Johnson, T.; Thorleifsson, G.; Luan, J.; Donnelly, L.A.; Kanoni, S.; et al. The genetics of blood pressure regulation and its target organs from association studies in 342,415 individuals. *Nat. Genet.* **2016**, *48*, 1171–1184. [CrossRef] [PubMed]

11. Hoffmann, T.J.; Ehret, G.B.; Nandakumar, P.; Ranatunga, D.; Schaefer, C.; Kwok, P.Y.; Iribarren, C.; Chakravarti, A.; Risch, N. Genome-wide association analyses using electronic health records identify new loci influencing blood pressure variation. *Nat. Genet.* **2017**, *49*, 54–64. [CrossRef] [PubMed]

12. The Wellcome Trust Case Control, C. Genome-wide association study of 14,000 cases of seven common diseases and 3000 shared controls. *Nature* **2007**, *447*, 661.

13. Seidel, E.; Scholl, U.I. Genetic mechanisms of human hypertension and their implications for blood pressure physiology. *Physiol. Genom.* **2017**, *49*, 630–652. [CrossRef] [PubMed]

14. Civelek, M.; Lusis, A.J. Systems genetics approaches to understand complex traits. *Nat. Rev. Genet.* **2014**, *15*, 34–48. [CrossRef] [PubMed]

15. Fehrmann, R.S.; Jansen, R.C.; Veldink, J.H.; Westra, H.J.; Arends, D.; Bonder, M.J.; Fu, J.; Deelen, P.; Groen, H.J.; Smolonska, A.; et al. Trans-eQTLs reveal that independent genetic variants associated with a complex phenotype converge on intermediate genes, with a major role for the HLA. *PLoS Genet.* **2011**, *7*, e1002197. [CrossRef] [PubMed]

16. Romanoski, C.E.; Lee, S.; Kim, M.J.; Ingram-Drake, L.; Plaisier, C.L.; Yordanova, R.; Tilford, C.; Guan, B.; He, A.; Gargalovic, P.S.; et al. Systems genetics analysis of gene-by-environment interactions in human cells. *Am. J. Hum. Genet.* **2010**, *86*, 399–410. [CrossRef] [PubMed]

17. Langley, S.R.; Bottolo, L.; Kunes, J.; Zicha, J.; Zidek, V.; Hubner, N.; Cook, S.A.; Pravenec, M.; Aitman, T.J.; Petretto, E. Systems-level approaches reveal conservation of trans-regulated genes in the rat and genetic determinants of blood pressure in humans. *Cardiovasc. Res.* **2013**, *97*, 653–665. [CrossRef] [PubMed]

18. Langfelder, P.; Horvath, S. WGCNA: An R package for weighted correlation network analysis. *BMC Bioinform.* **2008**, *9*, 559. [CrossRef] [PubMed]

19. Burrello, J.; Monticone, S.; Buffolo, F.; Tetti, M.; Veglio, F.; Williams, T.A.; Mulatero, P. Is There a Role for Genomics in the Management of Hypertension? *Int. J. Mol. Sci.* **2017**, *18*, 1131. [CrossRef] [PubMed]

20. Doris, P.A. The genetics of blood pressure and hypertension: the role of rare variation. *Cardiovasc. Ther.* **2011**, *29*, 37–45. [CrossRef] [PubMed]

21. Munroe, P.B.; Barnes, M.R.; Caulfield, M.J. Advances in blood pressure genomics. *Circ. Res.* **2013**, *112*, 1365–1379. [CrossRef] [PubMed]

22. Jeppesen, J.; Berg, N.D.; Torp-Pedersen, C.; Hansen, T.W.; Linneberg, A.; Fenger, M. Fibrillin-1 genotype and risk of prevalent hypertension: a study in two independent populations. *Blood Press.* **2012**, *21*, 273–280. [CrossRef] [PubMed]

23. Ridker, P.M.; Hennekens, C.H.; Buring, J.E.; Rifai, N. C-reactive protein and other markers of inflammation in the prediction of cardiovascular disease in women. *N. Engl. J. Med.* **2000**, *342*, 836–843. [CrossRef] [PubMed]

24. Karaman, E.; Urhan Kucuk, M.; Bayramoglu, A.; Uzun Gocmen, S.; Ercan, S.; Guler, H.I.; Kucukkaya, Y.; Erden, S. Investigation of relationship between IL-6 gene variants and hypertension in Turkish population. *Cytotechnology* **2015**, *67*, 947–954. [CrossRef] [PubMed]

25. Gao, S.P.; Pan, M.; Chen, C.; Ge, L.J.; Jiang, M.H.; Luan, H.; Zheng, J.G.; Deng, X.T.; Pan, H.Y.; Zhu, J.H. The G to A polymorphism at-597 of the interleukin-6 gene is extremely rare in southern Han Chinese. *Cytokine* **2011**, *55*, 1–3. [CrossRef] [PubMed]

26. Villa, F.; Carrizzo, A.; Spinelli, C.C.; Ferrario, A.; Malovini, A.; Maciag, A.; Damato, A.; Auricchio, A.; Spinetti, G.; Sangalli, E.; et al. Genetic Analysis Reveals a Longevity-Associated Protein Modulating Endothelial Function and Angiogenesis. *Circ. Res.* **2015**, *117*, 333–345. [CrossRef] [PubMed]

27. Vecchione, C.; Villa, F.; Carrizzo, A.; Spinelli, C.C.; Damato, A.; Ambrosio, M.; Ferrario, A.; Madonna, M.; Uccellatore, A.; Lupini, S.; et al. A rare genetic variant of BPIFB4 predisposes to high blood pressure via impairment of nitric oxide signaling. *Sci. Rep.* **2017**, *7*, 9706. [CrossRef] [PubMed]

28. Sung, Y.J.; Basson, J.; Cheng, N.; Nguyen, K.D.; Nandakumar, P.; Hunt, S.C.; Arnett, D.K.; Davila-Roman, V.G.; Rao, D.C.; Chakravarti, A. The role of rare variants in systolic blood pressure: analysis of ExomeChip data in HyperGEN African Americans. *Hum. Hered.* **2015**, *79*, 20–27. [CrossRef] [PubMed]

29. Tran, N.T.; Aslibekyan, S.; Tiwari, H.K.; Zhi, D.; Sung, Y.J.; Hunt, S.C.; Rao, D.C.; Broeckel, U.; Judd, S.E.; Muntner, P.; et al. PCSK9 variation and association with blood pressure in African Americans: preliminary findings from the HyperGEN and REGARDS studies. *Front. Genet.* **2015**, *6*, 136. [CrossRef] [PubMed]

30. Yasukochi, Y.; Sakuma, J.; Takeuchi, I.; Kato, K.; Oguri, M.; Fujimaki, T.; Horibe, H.; Yamada, Y. Longitudinal exome-wide association study to identify genetic susceptibility loci for hypertension in a Japanese population. *Exp. Mol. Med.* **2017**, *49*, e409. [CrossRef] [PubMed]

31. Ohlsson, T.; Lindgren, A.; Engstrom, G.; Jern, C.; Melander, O. A stop-codon of the phosphodiesterase 11A gene is associated with elevated blood pressure and measures of obesity. *J. Hypertens.* **2016**, *34*, 445–451. [CrossRef] [PubMed]

32. Surendran, P.; Drenos, F.; Young, R.; Warren, H.; Cook, J.P.; Manning, A.K.; Grarup, N.; Sim, X.; Barnes, D.R.; Witkowska, K.; et al. Trans-ancestry meta-analyses identify rare and common variants associated with blood pressure and hypertension. *Nat. Genet.* **2016**, *48*, 1151–1161. [CrossRef] [PubMed]

33. Liu, C.; Kraja, A.T.; Smith, J.A.; Brody, J.A.; Franceschini, N.; Bis, J.C.; Rice, K.; Morrison, A.C.; Lu, Y.; Weiss, S.; et al. Meta-analysis identifies common and rare variants influencing blood pressure and overlapping with metabolic trait loci. *Nat. Genet.* **2016**, *48*, 1162–1170. [CrossRef] [PubMed]

34. Nandakumar, P.; Lee, D.; Richard, M.A.; Tekola-Ayele, F.; Tayo, B.O.; Ware, E.; Sung, Y.J.; Salako, B.; Ogunniyi, A.; Gu, C.C.; et al. Rare coding variants associated with blood pressure variation in 15,914 individuals of African ancestry. *J. Hypertens.* **2017**, *35*, 1381–1389. [CrossRef] [PubMed]

35. He, K.Y.; Wang, H.; Cade, B.E.; Nandakumar, P.; Giri, A.; Ware, E.B.; Haessler, J.; Liang, J.; Smith, J.A.; Franceschini, N.; et al. Rare variants in fox-1 homolog A (RBFOX1) are associated with lower blood pressure. *PLoS Genet.* **2017**, *13*, e1006678. [CrossRef] [PubMed]

36. Kraja, A.T.; Cook, J.P.; Warren, H.R.; Surendran, P.; Liu, C.; Evangelou, E.; Manning, A.K.; Grarup, N.; Drenos, F.; Sim, X.; et al. New Blood Pressure-Associated Loci Identified in Meta-Analyses of 475,000 Individuals. *Circ. Cardiovasc. Genet.* **2017**, *10*. [CrossRef] [PubMed]

37. Warren, H.R.; Evangelou, E.; Cabrera, C.P.; Gao, H.; Ren, M.; Mifsud, B.; Ntalla, I.; Surendran, P.; Liu, C.; Cook, J.P.; et al. Genome-wide association analysis identifies novel blood pressure loci and offers biological insights into cardiovascular risk. *Nat. Genet.* **2017**, *49*, 403–415. [CrossRef] [PubMed]

38. Kim, Y.K.; Hwang, M.Y.; Kim, Y.J.; Moon, S.; Han, S.; Kim, B.J. Evaluation of pleiotropic effects among common genetic loci identified for cardio-metabolic traits in a Korean population. *Cardiovasc. Diabetol.* **2016**, *15*, 20. [CrossRef] [PubMed]

39. Grove, M.L.; Yu, B.; Cochran, B.J.; Haritunians, T.; Bis, J.C.; Taylor, K.D.; Hansen, M.; Borecki, I.B.; Cupples, L.A.; Fornage, M.; et al. Best Practices and Joint Calling of the HumanExome BeadChip: The CHARGE Consortium. *PLoS ONE* **2013**, *8*, e68095. [CrossRef] [PubMed]

40. Sharotri, V.; Collier, D.M.; Olson, D.R.; Zhou, R.; Snyder, P.M. Regulation of epithelial sodium channel trafficking by proprotein convertase subtilisin/kexin type 9 (PCSK9). *J. Biol. Chem.* **2012**, *287*, 19266–19274. [CrossRef] [PubMed]

41. Riserus, U.; Arnlov, J.; Berglund, L. Long-term predictors of insulin resistance: role of lifestyle and metabolic factors in middle-aged men. *Diabetes Care* **2007**, *30*, 2928–2933. [CrossRef] [PubMed]

42. Horvath, A.; Boikos, S.; Giatzakis, C.; Robinson-White, A.; Groussin, L.; Griffin, K.J.; Stein, E.; Levine, E.; Delimpasi, G.; Hsiao, H.P.; et al. A genome-wide scan identifies mutations in the gene encoding phosphodiesterase 11A4 (PDE11A) in individuals with adrenocortical hyperplasia. *Nat. Genet.* **2006**, *38*, 794–800. [CrossRef] [PubMed]

43. Manolio, T.A.; Collins, F.S.; Cox, N.J.; Goldstein, D.B.; Hindorff, L.A.; Hunter, D.J.; McCarthy, M.I.; Ramos, E.M.; Cardon, L.R.; Chakravarti, A.; et al. Finding the missing heritability of complex diseases. *Nature* **2009**, *461*, 747–753. [CrossRef] [PubMed]

44. Caulfield, M.; Munroe, P.; Pembroke, J.; Samani, N.; Dominiczak, A.; Brown, M.; Benjamin, N.; Webster, J.; Ratcliffe, P.; O'Shea, S.; et al. Genome-wide mapping of human loci for essential hypertension. *Lancet* **2003**, *361*, 2118–2123. [CrossRef]

45. Province, M.A.; Kardia, S.L.; Ranade, K.; Rao, D.C.; Thiel, B.A.; Cooper, R.S.; Risch, N.; Turner, S.T.; Cox, D.R.; Hunt, S.C.; et al. A meta-analysis of genome-wide linkage scans for hypertension: the National Heart, Lung and Blood Institute Family Blood Pressure Program. *Am. J. Hypertens.* **2003**, *16*, 144–147. [CrossRef]

46. Wu, X.; Kan, D.; Province, M.; Quertermous, T.; Rao, D.C.; Chang, C.; Mosley, T.H.; Curb, D.; Boerwinkle, E.; Cooper, R.S. An updated meta-analysis of genome scans for hypertension and blood pressure in the NHLBI Family Blood Pressure Program (FBPP). *Am. J. Hypertens.* **2006**, *19*, 122–127. [CrossRef] [PubMed]

47. Elliott, P.; Peakman, T.C. The UK Biobank sample handling and storage protocol for the collection, processing and archiving of human blood and urine. *Int. J. Epidemiol.* **2008**, *37*, 234–244. [CrossRef] [PubMed]

48. Pazoki, R.; Dehghan, A.; Evangelou, E.; Warren, H.; Gao, H.; Caulfield, M.; Elliott, P.; Tzoulaki, I. Genetic Predisposition to High Blood Pressure and Lifestyle Factors: Associations With Midlife Blood Pressure Levels and Cardiovascular Events. *Circulation* **2018**, *137*, 653–661. [CrossRef] [PubMed]

49. Hoffmann, T.J.; Zhan, Y.; Kvale, M.N.; Hesselson, S.E.; Gollub, J.; Iribarren, C.; Lu, Y.; Mei, G.; Purdy, M.M.; Quesenberry, C.; et al. Design and coverage of high throughput genotyping arrays optimized for individuals of East Asian, African American, and Latino race/ethnicity using imputation and a novel hybrid SNP selection algorithm. *Genomics* **2011**, *98*, 422–430. [CrossRef] [PubMed]

50. Okuda, T.; Fujioka, Y.; Kamide, K.; Kawano, Y.; Goto, Y.; Yoshimasa, Y.; Tomoike, H.; Iwai, N.; Hanai, S.; Miyata, T. Verification of 525 coding SNPs in 179 hypertension candidate genes in the Japanese population: identification of 159 SNPs in 93 genes. *J. Hum. Genet.* **2002**, *47*, 387–394. [CrossRef] [PubMed]

51. Atlas, S.A. The renin-angiotensin aldosterone system: pathophysiological role and pharmacologic inhibition. *J. Manag. Care Pharm.* **2007**, *13*, 9–20. [CrossRef] [PubMed]

52. Scurrah, K.J.; Lamantia, A.; Ellis, J.A.; Harrap, S.B. Familial Analysis of Epistatic and Sex-Dependent Association of Genes of the Renin-Angiotensin-Aldosterone System and Blood Pressure. *Circ. Cardiovasc. Genet.* **2017**, *10*, e001595. [CrossRef] [PubMed]

53. Kelly, T.N.; Li, C.; Hixson, J.E.; Gu, D.; Rao, D.C.; Huang, J.; Rice, T.K.; Chen, J.; Cao, J.; Li, J.; et al. Resequencing Study Identifies Rare Renin-Angiotensin-Aldosterone System Variants Associated with Blood Pressure Salt-Sensitivity: The GenSalt Study. *Am. J. Hypertens.* **2017**, *30*, 495–501. [CrossRef] [PubMed]

54. Rossier, B.C. Epithelial sodium channel (ENaC) and the control of blood pressure. *Curr. Opin. Pharmacol.* **2014**, *15*, 33–46. [CrossRef] [PubMed]

55. Rossier, B.C.; Baker, M.E.; Studer, R.A. Epithelial sodium transport and its control by aldosterone: The story of our internal environment revisited. *Physiol. Rev.* **2015**, *95*, 297–340. [CrossRef] [PubMed]

56. Gu, X.; Gu, D.; He, J.; Rao, D.C.; Hixson, J.E.; Chen, J.; Li, J.; Huang, J.; Wu, X.; Rice, T.K.; et al. Resequencing Epithelial Sodium Channel Genes Identifies Rare Variants Associated with Blood Pressure Salt-Sensitivity: The GenSalt Study. *Am. J. Hypertens.* **2017**, *31*, 205–211. [CrossRef] [PubMed]

57. Joshi, M.B.; Philippova, M.; Ivanov, D.; Allenspach, R.; Erne, P.; Resink, T.J. T-cadherin protects endothelial cells from oxidative stress-induced apoptosis. *FASEB J.* **2005**, *19*, 1737–1739. [CrossRef] [PubMed]

58. Ranscht, B.; Dours-Zimmermann, M.T. T-Cadherin, a Novel Cadherin Cell-Adhesion Molecule in the Nervous-System Lacks the Conserved Cytoplasmic Region. *Neuron* **1991**, *7*, 391–402. [CrossRef]

59. Philippova, M.; Joshi, M.B.; Kyriakakis, E.; Pfaff, D.; Erne, P.; Resink, T.J. A guide and guard: The many faces of T-cadherin. *Cell. Signal.* **2009**, *21*, 1035–1044. [CrossRef] [PubMed]

60. Putku, M.; Kals, M.; Inno, R.; Kasela, S.; Org, E.; Kozich, V.; Milani, L.; Laan, M. CDH13 promoter SNPs with pleiotropic effect on cardiometabolic parameters represent methylation QTLs. *Hum. Genet.* **2015**, *134*, 291–303. [CrossRef] [PubMed]

61. Datta, A.S.; Zhang, Y.; Zhang, L.; Biswas, S. Association of rare haplotypes on ULK4 and MAP4 genes with hypertension. *BMC Proc.* **2016**, *10*, 363–369. [CrossRef] [PubMed]

62. Ji, W.; Foo, J.N.; O'Roak, B.J.; Zhao, H.; Larson, M.G.; Simon, D.B.; Newton-Cheh, C.; State, M.W.; Levy, D.; Lifton, R.P. Rare independent mutations in renal salt handling genes contribute to blood pressure variation. *Nat. Genet.* **2008**, *40*, 592–599. [CrossRef] [PubMed]

63. Morrison, A.C.; Bis, J.C.; Hwang, S.J.; Ehret, G.B.; Lumley, T.; Rice, K.; Muzny, D.; Gibbs, R.A.; Boerwinkle, E.; Psaty, B.M.; et al. Sequence analysis of six blood pressure candidate regions in 4178 individuals: the Cohorts for Heart and Aging Research in Genomic Epidemiology (CHARGE) targeted sequencing study. *PLoS ONE* **2014**, *9*, e109155. [CrossRef] [PubMed]

64. Salvi, E.; Kuznetsova, T.; Thijs, L.; Lupoli, S.; Stolarz-Skrzypek, K.; D'Avila, F.; Tikhonoff, V.; De Astis, S.; Barcella, M.; Seidlerova, J.; et al. Target sequencing, cell experiments, and a population study establish endothelial nitric oxide synthase (eNOS) gene as hypertension susceptibility gene. *Hypertension* **2013**, *62*, 844–852. [CrossRef] [PubMed]

65. Wang, W.; Wei, Z. Collapsing singletons may boost signal for associating rare variants in sequencing study. *BMC Proc.* **2014**, *8*, S50. [CrossRef] [PubMed]

66. Yu, B.; Pulit, S.L.; Hwang, S.J.; Brody, J.A.; Amin, N.; Auer, P.L.; Bis, J.C.; Boerwinkle, E.; Burke, G.L.; Chakravarti, A.; et al. Rare Exome Sequence Variants in CLCN6 Reduce Blood Pressure Levels and Hypertension Risk. *Circ. Cardiovasc. Genet.* **2016**, *9*, 64–70. [CrossRef] [PubMed]

67. Zhao, X.; Sha, Q.; Zhang, S.; Wang, X. Testing optimally weighted combination of variants for hypertension. *BMC Proc.* **2014**, *8*, S59. [CrossRef] [PubMed]

68. Lin, P.-L.; Tsai, W.-Y.; Chung, R.-H. A combined association test for rare variants using family and case-control data. *BMC Proc.* **2016**, *10*, 215–219. [CrossRef] [PubMed]

69. Tong, X.; Wei, C.; Lu, Q. Genome-wide joint analysis of single-nucleotide variant sets and gene expression for hypertension and related phenotypes. *BMC Proc.* **2016**, *10*, 125–129. [CrossRef] [PubMed]

70. Sun, R.; Weng, H.Y.; Hu, I.C.; Guo, J.F.; Wu, W.K.K.; Zee, B.C.Y.; Wang, M.H. A W-test collapsing method for rare-variant association testing in exome sequencing data. *Genet. Epidemiol.* **2016**, *40*, 591–596. [CrossRef] [PubMed]

71. Lu, A.T.; Cantor, R.M. Identifying rare-variant associations in parent-child trios using a Gaussian support vector machine. *BMC Proc.* **2014**, *8*, S98. [CrossRef] [PubMed]

72. Derkach, A.; Lawless, J.F.; Merico, D.; Paterson, A.D.; Sun, L. Evaluation of gene-based association tests for analyzing rare variants using Genetic Analysis Workshop 18 data. *BMC Proc.* **2014**, *8*, S9. [CrossRef] [PubMed]

73. Cao, S.; Qin, H.; Deng, H.W.; Wang, Y.P. A unified sparse representation for sequence variant identification for complex traits. *Genet. Epidemiol.* **2014**, *38*, 671–679. [CrossRef] [PubMed]

74. Lin, H.; Wang, M.; Brody, J.A.; Bis, J.C.; Dupuis, J.; Lumley, T.; McKnight, B.; Rice, K.M.; Sitlani, C.M.; Reid, J.G.; et al. Strategies to design and analyze targeted sequencing data: cohorts for Heart and Aging Research in Genomic Epidemiology (CHARGE) Consortium Targeted Sequencing Study. *Circ. Cardiovasc. Genet.* **2014**, *7*, 335–343. [CrossRef] [PubMed]

75. Salvi, E.; Kutalik, Z.; Glorioso, N.; Benaglio, P.; Frau, F.; Kuznetsova, T.; Arima, H.; Hoggart, C.; Tichet, J.; Nikitin, Y.P.; et al. Genomewide association study using a high-density single nucleotide polymorphism array and case-control design identifies a novel essential hypertension susceptibility locus in the promoter region of endothelial NO synthase. *Hypertension* **2012**, *59*, 248–255. [CrossRef] [PubMed]

76. Choi, M.; Scholl, U.I.; Ji, W.; Liu, T.; Tikhonova, I.R.; Zumbo, P.; Nayir, A.; Bakkaloglu, A.; Ozen, S.; Sanjad, S.; et al. Genetic diagnosis by whole exome capture and massively parallel DNA sequencing. *Proc. Natl. Acad. Sci. USA* **2009**, *106*, 19096–19101. [CrossRef] [PubMed]

77. Wain, L.V.; Verwoert, G.C.; O'Reilly, P.F.; Shi, G.; Johnson, T.; Johnson, A.D.; Bochud, M.; Rice, K.M.; Henneman, P.; Smith, A.V.; et al. Genome-wide association study identifies six new loci influencing pulse pressure and mean arterial pressure. *Nat. Genet.* **2011**, *43*, 1005–1012. [CrossRef] [PubMed]

78. Blangero, J.; Teslovich, T.M.; Sim, X.; Almeida, M.A.; Jun, G.; Dyer, T.D.; Johnson, M.; Peralta, J.M.; Manning, A.; Wood, A.R.; et al. Omics-squared: human genomic, transcriptomic and phenotypic data for genetic analysis workshop 19. *BMC Proc.* **2016**, *10*, 71–77. [CrossRef] [PubMed]

79. Higasa, K.; Ogawa, A.; Terao, C.; Shimizu, M.; Kosugi, S.; Yamada, R.; Date, H.; Matsubara, H.; Matsuda, F. A burden of rare variants in BMPR2 and KCNK3 contributes to a risk of familial pulmonary arterial hypertension. *BMC Pulm. Med.* **2017**, *17*, 57. [CrossRef] [PubMed]

80. Lee, S.; Wu, M.C.; Lin, X. Optimal tests for rare variant effects in sequencing association studies. *Biostatistics* **2012**, *13*, 762–775. [CrossRef] [PubMed]

81. Morgenthaler, S.; Thilly, W.G. A strategy to discover genes that carry multi-allelic or mono-allelic risk for common diseases: A cohort allelic sums test (CAST). *Mutat. Res. Fund. Mol. Mech. Mutagen.* **2007**, *615*, 28–56. [CrossRef] [PubMed]

82. Bansal, V.; Libiger, O.; Torkamani, A.; Schork, N.J. Statistical analysis strategies for association studies involving rare variants. *Nat. Rev. Genet.* **2010**, *11*, 773–785. [CrossRef] [PubMed]

83. Wu, M.C.; Lee, S.; Cai, T.; Li, Y.; Boehnke, M.; Lin, X. Rare-variant association testing for sequencing data with the sequence kernel association test. *Am. J. Hum. Genet.* **2011**, *89*, 82–93. [CrossRef] [PubMed]

84. Madsen, B.E.; Browning, S.R. A Groupwise Association Test for Rare Mutations Using a Weighted Sum Statistic. *PLoS Genet.* **2009**, *5*, e1000384. [CrossRef] [PubMed]

85. Morris, A.P.; Zeggini, E. An Evaluation of Statistical Approaches to Rare Variant Analysis in Genetic Association Studies. *Genet. Epidemiol.* **2010**, *34*, 188–193. [CrossRef] [PubMed]

86. Basu, S.; Pan, W. Comparison of Statistical Tests for Disease Association With Rare Variants. *Genet. Epidemiol.* **2011**, *35*, 606–619. [CrossRef] [PubMed]

87. Neale, B.M.; Rivas, M.A.; Voight, B.F.; Altshuler, D.; Devlin, B.; Orho-Melander, M.; Kathiresan, S.; Purcell, S.M.; Roeder, K.; Daly, M.J. Testing for an Unusual Distribution of Rare Variants. *PLoS Genet.* **2011**, *7*, e1001322. [CrossRef] [PubMed]

88. Li, B.; Leal, S.M. Methods for Detecting Associations with Rare Variants for Common Diseases: Application to Analysis of Sequence Data. *Am. J. Hum. Genet.* **2008**, *83*, 311–321. [CrossRef] [PubMed]

89. De, G.; Yip, W.K.; Ionita-Laza, I.; Laird, N. Rare Variant Analysis for Family-Based Design. *PLoS ONE* **2013**, *8*, e48495. [CrossRef] [PubMed]

90. Sha, Q.Y.; Wang, X.X.; Wang, X.L.; Zhang, S.L. Detecting Association of Rare and Common Variants by Testing an Optimally Weighted Combination of Variants. *Genet. Epidemiol.* **2012**, *36*, 561–571. [CrossRef] [PubMed]

91. Chen, H.; Meigs, J.B.; Dupuis, J. Sequence Kernel Association Test for Quantitative Traits in Family Samples. *Genet. Epidemiol.* **2013**, *37*, 196–204. [CrossRef] [PubMed]

92. Wang, X.; Zhao, X.; Zhou, J. Testing rare variants for hypertension using family-based tests with different weighting schemes. *BMC Proc.* **2016**, *10*, 233–237. [CrossRef] [PubMed]

93. Ionita-Laza, I.; Lee, S.; Makarov, V.; Buxbaum, J.D.; Lin, X.H. General Class of Family-based Association Tests for Sequence Data, and Comparisons with Population-based Association Tests. *Genet. Epidemiol.* **2012**, *36*, 720.

94. Lee, S.; Emond, M.J.; Bamshad, M.J.; Barnes, K.C.; Rieder, M.J.; Nickerson, D.A.; Christiani, D.C.; Wurfel, M.M.; Lin, X. Optimal unified approach for rare-variant association testing with application to small-sample case-control whole-exome sequencing studies. *Am. J. Hum. Genet.* **2012**, *91*, 224–237. [CrossRef] [PubMed]

95. Chung, R.H.; Schmidt, M.A.; Morris, R.W.; Martin, E.R. CAPL: A Novel Association Test Using Case-Control and Family Data and Accounting for Population Stratification. *Genet. Epidemiol.* **2010**, *34*, 747–755. [CrossRef] [PubMed]

96. Wang, K. Boosting the Power of the Sequence Kernel Association Test by Properly Estimating Its Null Distribution. *Am. J. Hum. Genet.* **2016**, *99*, 104–114. [CrossRef] [PubMed]

97. Daye, Z.J.; Li, H.; Wei, Z. A powerful test for multiple rare variants association studies that incorporates sequencing qualities. *Nucleic Acids Res.* **2012**, *40*, e60. [CrossRef] [PubMed]

98. Pan, W. Asymptotic tests of association with multiple SNPs in linkage disequilibrium. *Genet. Epidemiol.* **2009**, *33*, 497–507. [CrossRef] [PubMed]

99. Pritchard, J.K. Are rare variants responsible for susceptibility to complex diseases? *Am. J. Hum. Genet.* **2001**, *69*, 124–137. [CrossRef] [PubMed]

100. Atanur, S.S.; Diaz, A.G.; Maratou, K.; Sarkis, A.; Rotival, M.; Game, L.; Tschannen, M.R.; Kaisaki, P.J.; Otto, G.W.; Ma, M.C.; et al. Genome sequencing reveals loci under artificial selection that underlie disease phenotypes in the laboratory rat. *Cell* **2013**, *154*, 691–703. [CrossRef] [PubMed]

101. Feng, M.J.; Deerhake, M.E.; Keating, R.; Thaisz, J.; Xu, L.F.; Tsaih, S.W.; Smith, R.; Ishige, T.; Sugiyama, F.; Churchill, G.A.; et al. Genetic Analysis of Blood Pressure in 8 Mouse Intercross Populations. *Hypertension* **2009**, *54*, 802–809. [CrossRef] [PubMed]
102. Pravenec, M.; Churchill, P.C.; Churchill, M.C.; Viklicky, O.; Kazdova, L.; Aitman, T.J.; Petretto, E.; Hubner, N.; Wallace, C.A.; Zimdahl, H.; et al. Identification of renal Cd36 as a determinant of blood pressure and risk for hypertension. *Nat. Genet.* **2008**, *40*, 952–954. [CrossRef] [PubMed]
103. Crackower, M.A.; Sarao, R.; Oudit, G.Y.; Yagil, C.; Kozieradzki, I.; Scanga, S.E.; Oliveira-dos-Santos, A.J.; da Costa, J.; Zhang, L.Y.; Pei, Y.; et al. Angiotensin-converting enzyme 2 is an essential regulator of heart function. *Nature* **2002**, *417*, 822–828. [CrossRef] [PubMed]
104. Zimdahl, H.; Kreider, T.; Gosele, C.; Ganten, D.; Hubner, N. Conserved synteny in rat and mouse for a blood pressure QTL on human chromosome 17. *Hypertension* **2002**, *39*, 1050–1052. [CrossRef] [PubMed]
105. Young, J.H.; Chang, Y.P.C.; Kim, J.D.O.; Chretien, J.P.; Klag, M.J.; Levine, M.A.; Ruff, C.B.; Wang, N.Y.; Chakravarti, A. Differential susceptibility to hypertension is due to selection during the out-of-Africa expansion. *PLoS Genet.* **2005**, *1*, e82. [CrossRef] [PubMed]

International Journal of
Molecular Sciences

Review

The Low-Renin Hypertension Phenotype: Genetics and the Role of the Mineralocorticoid Receptor

Rene Baudrand [1] and Anand Vaidya [2,*]

[1] Department of Endocrinology, Endocrine Hypertension and Adrenal Disease Program, School of Medicine, Pontificia Universidad Catolica De Chile, Santiago 8330074, Chile; baudrandrene@gmail.com
[2] Department of Medicine, Division of Endocrinology Diabetes and Hypertension, Center for Adrenal Disorders, Brigham and Women's Hospital, Harvard Medical School, Boston, MA 02115, USA
* Correspondence: anandvaidya@bwh.harvard.edu

Received: 19 January 2018; Accepted: 8 February 2018; Published: 11 February 2018

Abstract: A substantial proportion of patients with hypertension have a low or suppressed renin. This phenotype of low-renin hypertension (LRH) may be the manifestation of inherited genetic syndromes, acquired somatic mutations, or environmental exposures. Activation of the mineralocorticoid receptor is a common final mechanism for the development of LRH. Classically, the individual causes of LRH have been considered to be rare diseases; however, recent advances suggest that there are milder and "non-classical" variants of many LRH-inducing conditions. In this regard, our understanding of the underlying genetics and mechanisms accounting for LRH, and therefore, potentially the pathogenesis of a large subset of essential hypertension, is evolving. This review will discuss the potential causes of LRH, with a focus on implicated genetic mechanisms, the expanding recognition of non-classical variants of conditions that induce LRH, and the role of the mineralocorticoid receptor in determining this phenotype.

Keywords: renin; low-renin; hypertension; mineralocorticoid receptor; genetics; aldosterone

1. Introduction

The renin–angiotensin–aldosterone system plays a crucial role in volume, sodium, and potassium homeostasis. When renal hypofiltration is sensed, this hormonal system is activated via the secretion of renin, which catalyzes the generation of angiotensin I that is subsequently modified to generate angiotensin II. The potent vasoconstrictive properties of angiotensin II, in addition to its ability to stimulate the release of vasopressin and adrenal aldosterone secretion, help maintain arterial blood pressure and restoration of intravascular volume. Thus, physiologic activation of the renin–angiotensin–aldosterone system is characterized as a renin-dependent aldosteronism and serves to maintain blood pressure and volume in terrestrial mammals that evolved in a milieu of scarce dietary sodium intake. Despite this highly evolved physiology, low-renin hypertension (LRH) is currently a prevalent biochemical phenotype described in up to 30% of hypertensives, depending on age and race [1,2]. LRH is characterized by the physiologic suppression of renin, often in the context of intravascular volume expansion; however, there are many potential pathophysiological events that can result in hypertension with a low-renin phenotype that will be discussed in this review.

LRH has been described and investigated for nearly 50 years [3,4]. Shortly after Jerome Conn first described primary aldosteronism as a condition of aldosterone excess independent of renin [5], a phenotype of low renin activity in hypertension without overt hyperaldosteronism was described [3,4]. This phenotype was considered unique from primary aldosteronism and termed LRH, and subsequent studies described it as a condition more prevalent in individuals of African descent and elderly populations, who are also prone to salt-sensitive hypertension [6]. In subsequent decades, it was speculated that LRH might represent a heterogeneous mixture of etiologies that could include states of excessive mineralocorticoid receptor (MR) activation, as initially supported by LRH subjects manifesting low salivary sodium-to-potassium ratios and marked decreases in blood pressure in response to spironolactone and aminoglutethimide (an adrenocortical steroid synthesis inhibitor) [7–9].

The potential role of MR activation in LRH is important, since LRH and primary aldosteronism are similarly associated with higher risk for cardiometabolic events and death [10,11], and the availability of oral MR antagonists permits a potential targeted therapy. Further, the LRH phenotype displays familial aggregation, where several polymorphisms and novel genes have been described [12]. In one study, family membership explained 35% of variance in renin levels, far beyond the classic low frequency monogenic causes [12]. The composite of evidence suggests that the LRH is likely a heterogeneous admixture of disease states that all converge on a phenotype of suppressed renin activity and high blood pressure. These disease states include primary aldosteronism (PA), as well as conditions that manifest with low aldosterone levels, such as endogenous hypercortisolism, the syndrome of apparent mineralocorticoid excess (AME), atypical forms of congenital adrenal hyperplasia, and alterations in the activity of the mineralocorticoid receptor or epithelial sodium channel (Liddle syndrome). Consistent with this differential diagnosis, LRH has been shown to exhibit a bimodal distribution of aldosterone levels in population-based studies, supporting the existence of two broad categories of LRH: those with suppressed aldosterone and those with normal or elevated aldosterone [13]. New evidence suggests that beyond these general etiologic categorizations, diseases such as PA, hypercortisolism, AME, and perhaps even Liddle syndrome, may exist across a phenotypic continuum that extends from their overt and "classical" presentations to milder and "non-classical" phenotypes; the recognition of these expanding phenotypic continuums may improve our understanding of the pathogenesis and treatment of LRH and essential hypertension. Finally, environmental and nutritional factors, such as obesity, diabetes, and especially high dietary sodium intake, also play an important role in the development of the LRH phenotype. Putative mechanisms for these links include, new MR interactions, such as ligand independent activation of MR by sodium-RAC1 [14] and epigenetic modifications of LRH-related genes by DNA methylation (e.g., *HSD11B2* or adducin gene), histone modifications (e.g., epithelial sodium channel gene) or noncoding RNA (e.g., renin-angiotensin-aldosterone pathway genes) [15].

Herein, we will review known conditions that manifest with a phenotype of LRH, while focusing especially on postulated genetic mechanisms and the role of excessive MR activation.

2. Primary Aldosteronism

The most prevalent cause of LRH is primary aldosteronism (PA) [10,16]. PA is characterized by hyperaldosteronism that is independent of renin and angiotensin II (thus renin-independent aldosteronism) that results in excessive MR activation, increases intra-vascular volume and blood pressure, and results in renal, vascular, and cardiac disease, and higher mortality [10,11,16–19].

PA is considered the most common form of endocrine hypertension, with an estimated prevalence of 5–10% in the general hypertensive population, at least 6% in the primary care population, and up to 20% in the setting of resistant hypertension [11,17,20]. Since Conn's initial description of the classical PA disease phenotype over 50 years ago, the understanding of the severity spectrum of PA and underlying genetics has greatly expanded [5,21]. First, human studies have shown that there is a broad spectrum of autonomous and renin-independent aldosteronism and MR activation; PA is not only a disease reserved for those with severe and resistant hypertension, rather can be detected in mild to moderate hypertension and also in normotension [20,22–24]. Normotensive individuals with higher aldosterone levels have a higher risk for developing hypertension, an association that is driven by normotensives exhibiting a PA phenotype: renin suppression with increasingly inappropriate aldosterone secretion [22–25]. Thus, it is becoming clearer that dysregulated autonomous aldosterone secretion that is independent of renin, even when it does not meet the classical definitions of overt PA, exists across a large continuum, and therefore, our strict categorization of PA may handicap clinical care by placing focus on only the most severe cases at the expense of ignoring milder disease [23,24,26,27].

Although clinical practice recommendations focus on defining PA using categorical thresholds [16,18], expert opinion is increasingly warning that "the strict definition of primary aldosteronism is no longer tenable," and calling to "recognize the true prevalence of primary aldosteronism to include dysregulated aldosterone secretion and inappropriate aldosterone production" [28]. Second, excessive MR activation in PA contributes to significant cardiovascular and metabolic diseases, independent of blood pressure, such as diabetes and metabolic syndrome, stroke, myocardial infarction, left ventricular hypertrophy, atrial fibrillation, heart failure, and death [11,16,19,29–31]. Collectively, these two important observations have made it clear that recognizing and treating PA as early as possible is critical to prevent long-term adverse outcomes. Finally, our understanding of the pathogenesis of PA has dramatically improved with new genetic and histopathologic discoveries that have shed light on the mechanisms that might underlie PA. These advances will be discussed in more detail below.

The Endocrine Society clinical practice guidelines recommend identifying overt cases of PA by measuring the aldosterone-to-renin ratio (ARR) based on the clinical detection of severe or resistant hypertension, hypokalemia, an adrenal nodule, sleep apnea, or a family history of PA or early cardiovascular disease [16]. The most widely used cut-off for an aldosterone concentration is at least 15 ng/dL (and less frequently 10 ng/dL) with an ARR of at least 30 ng/dL per ng/mL/h [16]. This clinical approach lowers the risk for false positive screening results and, in general, is designed to detect overt and severe cases of PA. Alternatively, to recognize milder forms of PA, and to maximize early case detection in order to mitigate future cardiometabolic disease, more permissive screening criteria have also been proposed: a suppressed renin activity in the context of non-suppressed aldosterone (>6–9 ng/dL) consistent with an ARR >20 [16]. This latter approach may detect milder cases of PA, but will increase the risk of false-positive screening results, and consequently, potentially more costly and/or invasive medical testing. The absence of a single diagnostic criterion is largely propagated by the lack of a histopathologic gold standard for PA diagnosis.

Given the high prevalence of PA, and particularly, the more recent recognition that milder forms of PA may be common even when there are is no radiographic evidence of adrenal neoplasia, a key issue is to understand what may underlie the pathogenesis of PA. The use of specific CYP11B2 antibodies has revealed the presence of aldosterone producing cell clusters (APCCs) in a remarkable proportion of morphologically normal adrenal glands [32,33]. APCCs have been described as non-neoplastic foci of CYP11B2 staining that can co-exist adjacent to aldosterone-producing adenomas, extend into zona fascisculata, and often harbor known somatic mutations in aldosterone driver genes [32,34,35]. Collectively, this evidence strongly suggests that APCCs may represent a common aldosterone-secretory abnormality, which can be detected in normotensives with normal adrenal glands [36], may increase in number with older age, and may thus be a pre-neoplastic and age-dependent precursor to more overt PA [37].

Along with the finding of APCCs, our understanding of the genetics of PA has undergone substantial revision in the past decade. These advances have been described in detail in recent reviews, and therefore, we will summarize them briefly herein. The first described inheritable form of PA is glucocorticoid remediable aldosteronism (GRA), also known as familial hyperaldosteronism type I (FH-I). The estimated prevalence of GRA is only 1% of PA subjects and is secondary to a chimeric gene with recombination of 11β-hydroxylase (*CYP11B1*) and aldosterone synthase (*CYP11B2*) genes [17,38]. The chimeric fusion results in regulation of aldosterone synthesis by adrenocorticotropic hormone (ACTH) and can be mitigated by suppressing ACTH with glucocorticoids. GRA should be considered in patients with early-onset hypertension and family history of PA or early-onset cerebrovascular accidents [17]. GRA has considerable variation in its clinical presentation, some families do not manifest with the classic hypertension and hypokalemia, and therefore, the gold standard diagnosis is sequencing to confirm the chimeric *CYP11B2/CYP11B1* gene [39].

Familial Hyperaldosteronism type II (FH-II) has been described in 6% of PA cases but is by definition a familial form of PA with yet unknown genetic loci [17]. Whether the discovery of newer inheritable forms of PA (below) results in a reclassification of FH-II remains to be seen.

Most recently, a new Familial Hyperaldosteronism named type III (FHIII), was described secondary to a gain-of-function germline mutation in the *KCNJ5* gene [40,41]. The *KCNJ5* gene mutation results in loss of potassium selectivity in a zona glomerulosa potassium channel, consequent increased influx of sodium resulting in a higher cell membrane potential and lower depolarization threshold, and therefore increased aldosterone synthesis and secretion [17,38,41]. Interestingly, although germline mutations in *KCNJ5* are rare, somatic mutations of *KCNJ5* have been described in nearly half of aldosterone-producing adenomas [38]. In addition, somatic (and rarely germline) mutations in other zona glomerulosa channels that result in increased cell membrane potential and decreased depolarization thresholds have been described in *CACNA1D* gene (codes for calcium channel, voltage-dependent, L-type, α-1d subunit), *CACNA1H* gene (codes for T-type voltage dependent calcium channel Cav3.2) [42], while somatic mutations have been found in the *ATP1A1* gene (Na$^+$, K$^+$-ATPase), and *ATP2B3* gene (calcium transporting ATPase 3) [38,43].

Treatment of PA should be tailored according to the severity of disease, age of the patient, anatomic type of disease (unilateral adenoma versus bilateral hyperplasia) and desire for surgery. Laparoscopic surgery is the recommended and ideal therapeutic intervention if PA is unilateral since it can cure aldosterone excess and improve long-term cardiometabolic outcomes and blood pressure control [16,19,44]. In most cases of PA, the source of autonomous aldosterone is bilateral or surgery is not pursued due to other complicating factors, and therefore, medical therapy with MR antagonists (such as spironolactone and eplerenone) is recommended [16]. Although medical therapy is often assumed to be equally efficacious to surgical therapy if blood pressure is normalized, a recent study suggested that this assumption may not be correct. In this large cohort study, patients with PA treated with MR antagonists had 2–3 times higher risk for incident cardiovascular events and death, when compared to age-matched essential hypertensives, even though they had similarly controlled (and relatively normalized) blood pressure while on MR antagonist therapy [19]. In contrast, PA patients who developed a longitudinal increase in renin activity while being treated with MR antagonists had the same risk for incident cardiovascular events and death as patients with essential hypertension, suggesting that the excess risk in medically treated PA may be due to inadequate MR blockade as manifested by a persistently suppressed renin. Future prospective studies will be needed to determine the optimal approach for medical therapy in PA, and how it compares with surgical adrenalectomy [19].

3. Hypercortisolism

Endogenous hypercortisolism, with or without the overt manifestations of Cushing syndrome, can result in chronic stimulation of the glucocorticoid receptor and also potentially the MR, with consequent development of hypertension, insulin resistance, diabetes, and cardiovascular disease and mortality [45–49].

There are several potential mechanisms for developing the LRH phenotype with hypercortisolism. Activation of the glucocorticoid receptor by excess cortisol can induce a direct vasopressor effect and elevations in blood pressure [50]. However, cortisol-mediated activation of the renal MR can also play a role in developing LRH. Cortisol and aldosterone are similarly potent MR agonists, but cortisol is inactivated to cortisone by 11β-hydroxysteroid dehydrogenase Type 2 (11βHSD2), thereby "protecting" the renal MR from abundant cortisol stimulation and permitting a high-affinity aldosterone-MR interaction. However, in states of severe hypercortisolism, excess cortisol can overwhelm 11βHSD2 activity and result in direct cortisol-mediated MR activation and subsequent intravascular volume expansion with suppression of renin and aldosterone [50]. Thus, the LRH phenotype with hypercortisolism is unique from PA in that it is a hyporeninemic hypoaldosteronism manifesting with Cushing syndrome and apparent MR overactivation (i.e., hypertension, hypokalemia and increased kaliuresis) [47,51,52]. Studies have also shown that chronic hypercortisolism can also result in activation of ENaC and increases in angiotensinogen [50,51,53].

Endogenous hypercortisolism is most commonly due to a benign pituitary ACTH secreting tumor (Cushing disease), and less frequently due to benign or malignant adrenal tumors (ACTH-independent hypercortisolism), and non-pituitary ACTH-secreting tumor (ectopic ACTH secretion). Current guidelines recommend that the diagnosis of overt hypercortisolism be confirmed using two distinct tests: elevated late night salivary cortisols, elevated 24 h urinary free cortisols, and/or incomplete suppression of cortisol following overnight dexamethasone [47]. Beyond overt hypercortisolism, the concept of "subclinical hypercortisolism" or "autonomous cortisol secretion" in association with adrenocortical adenomas, whereby there is excess cortisol without the hallmark signs of Cushing syndrome, is being recognized as a prevalent phenotype that is associated with higher risk for cardiometabolic disease [54]. For example, studies using mass spectrometry to analyze steroid metabolites have shown that even apparently nonfunctional adrenal adenomas secrete higher concentrations of glucocorticoids when compared to patients with no adenomas [55]; therefore, it is possible that many or most adrenocortical neoplasms secrete at least miniscule amounts of glucocorticoid, and that the categorization of "nonfunctional adenoma" may be misleading or a misnomer. In parallel, cohort studies in patients with incidentally discovered benign adrenocortical tumors have observed that patients with subclinical hypercortisolism have a higher risk for incident cardiovascular disease and death when compared to those with "nonfunctional" tumors [49]. Further, patients with apparently "nonfunctional" adrenocortical tumors have a higher risk for incident diabetes, a risk that was related to the degree of autonomous cortisol secretion within the normal range (normal defined as cortisol < 1.8 mcg/dL following overnight 1 mg dexamethasone suppression), compared with patients with no adrenal tumors [48]. Thus, as with PA, there is increased recognition that adrenal hypercortisolism is not a categorical phenotype, rather exists over an expanded continuum ranging from mild and subclinical to overt Cushing syndrome, with a parallel risk profile of increasing cardiometabolic risk. Future and ongoing studies will be necessary to determine when intervention to treat the autonomous cortisol is indicated with respect to abrogating incident adverse outcomes.

The genetics of hypercortisolism, both related to pituitary and adrenals tumors, has undergone dramatic change in the recent decade. Concerning pituitary tumors, germline mutations that cause Cushing disease have been previously ascribed to *MEN1* gene mutations and *AIP* mutations in Familial isolated pituitary adenomas (FIPA). More recently, next generation sequencing of pituitary adenomas identified somatic driver mutations in ubiquitin-specific protease 8 (*USP8*), leading to ACTH excess and hypercortisolism [56]. In adrenal hypercortisolism, novel genes have also been described beyond the classic germline mutations of the Carney complex (*PRKAR1A* gene, regulatory subunit of protein kinase) or McCune-Albright syndrome (activating somatic mutations in *GNAS1* oncogene) [56]. Germline mutations in *ARMC5* (Armadillo repeat-containing protein 5) have been identified in familial cases and are present in approximately 50% of sporadic cases of macronodular adrenal hyperplasia [57]. In addition, somatic events in *ARMC5* and *PRKACA* (encodes for catalytic subunit α of protein kinase A) are frequently observed in cortisol-producing adrenal adenomas [38,58,59].

4. Apparent Mineralocorticoid Excess Syndrome

The syndrome of apparent mineralocorticoid excess (AME) is a rare disease, first described in the late 1970s, as a syndrome of severe pediatric LRH [60]. The AME syndrome is an autosomal recessive condition due to loss of function mutations in 11βHSD2. The insufficient activity of 11βHSD2 permits normal cortisol concentrations to activate the renal MR, resulting in a syndrome of MR-mediated LRH, low aldosterone, hypokalemia, alkalosis, and usually failure to thrive and poor weight gain.

The diagnosis of AME is usually suspected in the setting on non-aldosterone dependent LRH with classic features of MR activation and confirmed by a high cortisol/cortisone (F/E) ratio in the serum or urine, and/or genetic sequencing of 11βHSD2 [61,62]. These infrequent cases of classic AME are treated with low dose dexamethasone to suppress endogenous ACTH and cortisol (since dexamethasone is not metabolized by 11βHSD2) in combination with an MR antagonist, and in extreme cases, renal transplantation [63].

Although the classic AME syndrome is rare, recent research suggests that the spectrum of cortisol-mediated MR activation may be more expansive than currently recognized in that milder forms of AME may be common. For instance, Ulick et al. described a milder version of AME and named it the type 2 variant, caused by a decrease in the cortisol clearance rate but not related to cortisone conversion [64]. More recently, cross-sectional human studies have shown that both cortisone levels and renin activity decline with older age, suggesting a potential age-dependent decline in the activity of 11βHSD2 [65]. Since higher F/E ratio and the low-renin phenotype have been correlated with higher blood pressure in both adults and children [66,67], new evidence suggests that a proportion of LRH may be explained by a less severe, or "non-classical", phenotype of AME that may respond to MR antagonists [68]. Interestingly, milder phenotypes of AME may be explained by less severe inactivating mutations or heterozygosity with partial activity of 11βHSD2, but also by consumption of exogenous inhibitors of 11βHSD2such as licorice or grapefruit [69]. In a recent study, we observed that lower cortisone levels (in combination with higher F/E ratio) were strongly associated with higher MR activity (lower renin activity and higher urinary potassium excretion) in patients suspected to have mild or non-classical AME (Tapia-Castillo, Baudrand, Vaidya, et al. personal communication). The summary of available data suggest that beyond the rare classical phenotype of AME, milder forms of non-classical AME may contribute to LRH and may be best detected by recognition of high F/E ratio in combination with low cortisone levels.

5. Congenital Adrenal Hyperplasia

Congenital adrenal hyperplasia (CAH) is one of the most common autosomal recessive disorders caused by mutations in genes involved in cortisol biosynthesis enzymes. More than 90% of described cases of CAH are 21-hydroxylase deficiency. Less common, and pertinent to the LRH phenotype, are CAH syndromes due to 11β-hydroxylase [P450c11β] and 17α-hydroxylase [P450c17] deficiency.

CAH caused by steroid 11β-hydroxylase deficiency is considered a rare recessive disorder, with an overall frequency of 1/100,000 live births. Unlike 21-hydroxylase CAH that is more frequent in Europe and North America, 11β-hydroxylase deficiency is more frequently described in the Middle East and Africa [70]. The gene *CYP11B1* encodes 11β-hydroxylase that catalyzes the conversion of 11-deoxycortisol and 11-deoxycorticosterone to cortisol and corticosterone in the zona fasciculata. Thus, from a clinical perspective, deficiency of 11β-hydroxylase, results in low levels of cortisol and high levels of 11-deoxycortisol and 11-deoxycorticosterone (DOC) and a shunting of metabolites into the androgen synthesis pathway. Depending on the severity of the mutations and 11β-hydroxylase deficiency, patients may present with signs of hyperandrogenism (masculinization of genitalia, hirsutism, premature bone maturation and precocious puberty) and excessive MR activation due to increased DOC (hypertension and hypokalemia) [70]. The increased activation of the MR leads to sodium reabsorption, increased intravascular volume expansion, and consequently LRH. The diagnosis can be confirmed via elevated DOC and androgen levels (with or without cosyntropin stimulation) and/or genetic sequencing of 11β-hydroxylase. Treatment involves the use of MR antagonists in addition to glucocorticoids in order to suppress ACTH.

17α-hydroxylase deficiency is another rare form of CAH and results as a consequence of inactivating mutations in the *CYP17A1* gene that regulates steroid 17-hydroxylation followed by the 17,20-lyase reactions. The most common presentation of 17α-hydroxylase deficiency is an adolescent female with absence of secondary sexual characteristics, amenorrhea, and LRH [71]. The 17α-hydroxylase deficiency phenotype is explained by impaired steroidogenesis in both the adrenals and the gonads: abnormal cortisol synthesis, elevation of DOC and resultant MR-mediated LRH, and impaired sex-steroid synthesis. These patients typically display low cortisol, but high levels of corticosterone, so adrenal crises are rarelyobserved. LRH is due to high levels of DOC that result in MR activation, hypertension, and hypokalemia [72], and treatment involves MR antagonists to normalize blood pressure and replacement of sex steroids [72].

6. Liddle Syndrome

The Liddle syndrome is a rare autosomal dominant monogenic disease caused by gain-of-function mutations of subunits of the epithelial sodium channel (ENaC). Since increased luminal ENaC in the distal nephron is the common final mechanism for MR-mediated sodium reabsorption, Liddle syndrome resembles states of mineralocorticoid excess. The classical clinical presentation is severe hypertension in a young patient, with hypokalemia and metabolic alkalosis, in the setting of suppressed renin; however, unlike PA, aldosterone levels are low or undetectable and the syndrome does not improve with MR antagonist therapy [73–76]. The diagnosis of Liddle syndrome can be confirmed by genetic sequencing of the *SCNN1B* and *SCNN1G* genes, which encode for the β and γ subunits of ENaC [75], or the *SCNN1A* gene [77], and the treatment of choice are ENaC inhibitors such as amiloride or triamterene [74].

In a similar theme to the aforementioned discussions on PA, hypercortisolism, and AME, Liddle syndrome may exist across a more extensive spectrum than currently recognized. "Classical" Liddle syndrome is rare and manifests with a severe phenotype; however, recent studies have suggested that a substantial proportion of LRH patients also have a low serum aldosterone [13,24]. Although there are potentially many mechanistic explanations for a phenotype of hypertension with low renin and aldosterone, a recent clinical trial demonstrated that treating African patients with a low-renin and aldosterone phenotype using ENaC inhibitors resulted in the most effective control of blood pressure [78], a finding that extended genetic studies demonstrating a high prevalence of *SCNN1B* variants in individuals of African descent with LRH [79].

7. Glucocorticoid Resistance Syndrome

Inactivating mutations of the glucocorticoid receptor gene (*NR3C1*) in chromosome 5q31–q32 cause familial glucocorticoid resistance [80–82]. All severe described cases are secondary to mutations that affect the function of either the ligand binding or the DNA binding domain. This is a rare syndrome, inherited either as an autosomal recessive or dominant disease, that is characterized by the classical biochemical pattern seen in endocrine receptor resistance syndromes: hypercortisolism and increased ACTH but without the classic clinical features of Cushing's syndrome [81]. Although the increased ACTH and cortisol are necessary to try and overcome glucocorticoid receptor resistance, the undesired effects are ACTH-induced hypersecretion of adrenal mineralocorticoids and androgens, and hypercortisolism-mediated renal-MR activation. Thus, the classic phenotype of glucocorticoid resistance is LRH, hypokalemia, hirsutism in females, alopecia in males, acne, precocious puberty and menstrual irregularities, in addition to chronic fatigue and malaise as a consequence of relative glucocorticoid deficiency [83]. The treatment is usually overnight low dose dexamethasone to suppress ACTH, and thus improving ACTH-induced hypercortisolism, excess mineralocorticoids, and hyperandrogenism. The addition of spironolactone can further help to control hypertension and hirsutism in females as well. Interestingly, haploinsufficiency of the GR receptor can also present with LRH due to MR activation by elevated cortisol rather than increased DOC as observed in classical glucocorticoid resistance [84].

8. Gordon Syndrome

Gordon Syndrome (GS) was described in the 1960s and is a very rare familial hypertension syndrome that presents with low renin and hyperkalemia. GS is considered to have autosomal dominant inheritance, although new molecular studies suggest some recessive cases [85]. Mutations in *WNK1*, *WNK4*, *CUL3* and *KLHL3* genes have all been identified in GS [85], due their interaction with the thiazide sensitive Na/Cl cotransporter (NCC) in the distal nephron, responsible for sodium reabsorption, but also reducing cell-surface expression of renal outer medullary potassium channel (ROMK), explaining the lower potassium excretion in the collecting ducts [86].

Subjects with GS have suppressed renin levels consistent with their high volume state, with relatively low aldosterone levels (that may still fall in the reference range) despite their hyperkalemia, that can be occasionally severe and with periodic paralysis [86]. Both hypertension and hyperkalemia can be reversed by dietary sodium restriction and/or low doses of thiazide diuretics.

9. MR Activating Mutations

This is a novel entity secondary to a gain of function mutation, by a substitution of leucine for serine at codon 810 (abbreviated as S810L), in the MR gene. The S810L mutation was recently described and is responsible for an MR-mediated LRH that has also been described as worsening during pregnancy [87]. The worsening in pregnancy is explained by the activation of the mutant S810L MR by progesterone, where progesterone typically antagonizes wild-type MR. In males and non-pregnant females, cortisone and 11-dehydrocorticosterone (cortisol and corticosterone metabolites respectively) can activate the mutant MR and result in increased sodium reabsorption [87]. Spironolactone is surprisingly ineffective since it has agonist properties on the mutant MR and can paradoxically increase MR activation [88]. Recently, Amazit et al. described that in addition to amiloride to inhibit ENaC, the novel and potent nonsteroidal selective MR antagonist finerenone has MR antagonistic properties and may therefore represent a useful treatment option [89].

10. Proposed Approach to the Patient with Low-Renin Hypertension

Prior to considering an elaborate differential diagnosis for LRH, it is important to confirm the phenotype. First, factors than can reversibly suppress renin include, increased intra-vascular volume, certain medications (see below), supine posture, menstrual phase and high dietary sodium intake, and therefore must all be considered [16,90,91]. The diagnosis of LRH inherently assumes a chronic phenotype, and therefore, it is important to ensure that a single measure of low renin was not induced by a confounding or transient factor. Second, it is important that renin levels are measured in a reliable laboratory using a validated technique. Renin activity assays are increasingly being measured via LC-MS/MS and the global forecast suggests that commercial renin activity assays may be largely replaced by assays of renin concentration [92]. To avoid misinterpretation, the most ideal confirmation of LRH entails measurement of renin in a seated posture, while on unrestricted dietary sodium intake, during the follicular phase of the menstrual cycle (in women), and ideally without the confounding effects of MR antagonists, β-blockers, diuretics, angiotensin-converting enzyme inhibitors, or angiotensin receptor blockers [16,18,93]. Although most patients in industrialized countries consume sufficient sodium to maintain a dietary sodium balance of greater than 100–150 mmol/24 h, it has been shown that severe dietary sodium restriction (<40 mmol/24 h) can raise renin even among patients with overt PA; thus, emphasizing the importance of recommending unrestricted dietary sodium intake when phenotyping renin status, and consideration of confirming 24 h urinary sodium balance [94].

The exact definition of what constitutes low renin is not well established. Suppression of renin below the limit of detection is probably the most commonly used definition, however, in non-clinical research settings, more elaborate physiological maneuvers have been employed to characterize a phenotype of low renin or a renin that does not adequately stimulate to physiological provocation [4,27,94–96]. Once there is confidence that the phenotype represents LRH, the examination of aldosterone levels in the context of renin suppression can provide insight into potential underlying mechanisms of disease (Figure 1). An inappropriately high aldosterone level narrows the differential diagnosis to PA. The most severely elevated aldosterone levels (>15–20 ng/dL) require little confirmation, and similarly, even levels > 10 ng/dL are highly suggestive among patients with LRH that can be confirmed using a variety of confirmatory tests (sodium or fludrocortisone suppression or captopril challenge). However, milder forms of PA are common and may not exhibit marked elevations in aldosterone or blood pressure [22–25], and therefore, these non-classical instances of PA may represent missed opportunities at mitigating the age-dependent cardiometabolic consequences of autonomous aldosterone secretion [10,11]. Future studies, and a renewed commitment to better understanding the true prevalence and severity spectrum of PA, will be needed to identify new and efficient methods for identifying non-classical PA, including novel biomarkers of MR activation [97].

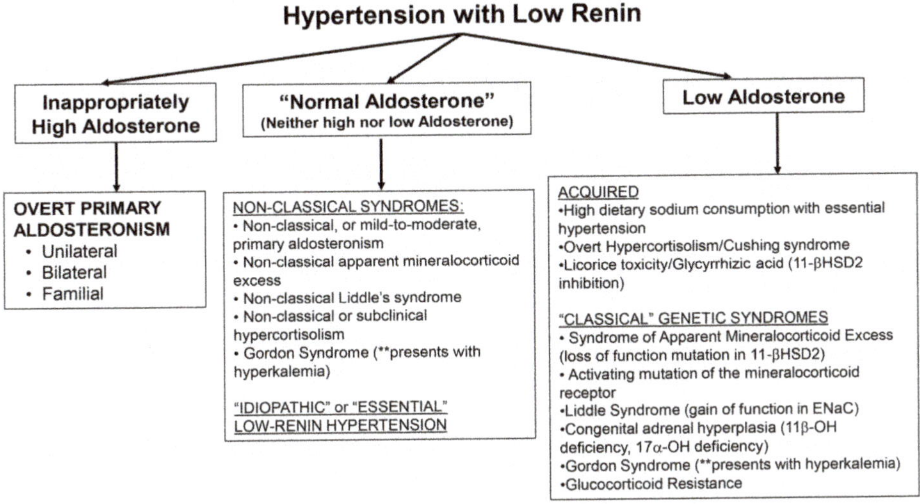

Figure 1. A proposed approach to the patient with low-renin hypertension. ENaC = epithelial sodium channel; 11-βHSD2 = 11-β hydroxysteroid dehydrogenase type 2; 11β-OH = 11-β hydroxylase; 17α-OH = 17-α hydroxylase.

A phenotype of LRH whereby aldosterone levels are clearly suppressed or very low generates a broad consideration of underlying mechanisms. Acquired or non-genetic etiologies include essential (or idiopathic) hypertension in the context of high dietary sodium intake, as well as severe hypercortisolism or reversible inhibition of 11BHSD2 (Figure 1). The classical genetic syndromes that result in a low renin and aldosterone phenotype are rare, and include classical AME, activating mutations in the MR or ENaC (Liddle syndrome), atypical forms of CAH, Gordon syndrome, and glucocorticoid resistance syndrome. The default explanation for LRH with low aldosterone, once these infrequent and classical diseases have been excluded, is essential hypertension with high dietary sodium consumption.

Finally, a great deal of progress has been made in identifying "non-classical" phenotypes that represent a milder continuum of classical disease states. These non-classical conditions should be considered when aldosterone levels are normal or not in the extremes: neither markedly elevated nor remarkably suppressed. Mild and moderate variants of PA, hypercortisolism, AME, and potentially even Liddle's syndrome, may be common and the early recognition and treatment of these phenotypes may prevent years of untreated hypertension and cardiovascular risk. Gordon's syndrome may also present with this phenotype. MR antagonists have already been shown to be the most effective add-on therapy in low-renin resistant hypertension [98], however, the adoption of MR antagonists as effective anti-hypertensives that can be used early in the treatment of hypertension using a renin- and aldosterone-based phenotyping approach has not been evaluated with sufficient evidence for widespread recommendation. Non-classical PA, hypercortisolism, and non-classical AME may all potentially respond well to MR antagonists, and therefore, future studies are needed to evaluate the efficacy of early initiation of MR antagonists in the treatment of LRH to formulate evidence-based clinical practice recommendations.

Acknowledgments: Rene Baudrand was funded by Fondo Nacional de Desarrollo Cientifico y Tecnologico (FONDECYT) 1150327, 1150437, 1160836, 1160695, and Corporacion de Fomento de la Produccion de Chile

(CORFO) 13CTI-21 526-P1. Anand Vaidya supported by the National Institutes of Diabetes and Digesttive and Kidney Diseases (NIDDK) of the National Institutes of Health (NIH) under award number R01DK107407, by grant 2015085 from the Doris Duke Charitable Foundation.

Conflicts of Interest: The authors declare no conflict of interest.

References

1. Mulatero, P.; Verhovez, A.; Morello, F.; Veglio, F. Diagnosis and treatment of low-renin hypertension. *Clin. Endocrinol.* **2007**, *67*, 324–334. [CrossRef] [PubMed]
2. Sagnella, G.A. Why is plasma renin activity lower in populations of african origin? *J. Hum. Hypertens.* **2001**, *15*, 17–25. [CrossRef] [PubMed]
3. Jose, A.; Crout, J.R.; Kaplan, N.M. Suppressed plasma renin activity in essential hypertension. Roles of plasma volume, blood pressure, and sympathetic nervous system. *Ann. Intern. Med.* **1970**, *72*, 9–16. [CrossRef] [PubMed]
4. Channick, B.J.; Adlin, E.V.; Marks, A.D. Suppressed plasma renin activity in hypertension. *Arch. Intern. Med.* **1969**, *123*, 131–140. [CrossRef] [PubMed]
5. Conn, J.W. Evolution of primary aldosteronism as a highly specific clinical entity. *J. Am. Med. Assoc.* **1960**, *172*, 1650–1653. [CrossRef] [PubMed]
6. Fisher, N.D.; Hurwitz, S.; Ferri, C.; Jeunemaitre, X.; Hollenberg, N.K.; Williams, G.H. Altered adrenal sensitivity to angiotensin II in low-renin essential hypertension. *Hypertension* **1999**, *34*, 388–394. [CrossRef] [PubMed]
7. Adlin, E.V.; Marks, A.D.; Channick, B.J. The salivary sodium/potassium ratio in hypertension: Relation to race and plasma renin activity. *Clin. Exp. Hypertens. A* **1982**, *4*, 1869–1880. [CrossRef] [PubMed]
8. Woods, J.W.; Liddle, G.W.; Michelakis, A.M.; Brill, A.B. Effect of an adrenal inhibitor in hypertensive patients with suppressed renin. *Arch. Intern. Med.* **1969**, *123*, 366–370. [CrossRef] [PubMed]
9. Adlin, E.V.; Marks, A.D.; Channick, B.J. Spironolactone and hydrochlorothiazide in essential hypertension. Blood pressure response and plasma renin activity. *Arch. Intern. Med.* **1972**, *130*, 855–858. [CrossRef] [PubMed]
10. Monticone, S.; Burrello, J.; Tizzani, D.; Bertello, C.; Viola, A.; Buffolo, F.; Gabetti, L.; Mengozzi, G.; Williams, T.A.; Rabbia, F.; et al. Prevalence and clinical manifestations of primary aldosteronism encountered in primary care practice. *J. Am. Coll. Cardiol.* **2017**, *69*, 1811–1820. [CrossRef] [PubMed]
11. Monticone, S.; D'Ascenzo, F.; Moretti, C.; Williams, T.A.; Veglio, F.; Gaita, F.; Mulatero, P. Cardiovascular events and target organ damage in primary aldosteronism compared with essential hypertension: A systematic review and meta-analysis. *Lancet Diabetes Endocrinol.* **2017**, *6*, 41–50. [CrossRef]
12. Fisher, N.D.; Hurwitz, S.; Jeunemaitre, X.; Hopkins, P.N.; Hollenberg, N.K.; Williams, G.H. Familial aggregation of low-renin hypertension. *Hypertension* **2002**, *39*, 914–918. [CrossRef] [PubMed]
13. Adlin, E.V.; Braitman, L.E.; Vasan, R.S. Bimodal aldosterone distribution in low-renin hypertension. *Am. J. Hypertens.* **2013**, *26*, 1076–1085. [CrossRef] [PubMed]
14. Kawarazaki, W.; Fujita, T. Aberrant Rac1-mineralocorticoid receptor pathways in salt-sensitive hypertension. *Clin. Exp. Pharmacol. Physiol.* **2013**, *40*, 929–936. [CrossRef] [PubMed]
15. Friso, S.; Carvajal, C.A.; Fardella, C.E.; Olivieri, O. Epigenetics and arterial hypertension: The challenge of emerging evidence. *Transl. Res.* **2015**, *165*, 154–165. [CrossRef] [PubMed]
16. Funder, J.W.; Carey, R.M.; Mantero, F.; Murad, M.H.; Reincke, M.; Shibata, H.; Stowasser, M.; Young, W.F., Jr. The management of primary aldosteronism: Case detection, diagnosis, and treatment: An endocrine society clinical practice guideline. *J. Clin. Endocrinol. Metab.* **2016**, *101*, 1889–1916. [CrossRef] [PubMed]
17. Vaidya, A.; Hamrahian, A.H.; Auchus, R.J. Genetics of primary aldosteronism. *Endocr. Pract.* **2015**, *21*, 400–405. [CrossRef] [PubMed]
18. Vaidya, A.; Malchoff, C.D.; Auchus, R.J.; Committee, A.A.S. An individualized approach to the evaluation and management of primary aldosteronism. *Endocr. Pract.* **2017**, *23*, 680–689. [CrossRef] [PubMed]
19. Hundemer, G.L.; Curhan, G.C.; Yozamp, N.; Wang, M.; Vaidya, A. Cardiometabolic outcomes and mortality in medically treated primary aldosteronism: A retrospective cohort study. *Lancet Diabetes Endocrinol.* **2017**, *6*, 51–59. [CrossRef]

20. Mosso, L.; Carvajal, C.; Gonzalez, A.; Barraza, A.; Avila, F.; Montero, J.; Huete, A.; Gederlini, A.; Fardella, C.E. Primary aldosteronism and hypertensive disease. *Hypertension* **2003**, *42*, 161–165. [CrossRef] [PubMed]

21. Buffolo, F.; Monticone, S.; Burrello, J.; Tetti, M.; Veglio, F.; Williams, T.A.; Mulatero, P. Is primary aldosteronism still largely unrecognized? *Horm. Metab. Res.* **2017**, *49*, 908–914. [CrossRef] [PubMed]

22. Markou, A.; Pappa, T.; Kaltsas, G.; Gouli, A.; Mitsakis, K.; Tsounas, P.; Prevoli, A.; Tsiavos, V.; Papanastasiou, L.; Zografos, G.; et al. Evidence of primary aldosteronism in a predominantly female cohort of normotensive individuals: A very high odds ratio for progression into arterial hypertension. *J. Clin. Endocrinol. Metab.* **2013**, *98*, 1409–1416. [CrossRef] [PubMed]

23. Baudrand, R.; Guarda, F.J.; Fardella, C.; Hundemer, G.; Brown, J.; Williams, G.; Vaidya, A. Continuum of renin-independent aldosteronism in normotension. *Hypertension* **2017**, *69*, 950–956. [CrossRef] [PubMed]

24. Brown, J.M.; Robinson-Cohen, C.; Luque-Fernandez, M.A.; Allison, M.A.; Baudrand, R.; Ix, J.H.; Kestenbaum, B.; de Boer, I.H.; Vaidya, A. The spectrum of subclinical primary aldosteronism and incident hypertension: A cohort study. *Ann. Intern. Med.* **2017**, *167*, 630–641. [CrossRef] [PubMed]

25. Vasan, R.S.; Evans, J.C.; Larson, M.G.; Wilson, P.W.; Meigs, J.B.; Rifai, N.; Benjamin, E.J.; Levy, D. Serum aldosterone and the incidence of hypertension in nonhypertensive persons. *N. Engl. J. Med.* **2004**, *351*, 33–41. [CrossRef] [PubMed]

26. Vaidya, A.; Underwood, P.C.; Hopkins, P.N.; Jeunemaitre, X.; Ferri, C.; Williams, G.H.; Adler, G.K. Abnormal aldosterone physiology and cardiometabolic risk factors. *Hypertension* **2013**, *61*, 886–893. [CrossRef] [PubMed]

27. Hundemer, G.L.; Baudrand, R.; Brown, J.M.; Curhan, G.; Williams, G.H.; Vaidya, A. Renin phenotypes characterize vascular disease, autonomous aldosteronism, and mineralocorticoid receptor activity. *J. Clin. Endocrinol. Metab.* **2017**, *102*, 1835–1843. [CrossRef] [PubMed]

28. Funder, J.W. Primary aldosteronism: The next five years. *Horm. Metab. Res.* **2017**, *49*, 977–983. [CrossRef] [PubMed]

29. Seccia, T.M.; Caroccia, B.; Adler, G.K.; Maiolino, G.; Cesari, M.; Rossi, G.P. Arterial hypertension, atrial fibrillation, and hyperaldosteronism: The triple trouble. *Hypertension* **2017**, *69*, 545–550. [CrossRef] [PubMed]

30. Mulatero, P.; Monticone, S.; Bertello, C.; Viola, A.; Tizzani, D.; Iannaccone, A.; Crudo, V.; Burrello, J.; Milan, A.; Rabbia, F.; et al. Long-term cardio- and cerebrovascular events in patients with primary aldosteronism. *J. Clin. Endocrinol. Metab.* **2013**, *98*, 4826–4833. [CrossRef] [PubMed]

31. Fallo, F.; Veglio, F.; Bertello, C.; Sonino, N.; Della Mea, P.; Ermani, M.; Rabbia, F.; Federspil, G.; Mulatero, P. Prevalence and characteristics of the metabolic syndrome in primary aldosteronism. *J. Clin. Endocrinol. Metab.* **2006**, *91*, 454–459. [CrossRef] [PubMed]

32. Nishimoto, K.; Nakagawa, K.; Li, D.; Kosaka, T.; Oya, M.; Mikami, S.; Shibata, H.; Itoh, H.; Mitani, F.; Yamazaki, T.; et al. Adrenocortical zonation in humans under normal and pathological conditions. *J. Clin. Endocrinol. Metab.* **2010**, *95*, 2296–2305. [CrossRef] [PubMed]

33. Nishimoto, K.; Seki, T.; Hayashi, Y.; Mikami, S.; Al-Eyd, G.; Nakagawa, K.; Morita, S.; Kosaka, T.; Oya, M.; Mitani, F.; et al. Human adrenocortical remodeling leading to aldosterone-producing cell cluster generation. *Int. J. Endocrinol.* **2016**, *2016*, 7834356. [CrossRef] [PubMed]

34. Nishimoto, K.; Tomlins, S.A.; Kuick, R.; Cani, A.K.; Giordano, T.J.; Hovelson, D.H.; Liu, C.J.; Sanjanwala, A.R.; Edwards, M.A.; Gomez-Sanchez, C.E.; et al. Aldosterone-stimulating somatic gene mutations are common in normal adrenal glands. *Proc. Natl. Acad. Sci. USA* **2015**, *112*, E4591–E4599. [CrossRef] [PubMed]

35. Omata, K.; Yamazaki, Y.; Nakamura, Y.; Anand, S.K.; Barletta, J.A.; Sasano, H.; Rainey, W.E.; Tomlins, S.A.; Vaidya, A. Genetic and histopathologic intertumor heterogeneity in primary aldosteronism. *J. Clin. Endocrinol. Metab.* **2017**, *102*, 1792–1796. [CrossRef] [PubMed]

36. Omata, K.; Anand, S.K.; Hovelson, D.H.; Liu, C.J.; Yamazaki, Y.; Nakamura, Y.; Ito, S.; Satoh, F.; Sasano, H.; Rainey, W.E.; et al. Aldosterone-producing cell clusters frequently harbor somatic mutations and accumulate with age in normal adrenals. *J. Endocr. Soc.* **2017**, *1*, 787–799. [CrossRef] [PubMed]

37. Nanba, K.; Vaidya, A.; Williams, G.H.; Zheng, I.; Else, T.; Rainey, W.E. Age-related autonomous aldosteronism. *Circulation* **2017**, *136*, 347–355. [CrossRef] [PubMed]

38. Zennaro, M.C.; Boulkroun, S.; Fernandes-Rosa, F. Genetic causes of functional adrenocortical adenomas. *Endocr. Rev.* **2017**, *38*, 516–537. [CrossRef] [PubMed]

39. Carvajal, C.A.; Campino, C.; Martinez-Aguayo, A.; Tichauer, J.E.; Bancalari, R.; Valdivia, C.; Trejo, P.; Aglony, M.; Baudrand, R.; Lagos, C.F.; et al. A new presentation of the chimeric *CYP11B1/CYP11B2* gene with low prevalence of primary aldosteronism and atypical gene segregation pattern. *Hypertension* **2012**, *59*, 85–91. [CrossRef] [PubMed]

40. Geller, D.S.; Zhang, J.; Wisgerhof, M.V.; Shackleton, C.; Kashgarian, M.; Lifton, R.P. A novel form of human mendelian hypertension featuring nonglucocorticoid-remediable aldosteronism. *J. Clin. Endocrinol. Metab.* **2008**, *93*, 3117–3123. [CrossRef] [PubMed]

41. Choi, M.; Scholl, U.I.; Yue, P.; Bjorklund, P.; Zhao, B.; Nelson-Williams, C.; Ji, W.; Cho, Y.; Patel, A.; Men, C.J.; et al. K$^+$ channel mutations in adrenal aldosterone-producing adenomas and hereditary hypertension. *Science* **2011**, *331*, 768–772. [CrossRef] [PubMed]

42. Daniil, G.; Fernandes-Rosa, F.L.; Chemin, J.; Blesneac, I.; Beltrand, J.; Polak, M.; Jeunemaitre, X.; Boulkroun, S.; Amar, L.; Strom, T.M.; et al. Cacna1h mutations are associated with different forms of primary aldosteronism. *EBioMedicine* **2016**, *13*, 225–236. [CrossRef] [PubMed]

43. Scholl, U.I.; Goh, G.; Stolting, G.; de Oliveira, R.C.; Choi, M.; Overton, J.D.; Fonseca, A.L.; Korah, R.; Starker, L.F.; Kunstman, J.W.; et al. Somatic and germline cacna1d calcium channel mutations in aldosterone-producing adenomas and primary aldosteronism. *Nat. Genet.* **2013**, *45*, 1050–1054. [CrossRef] [PubMed]

44. Williams, T.A.; Lenders, J.W.M.; Mulatero, P.; Burrello, J.; Rottenkolber, M.; Adolf, C.; Satoh, F.; Amar, L.; Quinkler, M.; Deinum, J.; et al. Outcomes after adrenalectomy for unilateral primary aldosteronism: An international consensus on outcome measures and analysis of remission rates in an international cohort. *Lancet Diabetes Endocrinol.* **2017**, *5*, 689–699. [CrossRef]

45. Lacroix, A.; Feelders, R.A.; Stratakis, C.A.; Nieman, L.K. Cushing's syndrome. *Lancet* **2015**, *386*, 913–927. [CrossRef]

46. Ntali, G.; Asimakopoulou, A.; Siamatras, T.; Komninos, J.; Vassiliadi, D.; Tzanela, M.; Tsagarakis, S.; Grossman, A.B.; Wass, J.A.; Karavitaki, N. Mortality in cushing's syndrome: Systematic analysis of a large series with prolonged follow-up. *Eur. J. Endocrinol.* **2013**, *169*, 715–723. [CrossRef] [PubMed]

47. Nieman, L.K.; Biller, B.M.; Findling, J.W.; Murad, M.H.; Newell-Price, J.; Savage, M.O.; Tabarin, A.; Endocrine, S. Treatment of cushing's syndrome: An endocrine society clinical practice guideline. *J. Clin. Endocrinol. Metab.* **2015**, *100*, 2807–2831. [CrossRef] [PubMed]

48. Lopez, D.; Luque-Fernandez, M.A.; Steele, A.; Adler, G.K.; Turchin, A.; Vaidya, A. "Nonfunctional" adrenal tumors and the risk for incident diabetes and cardiovascular outcomes: A cohort study. *Ann. Intern. Med.* **2016**, *165*, 533–542. [CrossRef] [PubMed]

49. Di Dalmazi, G.; Vicennati, V.; Garelli, S.; Casadio, E.; Rinaldi, E.; Giampalma, E.; Mosconi, C.; Golfieri, R.; Paccapelo, A.; Pagotto, U.; et al. Cardiovascular events and mortality in patients with adrenal incidentalomas that are either non-secreting or associated with intermediate phenotype or subclinical cushing's syndrome: A 15-year retrospective study. *Lancet Diabetes Endocrinol.* **2014**, *2*, 396–405. [CrossRef]

50. Isidori, A.M.; Graziadio, C.; Paragliola, R.M.; Cozzolino, A.; Ambrogio, A.G.; Colao, A.; Corsello, S.M.; Pivonello, R.; Group, A.B.C.S. The hypertension of cushing's syndrome: Controversies in the pathophysiology and focus on cardiovascular complications. *J. Hypertens.* **2015**, *33*, 44–60. [CrossRef] [PubMed]

51. Van der Pas, R.; van Esch, J.H.; de Bruin, C.; Danser, A.H.; Pereira, A.M.; Zelissen, P.M.; Netea-Maier, R.; Sprij-Mooij, D.M.; van den Berg-Garrelds, I.M.; van Schaik, R.H.; et al. Cushing's disease and hypertension: In vivo and in vitro study of the role of the renin-angiotensin-aldosterone system and effects of medical therapy. *Eur. J. Endocrinol.* **2014**, *170*, 181–191. [CrossRef] [PubMed]

52. Baudrand, R.; Vaidya, A. Cortisol dysregulation in obesity-related metabolic disorders. *Curr. Opin. Endocrinol. Diabetes Obes.* **2015**, *22*, 143–149. [CrossRef] [PubMed]

53. Bailey, M.A.; Mullins, J.J.; Kenyon, C.J. Mineralocorticoid and glucocorticoid receptors stimulate epithelial sodium channel activity in a mouse model of cushing syndrome. *Hypertension* **2009**, *54*, 890–896. [CrossRef] [PubMed]

54. Fassnacht, M.; Arlt, W.; Bancos, I.; Dralle, H.; Newell-Price, J.; Sahdev, A.; Tabarin, A.; Terzolo, M.; Tsagarakis, S.; Dekkers, O.M. Management of adrenal incidentalomas: European society of endocrinology clinical practice guideline in collaboration with the european network for the study of adrenal tumors. *Eur. J. Endocrinol.* **2016**, *175*, G1–G34. [CrossRef] [PubMed]

55. Arlt, W.; Biehl, M.; Taylor, A.E.; Hahner, S.; Libe, R.; Hughes, B.A.; Schneider, P.; Smith, D.J.; Stiekema, H.; Krone, N.; et al. Urine steroid metabolomics as a biomarker tool for detecting malignancy in adrenal tumors. *J. Clin. Endocrinol. Metab.* **2011**, *96*, 3775–3784. [CrossRef] [PubMed]

56. Albani, A.; Theodoropoulou, M.; Reincke, M. Genetics of cushing's disease. *Clin. Endocrinol.* **2018**, *88*, 3–12. [CrossRef] [PubMed]

57. Assie, G.; Libe, R.; Espiard, S.; Rizk-Rabin, M.; Guimier, A.; Luscap, W.; Barreau, O.; Lefevre, L.; Sibony, M.; Guignat, L.; et al. ARMC5 mutations in macronodular adrenal hyperplasia with cushing's syndrome. *N. Engl. J. Med.* **2013**, *369*, 2105–2114. [CrossRef] [PubMed]

58. Alencar, G.A.; Lerario, A.M.; Nishi, M.Y.; Mariani, B.M.; Almeida, M.Q.; Tremblay, J.; Hamet, P.; Bourdeau, I.; Zerbini, M.C.; Pereira, M.A.; et al. ARMC5 mutations are a frequent cause of primary macronodular adrenal hyperplasia. *J. Clin. Endocrinol. Metab.* **2014**, *99*, E1501–E1509. [CrossRef] [PubMed]

59. Bonnet-Serrano, F.; Bertherat, J. Genetics of tumors of the adrenal cortex. *Endocr. Relat. Cancer* **2017**. [CrossRef] [PubMed]

60. Ulick, S.; Levine, L.S.; Gunczler, P.; Zanconato, G.; Ramirez, L.C.; Rauh, W.; Rosler, A.; Bradlow, H.L.; New, M.I. A syndrome of apparent mineralocorticoid excess associated with defects in the peripheral metabolism of cortisol. *J. Clin. Endocrinol. Metab.* **1979**, *49*, 757–764. [CrossRef] [PubMed]

61. Funder, J.W. Apparent mineralocorticoid excess. *J. Steroid Biochem. Mol. Biol.* **2017**, *165*, 151–153. [CrossRef] [PubMed]

62. Carvajal, C.A.; Romero, D.G.; Mosso, L.M.; Gonzalez, A.A.; Campino, C.; Montero, J.; Fardella, C.E. Biochemical and genetic characterization of 11 β-hydroxysteroid dehydrogenase type 2 in low-renin essential hypertensives. *J. Hypertens.* **2005**, *23*, 71–77. [CrossRef] [PubMed]

63. Khattab, A.M.; Shackleton, C.H.; Hughes, B.A.; Bodalia, J.B.; New, M.I. Remission of hypertension and electrolyte abnormalities following renal transplantation in a patient with apparent mineralocorticoid excess well documented throughout childhood. *J. Pediatr. Endocrinol. Metab.* **2014**, *27*, 17–21. [CrossRef] [PubMed]

64. Ulick, S.; Chan, C.K.; Rao, K.N.; Edassery, J.; Mantero, F. A new form of the syndrome of apparent mineralocorticoid excess. *J. Steroid Biochem.* **1989**, *32*, 209–212. [CrossRef]

65. Campino, C.; Martinez-Aguayo, A.; Baudrand, R.; Carvajal, C.A.; Aglony, M.; Garcia, H.; Padilla, O.; Kalergis, A.M.; Fardella, C.E. Age-related changes in 11β-hydroxysteroid dehydrogenase type 2 activity in normotensive subjects. *Am. J. Hypertens.* **2013**, *26*, 481–487. [CrossRef] [PubMed]

66. Campino, C.; Carvajal, C.A.; Cornejo, J.; San Martin, B.; Olivieri, O.; Guidi, G.; Faccini, G.; Pasini, F.; Sateler, J.; Baudrand, R.; et al. 11β-hydroxysteroid dehydrogenase type-2 and type-1 (11β-HSD2 and 11β-HSD1) and 5β-reductase activities in the pathogenia of essential hypertension. *Endocrine* **2010**, *37*, 106–114. [CrossRef] [PubMed]

67. Martinez-Aguayo, A.; Campino, C.; Baudrand, R.; Carvajal, C.A.; Garcia, H.; Aglony, M.; Bancalari, R.; Garcia, L.; Loureiro, C.; Vecchiola, A.; et al. Cortisol/cortisone ratio and matrix metalloproteinase-9 activity are associated with pediatric primary hypertension. *J. Hypertens.* **2016**, *34*, 1808–1814. [CrossRef] [PubMed]

68. Tapia-Castillo, A.; Carvajal, C.A.; Allende, F.; Campino, C.; Fardella, C.E. Hypertensive patients that respond to aldosterone antagonists may have a nonclassical 11β-HSD2 deficiency. *Am. J. Hypertens.* **2017**, *30*, e6. [CrossRef] [PubMed]

69. Sabbadin, C.; Armanini, D. Syndromes that mimic an excess of mineralocorticoids. *High Blood Press. Cardiovasc. Prev.* **2016**, *23*, 231–235. [CrossRef] [PubMed]

70. Khattab, A.; Haider, S.; Kumar, A.; Dhawan, S.; Alam, D.; Romero, R.; Burns, J.; Li, D.; Estatico, J.; Rahi, S.; et al. Clinical, genetic, and structural basis of congenital adrenal hyperplasia due to 11β-hydroxylase deficiency. *Proc. Natl. Acad. Sci. USA* **2017**, *114*, E1933–E1940. [CrossRef] [PubMed]

71. Sahakitrungruang, T. Clinical and molecular review of atypical congenital adrenal hyperplasia. *Ann. Pediatr. Endocrinol. Metab.* **2015**, *20*, 1–7. [CrossRef] [PubMed]

72. Auchus, R.J. Steroid 17-hydroxylase and 17,20-lyase deficiencies, genetic and pharmacologic. *J. Steroid Biochem. Mol. Biol.* **2017**, *165*, 71–78. [CrossRef] [PubMed]

73. Monnens, L.; Levtchenko, E. Distinction between liddle syndrome and apparent mineralocorticoid excess. *Pediatr. Nephrol.* **2004**, *19*, 118–119. [CrossRef] [PubMed]

74. Nesterov, V.; Krueger, B.; Bertog, M.; Dahlmann, A.; Palmisano, R.; Korbmacher, C. In liddle syndrome, epithelial sodium channel is hyperactive mainly in the early part of the aldosterone-sensitive distal nephron. *Hypertension* **2016**, *67*, 1256–1262. [CrossRef] [PubMed]

75. Cui, Y.; Tong, A.; Jiang, J.; Wang, F.; Li, C. Liddle syndrome: Clinical and genetic profiles. *J. Clin. Hypertens. (Greenwich)* **2017**, *19*, 524–529. [CrossRef] [PubMed]

76. Pappachan, J.M.; Buch, H.N. Endocrine hypertension: A practical approach. *Adv. Exp. Med. Biol.* **2017**, *956*, 215–237. [PubMed]

77. Salih, M.; Gautschi, I.; van Bemmelen, M.X.; di Benedetto, M.; Brooks, A.S.; Lugtenberg, D.; Schild, L.; Hoorn, E.J. A missense mutation in the extracellular domain of alphaenac causes liddle syndrome. *J. Am. Soc. Nephrol.* **2017**, *28*, 3291–3299. [CrossRef] [PubMed]

78. Akintunde, A.; Nondi, J.; Gogo, K.; Jones, E.S.W.; Rayner, B.L.; Hackam, D.G.; Spence, J.D. Physiological phenotyping for personalized therapy of uncontrolled hypertension in africa. *Am. J. Hypertens.* **2017**, *30*, 923–930. [CrossRef] [PubMed]

79. Jones, E.S.; Spence, J.D.; McIntyre, A.D.; Nondi, J.; Gogo, K.; Akintunde, A.; Hackam, D.G.; Rayner, B.L. High frequency of variants of candidate genes in black africans with low renin-resistant hypertension. *Am. J. Hypertens.* **2017**, *30*, 478–483. [CrossRef] [PubMed]

80. Karl, M.; Lamberts, S.W.; Detera-Wadleigh, S.D.; Encio, I.J.; Stratakis, C.A.; Hurley, D.M.; Accili, D.; Chrousos, G.P. Familial glucocorticoid resistance caused by a splice site deletion in the human glucocorticoid receptor gene. *J. Clin. Endocrinol. Metab.* **1993**, *76*, 683–689. [PubMed]

81. Yang, N.; Ray, D.W.; Matthews, L.C. Current concepts in glucocorticoid resistance. *Steroids* **2012**, *77*, 1041–1049. [CrossRef] [PubMed]

82. Vitellius, G.; Fagart, J.; Delemer, B.; Amazit, L.; Ramos, N.; Bouligand, J.; le Billan, F.; Castinetti, F.; Guiochon-Mantel, A.; Trabado, S.; et al. Three novel heterozygous point mutations of NR3C1 causing glucocorticoid resistance. *Hum. Mutat.* **2016**, *37*, 794–803. [CrossRef] [PubMed]

83. Van Rossum, E.F.; Lamberts, S.W. Glucocorticoid resistance syndrome: A diagnostic and therapeutic approach. *Best Pract. Res. Clin. Endocrinol. Metab.* **2006**, *20*, 611–626. [CrossRef] [PubMed]

84. Bouligand, J.; Delemer, B.; Hecart, A.C.; Meduri, G.; Viengchareun, S.; Amazit, L.; Trabado, S.; Feve, B.; Guiochon-Mantel, A.; Young, J.; et al. Familial glucocorticoid receptor haploinsufficiency by non-sense mediated mrna decay, adrenal hyperplasia and apparent mineralocorticoid excess. *PLoS ONE* **2010**, *5*, e13563. [CrossRef] [PubMed]

85. Park, J.S.; Park, E.; Hyun, H.S.; Ahn, Y.H.; Kang, H.G.; Ha, I.S.; Cheong, H.I. Three cases of gordon syndrome with dominant KLHL3 mutations. *J. Pediatr. Endocrinol. Metab.* **2017**, *30*, 361–364. [CrossRef] [PubMed]

86. O'Shaughnessy, K.M. Gordon syndrome: A continuing story. *Pediatr. Nephrol.* **2015**, *30*, 1903–1908. [CrossRef] [PubMed]

87. Rafestin-Oblin, M.E.; Souque, A.; Bocchi, B.; Pinon, G.; Fagart, J.; Vandewalle, A. The severe form of hypertension caused by the activating S810L mutation in the mineralocorticoid receptor is cortisone related. *Endocrinology* **2003**, *144*, 528–533. [CrossRef] [PubMed]

88. Kang, S.H.; Seok, Y.M.; Song, M.J.; Lee, H.A.; Kurz, T.; Kim, I. Histone deacetylase inhibition attenuates cardiac hypertrophy and fibrosis through acetylation of mineralocorticoid receptor in spontaneously hypertensive rats. *Mol. Pharmacol.* **2015**, *87*, 782–791. [CrossRef] [PubMed]

89. Amazit, L.; Le Billan, F.; Kolkhof, P.; Lamribet, K.; Viengchareun, S.; Fay, M.R.; Khan, J.A.; Hillisch, A.; Lombes, M.; Rafestin-Oblin, M.E.; et al. Finerenone impedes aldosterone-dependent nuclear import of the mineralocorticoid receptor and prevents genomic recruitment of steroid receptor coactivator-1. *J. Biol. Chem.* **2015**, *290*, 21876–21889. [CrossRef] [PubMed]

90. Raizman, J.E.; Diamandis, E.P.; Holmes, D.; Stowasser, M.; Auchus, R.; Cavalier, E. A renin-ssance in primary aldosteronism testing: Obstacles and opportunities for screening, diagnosis, and management. *Clin. Chem.* **2015**, *61*, 1022–1027. [CrossRef] [PubMed]

91. Stowasser, M.; Ahmed, A.H.; Pimenta, E.; Taylor, P.J.; Gordon, R.D. Factors affecting the aldosterone/renin ratio. *Horm. Metab. Res.* **2012**, *44*, 170–176. [CrossRef] [PubMed]

92. Burrello, J.; Monticone, S.; Buffolo, F.; Lucchiari, M.; Tetti, M.; Rabbia, F.; Mengozzi, G.; Williams, T.A.; Veglio, F.; Mulatero, P. Diagnostic accuracy of aldosterone and renin measurement by chemiluminescent immunoassay and radioimmunoassay in primary aldosteronism. *J. Hypertens.* **2016**, *34*, 920–927. [CrossRef] [PubMed]

93. Ahmed, A.H.; Gordon, R.D.; Ward, G.; Wolley, M.; Kogovsek, C.; Stowasser, M. Should aldosterone suppression tests be conducted during a particular phase of the menstrual cycle, and, if so, which phase? Results of a preliminary study. *Clin. Endocrinol.* **2015**, *83*, 303–307. [CrossRef] [PubMed]

94. Baudrand, R.; Guarda, F.J.; Torrey, J.; Williams, G.; Vaidya, A. Dietary sodium restriction increases the risk of misinterpreting mild cases of primary aldosteronism. *J. Clin. Endocrinol. Metab.* **2016**, *101*, 3989–3996. [CrossRef] [PubMed]

95. Adlin, E.V. Subclinical primary aldosteronism. *Ann. Intern. Med.* **2017**, *167*, 673–674. [CrossRef] [PubMed]

96. Adlin, E.V. Letter: Plasma-renin and blood-pressure. *Lancet* **1975**, *305*, 699. [CrossRef]

97. Qi, Y.; Wang, X.; Rose, K.L.; MacDonald, W.H.; Zhang, B.; Schey, K.L.; Luther, J.M. Activation of the endogenous renin-angiotensin-aldosterone system or aldosterone administration increases urinary exosomal sodium channel excretion. *J. Am. Soc. Nephrol.* **2016**, *27*, 646–656. [CrossRef] [PubMed]

98. Williams, B.; MacDonald, T.M.; Morant, S.; Webb, D.J.; Sever, P.; McInnes, G.; Ford, I.; Cruickshank, J.K.; Caulfield, M.J.; Salsbury, J.; et al. Spironolactone versus placebo, bisoprolol, and doxazosin to determine the optimal treatment for drug-resistant hypertension (pathway-2): A randomised, double-blind, crossover trial. *Lancet* **2015**, *386*, 2059–2068. [CrossRef]

 International Journal of
Molecular Sciences

Review

The T2238C Human Atrial Natriuretic Peptide Molecular Variant and the Risk of Cardiovascular Diseases

Speranza Rubattu [1,2,*], Sebastiano Sciarretta [2,3], Simona Marchitti [2], Franca Bianchi [2], Maurizio Forte [2] and Massimo Volpe [1,2]

1   Department of Clinical and Molecular Medicine, School of Medicine and Psychology, Sapienza University of Rome, S. Andrea Hospital, via di Grottarossa 1035, 00189 Rome, Italy; massimo.volpe@uniroma1.it
2   Istituto di Ricovero e Cura a Carattere Scientifico (IRCCS) Neuromed, Località Camerelle, 86077 Pozzilli (Is), Italy; sebastiano.sciarretta@uniroma1.it (S.S.); simona.marchitti@neuromed.it (S.M.); franca.bianchi@neuromed.it (F.B.); maurizio.forte@neuromed.it (M.F.)
3   Department of Medical-Surgical Sciences and Biotechnologies, Sapienza University of Rome, 04100 Latina, Italy
*   Correspondence: rubattu.speranza@neuromed.it; Tel.: +39-06-3377-5979; Fax: +39-06-3377-5061

Received: 25 January 2018; Accepted: 10 February 2018; Published: 11 February 2018

Abstract: Atrial natriuretic peptide (ANP) is a cardiac hormone which plays important functions to maintain cardio-renal homeostasis. The peptide structure is highly conserved among species. However, a few gene variants are known to fall within the human ANP gene. The variant rs5065 (T2238C) exerts the most substantial effects. The T to C transition at the 2238 position of the gene (13–23% allele frequency in the general population) leads to the production of a 30-, instead of 28-, amino-acid-long α-carboxy-terminal peptide. In vitro, CC2238/αANP increases the levels of reactive oxygen species and causes endothelial damage, vascular smooth muscle cells contraction, and increased platelet aggregation. These effects are achieved through the deregulated activation of type C natriuretic peptide receptor, the consequent inhibition of adenylate cyclase activity, and the activation of Giα proteins. In vivo, endothelial dysfunction and increased platelet aggregation are present in human subjects carrying the C2238/αANP allele variant. Several studies documented an increased risk of stroke and of myocardial infarction in C2238/αANP carriers. Recently, an incomplete response to antiplatelet therapy in ischemic heart disease patients carrying the C2238/αANP variant and undergoing percutaneous coronary revascularization has been reported. In summary, the overall evidence supports the concept that T2238C/ANP is a cardiovascular genetic risk factor that needs to be taken into account in daily clinical practice.

Keywords: atrial natriuretic peptide; T2238C variant; endothelial dysfunction; smooth muscle cells contraction; platelet aggregation; epigenetics; cardiovascular diseases

---

1. Introduction

The atrial natriuretic peptide (ANP) belongs to the natriuretic peptide family along with brain (BNP) and C-type (CNP) natriuretic peptides [1]. It is mainly secreted by the atrial cardiomyocytes and exerts important regulatory functions for the maintenance of cardio-renal homeostasis. The latter is achieved through the activation of guanylyl cyclase (GC)-coupled receptor, mainly GC-A, and an increase of cyclic guanylate monophosphate (cGMP) levels [2]. ANP plays natriuretic, diuretic, and vasorelaxant effects [1,2]. In addition, it exerts antiproliferative, antifibrotic, antiangiogenetic functions and, consequently, it contributes to the cardiovascular remodeling process [3]. An extra-atrial expression of ANP has been reported [4].

The gene encoding ANP (*NPPA*) is located in the distal arm of chromosome 1 (1p36.2), in tandem with the gene encoding brain natriuretic peptide (BNP). It includes three exons and two introns [5]. The signal sequence is located on exon 1, whereas the coding sequence is located on exon 2; exon3 encodes the terminal tyrosine and the 3′ untranslated region.

Human ANP is synthetized as a pre-prohormone of 151 amino acids and it is subsequently cleaved to obtain a biologically active α-carboxy-terminal peptide along with the amino-terminal end. After removal of the signal peptide, the proANP$_{1-126}$ is released and stored into granules within the atrial cardiomyocytes. Before secretion, proANP$_{1-126}$ is processed by corin, a type II transmembrane serine protease [2], into the circulating forms of ANP$_{(1-98)}$ and ANP$_{(99-126)}$. Of note, the active corin protease is obtained through the cleavage of procorin by PCSK6 [6,7]. The major form of biological active ANP is the 28-amino acid carboxy-terminal peptide, ANP$_{(99-126)}$. On the other hand, further cleavage of the ANP$_{(1-98)}$ generates LANP (long-acting natriuretic peptide), the vessel dilator, and the kaliuretic hormone [8]. The primary structure of αANP is conserved across species, apart from few variations. In particular, an isoleucine is present at position 10 in rats, mice, and rabbits, whereas humans, dogs, and bovines have a methionine at this position.

Both the high degree of homology and the persistence in the phylogenetic scale support the key role of the ANP primary structure in order to perform its functions.

The relevance of the ANP primary structure to allow the regular biological activities of the peptide has also been underscored by the discovery of few gene variants that change the amino acid sequence and, consequently, the physiological properties of ANP [9].

The first evidence in this regard was obtained in an animal model of spontaneous hypertension and increased predisposition to cerebrovascular disease, the stroke-prone spontaneously hypertensive rat (SHRSP). In this model, the ANP gene was found to map at the peak of linkage of a stroke quantitative trait locus (QTL) [10] and to carry both a variant within the promoter sequence and a variant within the exon 2 sequence, with consequent alterations of peptide regulation, expression, and function [11–13]. The latter were associated with the higher stroke susceptibility of the strain [10–14].

The subsequent translation of the rat data to the human disease led to very interesting findings. In fact, with regard to the human gene, several variants were reported to fall within the promoter region, the coding, and the intronic parts, as well as within the 3′ end of the gene [9]. These gene variants were associated either with alterations in gene expression and peptide levels (rs5068, G664C) or with an abnormal peptide structure (rs5065). Their contribution to hypertension, coronary artery disease, atrial fibrillation, cardiac hypertrophy, heart failure, and the metabolic syndrome was explored in distinct studies with several remarkable results [9].

The present review article will discuss the role of the rs5065 (T2238C) ANP gene variant since it has revealed important biological and pathological effects and has emerged as a relevant risk factor for the development of cardiovascular diseases.

1.1. Clinical Evidence of the Pathological Role of the T2238C/ANP Molecular Variant

The T2238C coding variant falls within the exon 3 of the gene and it changes the native stop codon by elongation with two additional amino acids (two arginines). Therefore, the resultant carboxy-terminal peptide contains 30, instead of 28, amino acids.

The frequency of the C allele is about 13–14% in the general population. It has been reported that its frequency may increase up to 23% in populations affected by cardiovascular diseases such as coronary and cerebrovascular diseases [9]. Several case-control association studies were performed in the attempt to explore the pathological relevance of the T2238C transition. As a result, an association of the C2238 allele with an increased risk of both myocardial infarction and stroke was reported in different cohorts, including Caucasian and Asian populations [15–23]. Moreover, carriers of this variant allele showed an increased risk of recurrent ischemic stroke and of myocardial infarction [16,18]. Interestingly, an analysis performed in a Caucasian general population found that subjects carrying the C allele had an increased risk to develop stroke and myocardial infarction over a long-term follow up (nine years) [24]. Of note, the C allele carriers had higher BNP values and tended to show an ejection fraction of about 40%, suggesting that their heart may suffer from vascular stress and also from possible direct cardiac consequences (although no evidence of either dilated or hypertrophic cardiac disease was revealed in that study).

Of note, studies aimed at the identification of a contributory role of T2238C in atrial fibrillation turned out to be negative [25,26]. No clear evidence has ever been obtained with regard to the predisposition to develop hypertension in relation to the carrier status of the C2238 allele. A meta-analysis performed in this regard showed only a modest trend to protect from hypertension [27]. Interestingly, a pharmacogenomic type of approach led to the discovery that hypertensive patients carrying the C2238 allele variant had a greater response to a diuretic-based therapy (chlortalidone) with a significant reduction of their cardiovascular risk [28].

Although negative results about the association of the C allele with an increased risk of cardiovascular events were also reported [29,30], the major clinical evidence suggested that the 30-amino-acid-long αANP could exert biological effects different from those of the common αANP, with the production of deleterious consequences for the health status of both coronary and cerebral blood vessels, and with the promotion of lesser diuretic effect.

Therefore, based on these premises, we decided to pursue detailed molecular studies to identify the precise mechanisms underlying the negative impact of the T2238C/ANP gene variant on the cardiovascular system.

1.2. Discovery of the Molecular Mechanisms Underlying the Deleterious Effects of CC2238/αAnp

In the attempt to fully explore this issue, we undertook a series of in vitro studies in a few vascular cell lines starting with the endothelial cells. Human umbilical vein endothelial cells (HUVEC), once exposed to CC2238/αANP, showed an increased rate of oxidative stress, apoptosis, and necrosis along with reduced cell viability and reduced endothelial cell tube formation [31]. Moreover, CC2238/αANP stimulated the production of proteins involved in atherogenesis [31]. Cannone et al. demonstrated also that CC2238/αANP produced a significant increase of endothelial cell permeability [24]. In this cellular model, the exposure to CC2238/αANP produced a significant decrease of cyclic adenosine monophosphate (cAMP) levels, apart from the expected increase of cyclic guanosine monophosphate (cGMP) levels [32]. In fact, binding affinity studies revealed that CC2238/αANP, compared to the common αANP, has a higher binding affinity for the type C natriuretic peptide receptor (NPR-C) rather than for the type A receptor (NPR-A), and consequently it reduces the adenylate cyclase activity dependent on NPR-C [32].

This result represented a remarkable achievement since the NPR-C, devoid of GC activity and mainly known as the clearance receptor of NPs, has direct biological properties that are not fully understood yet [33]. A synthetic form of ANP (C-ANP$_{4-23}$) has been the only known agonist of NPR-C until the discovery of the functional properties of CC2238/αANP [33].

The experimental setup performed in endothelial cells identified a novel mechanism of vascular damage. In fact, the reduction of cAMP levels consequent to the activation of NPR-C, led, in turn, to reduced protein kinase A (PKA) activity and, consequently, to a reduced rate of protein kinase B (Akt) phosphorylation, which is a fundamental factor to maintain endothelial integrity and function [34]. In addition, an increased expression of nicotinamide adenine dinucleotide phosphate (NADPH) oxidase, generating reactive oxygen species (ROS) accumulation, observed in the presence of CC2238/αANP, could certainly contribute to the observed endothelial damage and dysfunction [31,32].

To further support our findings, the cAMP reduction upon exposure to CC2238/αANP was abrogated by the inhibition of NPR-C, as well as by the Giα inhibitor pertuxis toxin. More importantly, either the inhibition of NPR-C by antisense oligonucleotide or the selective NPR-C knockdown by RNA gene silencing, rescued CC2238/αANP-induced endothelial cell death [32].

Thus, in the pioneering studies performed in endothelial cells, the cAMP-PKA-Akt signalling pathway was revealed as the main one mediating the deleterious effects of CC2238/αANP.

Its stimulation appeared to be the consequence of NPR-C rather NPR-A activation, in a manner opposed to that of the regular ANP. The subsequent in vivo translation of these in vitro data, through the characterization of the forearm endothelial-dependent vasorelaxation, revealed that apparently healthy subjects carrying the C2238 allele had a significantly reduced endothelial function [32]. Of note, the reduced endothelial-dependent vasorelaxation was directly related to the number of the mutant alleles, being more severe in double mutant homozygotes. These results provided clear evidence of the clinical implications of the C2238/ANP allele as a hallmark of increased predisposition to human vascular diseases.

A subsequent series of in vitro studies was carried out in both umbilical and coronary smooth muscle cells in the attempt to unravel a potential link between this ANP variant and atherosclerosis [35]. Herein, the in vitro exposure to CC2238/αANP caused oxidative stress damage and increased cell migration and contraction. In this cell line, the role of a reduced cAMP-PKA-Akt axis as well as of the increased NADPH oxidase expression was fully confirmed in response to NPR-C activation. Moreover, we were able to dissect out a novel pathway of vascular damage dependent on PKA inhibition in smooth muscle cells. In fact, we discovered that, as a consequence of PKA inhibition, cAMP response element-binding protein (CREB) was also inhibited, and, further down, microRNA-21 (miR21) expression was reduced, with the expected modulation of miR21 molecular targets. Accordingly, phosphatase and tensin homolog (PTEN) and programmed cell death protein 4 (PDCD4) increased, whereas B-cell lymphoma 2 (Bcl2) decreased. As a result of the modulation of these deleterious mechanisms, a severely compromised vascular smooth muscle cell viability was observed in the presence of CC2238/αANP [35]. Thus, our set of data provided an original evidence for the existence of an epigenetic regulation mediating the deleterious effects of CC2238/αANP in vascular cells. The replacement of miR21 could restore smooth muscle cell viability and function in the presence of CC2238/αANP.

miR21 is known as a fundamental protective factor for maintaining viability and proliferation in several cell types and it has been found to be highly expressed in all main types of cardiovascular cells [36]. It plays important roles in vascular smooth muscle cell proliferation and apoptosis through the targeting of PTEN, Bcl2, and PDCD4 proteins [37–39]. Notably, the antiproliferative effect of wild type αANP has been previously shown to associate with a decrease of miR21 in aortic smooth muscle cells, whereas the overexpression of miR21 restored vascular smooth muscle cell proliferation [40]. In addition, our studies supported previous evidence that miR21 belongs to the Akt regulatory network in cardiovascular cells as an upstream factor, because of its ability to inhibit PTEN and to exert, through the consequent increased Akt activity, important effects on cell proliferation [37,41]. In fact, both PTEN and Akt represent key molecules for cell growth and survival, as well as for the development of many cardiovascular diseases [42]. Finally, these original results were able to support knowledge on the ability of the cAMP-PKA-CREB axis to regulate the expression of miRNAs in vascular cells [43,44].

The relevance of an epigenetic regulation underlying some of the vascular effects of CC2238/αANP was later on reinforced by additional findings on the role of microRNA-199a (miR199a) in the same cell line [45]. In fact, we searched for differential gene expression of atherosclerosis-related pathways in vascular smooth muscle cells exposed to either TT2238/αANP (wild type) or to the variant CC2238/αANP. The major finding was that CC2238/αANP induced Apolipoprotein E (ApoE) downregulation through NPR-C-dependent mechanisms involving the upregulation of miR199a-3p and miR199a-5p and the downregulation of DnaJ (Hsp40) homolog (DNAJA4). The reduced expression of ApoE by CC2238/αANP was associated with a significant increase of inflammation, apoptosis, and necrosis that were completely rescued by the exogenous administration of recombinant ApoE [45]. Of note, the upregulation of miR199a by NPR-C was mediated by a ROS-dependent increase of early growth response protein-1 (Egr-1) transcription factor. In fact, Egr-1 knockdown abolished the impact of CC2238/αANP on ApoE and miR199a. As expected, NPR-C knockdown rescued the ApoE levels [45].

This set of in vitro data provided the first original demonstration that CC2238/αANP, through NPR-C-dependent activation of Egr-1 and the consequent upregulation of miR199a, downregulates ApoE in vascular smooth muscle cells. ApoE is a 34 kDa protein that participates in the mobilization and distribution of cholesterol and of other lipids among various tissues of the body [46]. ApoE was previously found to be expressed in vascular smooth muscle cells where ApoE downregulation is positively correlated with typical markers of inflammation, such as small mother against decapentaplegic 4 (Smad4) [47] and nuclear factor kappa-light-chain-enhancer of activated B cells (NF-κB) [48]. Previous studies demonstrated that ApoE exerts also pleiotropic anti-atherosclerotic and anti-inflammatory cellular effects. It is known that ApoE deficiency alone promotes the development of aortic atherosclerotic plaques in mice [49]. In humans, where the lack of ApoE is rare, the risk of atherosclerosis is strongly associated with three common apoE isoforms in the order of APOE4>APOE3>APOE2 [46]. Furthermore, ApoE is known to play a role in the development of cardiovascular diseases [50]. Thus, our studies may suggest that CC2238/αANP induces vascular damage also through the inhibition of the above-described pleiotropic effects of ApoE. As a consequence, restoring ApoE levels could represent a potential therapeutic strategy to counteract the harmful effects of CC2238/αANP in subjects carriers of the variant peptide.

A third cellular element stimulated our interest, on the basis of the knowledge that platelets play a key role in the promotion of atherothrombotic events [51] and on previous evidence of a certain functional impact of ANP on platelet aggregation [52]. The hypothesis was that CC2238/αANP could exert pro-aggregant effects on platelets and therefore, through this mechanism, could also promote the higher occurrence of stroke and myocardial infarction in human subjects.

In fact, we discovered that CC2238/αANP favored platelet activation and aggregation in vitro through the activation of NPR-C and the reduction of cAMP levels [53]. It increased ROS production through the upregulation of Nox2, a major NADPH oxidase isoform that is abundant in platelets [54]. More importantly, a translation of the in vitro data to the human condition was achieved through the demonstration that patients with atrial fibrillation (who are more prone to thromboembolic events) had a higher level of platelet aggregation and activation and of oxidative stress if they were also carriers of the C2238/αANP allele [53]. On the basis of these observations, Nox2 inhibition may represent a potential intervention to decrease the cardiovascular risk in patients carrying the C2238/αANP allele. These findings were subsequently extended to ischemic heart disease with a study performed in patients undergoing percutaneous coronary revascularization and taking dual antiplatelet therapy. The study revealed that carriers of the C2238/ANP allele had a higher residual platelet reactivity when diabetic [55]. Since diabetes represents on its own a pathological condition characterized by an increased platelet aggregation [56], a close monitoring should be deserved to this cohort of patients undergoing coronary revascularization if they are carriers of the T2238C/ANP variant.

Altogether, the combined evidence of endothelial damage and dysfunction, of increased smooth muscle cells migration and contraction with decreased levels of the protective ApoE, and of increased platelet activation and aggregation outline a pathway that may favor atherosclerotic plaque formation as well as plaque instability up to thrombus formation, with the consequent cardiovascular acute event.

Obviously, CC2238/αANP should be interpreted as one of the factors contributing to the above-mentioned deleterious effects on the vessel wall and to the consequent clinical outcome.

2. Summary and Outlook

Dissecting out novel pathways of vascular damage dependent on a gene variant is an intriguing issue from both a scientific and a clinical point of view. In the case of T2238C/ANP, we demonstrated that the extended (30-amino acid) gene product acts through the activation of a specific receptor (the NPR-C receptor) to induce multiple deleterious mechanisms. In addition, the ANP variant targets several cell types belonging to both the vascular wall and the bloodstream and, by doing so, it increases the cardiovascular risk (Figure 1).

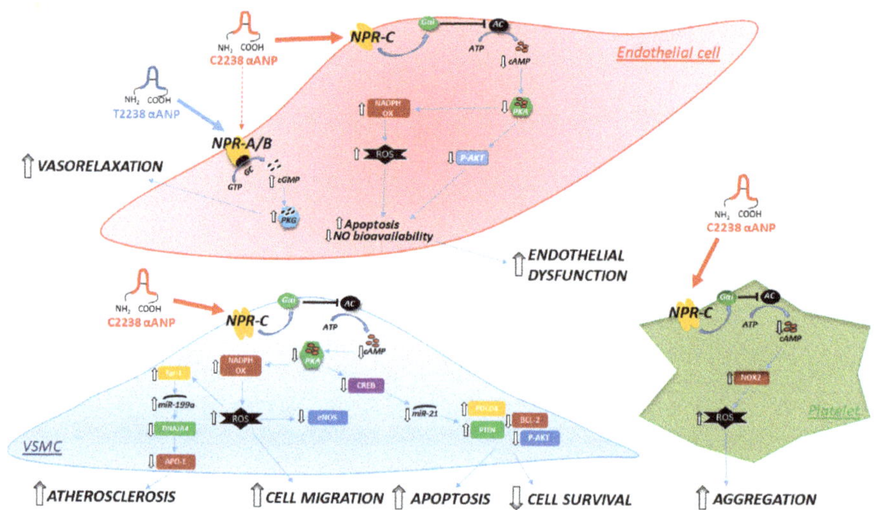

Figure 1. Schematic representation of the molecular signaling pathways underlying the effects of CC2238/αANP in endothelial cells, smooth muscle cells, and platelets. Abbreviations: AC, adenylate cyclase; Akt, protein kinase B; APO-E, apolipoprotein E; ATP, adenosine triphosphate; BCL-2, B-cell lymphoma 2; cAMP, cyclic adenosine monophosphate; cGMP, cyclic guanosine monophosphate; CREB, cAMP response element-binding protein; DNAJA4, DnaJ (Hsp40) homolog; Egr-1, early growth response protein-1; eNOS, endothelial nitric oxide synthase; GC, guanylate cyclase; GTP, guanosine triphosphate; Gαi, inhibitory G protein; miR-21, microRNA-21; miR-199, microRNA-199; NADPH OX, nicotinamide adenine dinucleotide phosphate-oxidase; NO, nitric oxide; NOX2, NADPH oxidase 2; NPR-A/B, natriuretic peptide type A/B receptors; NPR-C, natriuretic peptide type C receptor; P-AKT, phosphoprotein kinase B; PDCD4, programmed cell death protein 4; PKA, protein kinase A; PKG, protein kinase G; PTEN, phosphatase and tensin homolog; ROS, reactive oxygen species; VSMCs, vascular smooth muscle cells; αANP, carboxy-terminal atrial natriuretic peptide, wild-type (T2238) and mutant (C2238). Arrows indicate the positive stimulation within each pathway. The dotted arrow indicates lower affinity binding of C2238 for NPRA/B.

Interestingly, the findings describe for the first time the cAMP-PKA axis as the one driving all other signaling events stemming from CC2238/αANP, such as the CREB-miR21and the Egr-1-miR199a-ApoE pathways in smooth muscle cells, NADPH increase, and particularly Nox2 increase in platelets (Figure 1). In vivo, we obtained evidence of reduced endothelial-dependent vasorelaxation in apparently healthy subjects, of increased platelet activation and aggregation in a cohort of patients with atrial fibrillation, and of higher residual platelet aggregation in ischemic heart disease diabetic patients (following administration of dual antiplatelet therapy) if they were carrier of the C2238/αANP variant [55]. A previous pharmacogenomic study had already revealed the greater efficacy of diuretic therapy in hypertensive carriers of the C2238/ANP variant allele [28].

The findings provide original evidence in the field of the pathogenesis of vascular damage. We believe that the complex, multifaceted mechanisms explaining the negative impact of a molecular variant of ANP give even more support to the beneficial role played by the common form of ANP within the cardiovascular system. In fact, we can better support the view that the common form of ANP, acting through NPR-A activation and the consequent cGMP release, is a peptide promoting vascular health, angiogenesis, vascular smooth muscle cells relaxation, regular platelets activation and aggregation. On the other hand, the change in the peptide sequence leads to higher binding affinity for NPR-C and introduces several differences in the functions of the peptide, making C2238/ANP a "toxic" substance for the vascular wall and for the cardiovascular system generally [57].

A few final considerations are needed in regard to what we have learned about the T2238C/ANP variant. First of all, discovering a functional biological effect of a gene variant is not common. Furthermore, the translation of in vitro effects to in vivo condition is even less likely. Even when this achievement is obtained, the expected effect in vivo is usually small in the context of complex multifactorial diseases.

The case of CC2238/αANP is uncommon. In fact, apart from dissecting out its specific biological effects, the molecular studies allowed the discovery of a novel role of a known receptor (NPR-C) and of novel signaling pathways leading to vascular damage. More importantly, the deleterious impact of the alternative peptide can be detected in vivo, which is uncommon and likely to be of significant relevance to the physiological effects of ANP.

Therefore, the current mechanistic knowledge on C2238/αANP underscores a novel cardiovascular genetic risk factor that needs to be taken into account in daily clinical practice. Of note, the cost of the determination of the presence of the T2238C allele is affordable.

Further studies should assess whether T2238C allele characterization may add some value on top of the conventional risk factors and may have clinical and therapeutic relevance.

Moreover, some of the molecules outlined in our studies could be targeted by novel therapeutic approaches to reduce the cardiovascular risk in carriers of the C2238/ANP allele.

Acknowledgments: This work was supported by a grant from the Italian Ministry of Health and from the 5‰ grant.

Author Contributions: Speranza Rubattu prepared the article, Sebastiano Sciarretta, Simona Marchitti, Franca Bianchi, Maurizio Forte and Massimo Volpe revised it. Maurizio Forte prepared the figure.

Conflicts of Interest: The authors declare no conflicts of interest.

References

1. Levin, E.R.; Gardner, D.G.; Samson, W.K. Natriuretic peptides. *N. Engl. J. Med.* **1998**, *339*, 321–328. [PubMed]
2. Potter, L.R.; Yoder, A.R.; Flora, D.R.; Antos, L.K.; Dickey, D.M. Natriuretic peptides: Their structures, receptors, physiologic functions and therapeutic applications. *Handb. Exp. Pharmacol.* **2009**, *191*, 341–366.
3. Rubattu, S.; Sciarretta, S.; Valenti, V.; Stanzione, R.; Volpe, M. Natriuretic peptides: An update on bioactivity, potential therapeutic use and implication in cardiovascular diseases. *Am. J. Hypertens.* **2008**, *21*, 733–741. [CrossRef] [PubMed]
4. Gardner, D.G.; Deschepper, C.F.; Ganong, W.F.; Hane, S.; Fiddes, J.; Baxter, J.D.; Lewicki, J. Extra-atrial expression of the gene for atrial natriuretic factor. *Proc. Natl. Acad. Sci. USA* **1986**, *83*, 6697–6701. [CrossRef] [PubMed]
5. Nemer, M.; Chamberland, M.; Sirois, D.; Argentin, S.; Drouin, J.; Dixon, R.A.; Zivin, R.A.; Condra, J.H. Gene structure of human cardiac hormone precursor, pronatriodilatin. *Nature* **1984**, *312*, 654–656. [CrossRef] [PubMed]
6. Chen, S.; Cao, P.; Dong, N.; Peng, J.; Zhang, C.; Wang, H.; Zhou, T.; Yang, J.; Zhang, Y.; Martelli, E.E.; et al. PCSK6-mediated corin activation is essential for normal blood pressure. *Nat. Med.* **2015**, *21*, 1048–1053. [CrossRef] [PubMed]
7. Volpe, M.; Rubattu, S. Novel insights into the mechanisms regulating pro-atrial natriuretic peptide cleavage in the heart and blood pressure regulation. Proprotein Convertase Subtilisin/Kexin 6 is the corin activating enzyme. *Circ. Res.* **2016**, *118*, 196–198. [CrossRef] [PubMed]
8. Vesely, D.L. Atrial natriuretic hormones originating from the N-terminus of the atrial natriureticfactor prohormone. *Clin. Exp. Pharmacol. Physiol.* **1995**, *22*, 108–114. [CrossRef] [PubMed]
9. Rubattu, S.; Sciarretta, S.; Volpe, M. Atrial natriuretic peptide gene variants and circulating levels: Implications in cardiovascular diseases. *Clin. Sci.* **2014**, *127*, 1–13. [CrossRef] [PubMed]
10. Rubattu, S.; Volpe, M.; Kreutz, R.; Ganten, U.; Ganten, D.; Lindpaintner, K. Chromosomal mapping of genetic loci contributing to stroke in an animal model of a complex human disease. *Nat. Genet.* **1996**, *13*, 429–434. [CrossRef] [PubMed]

11. Rubattu, S.; Lee, M.A.; De Paolis, P.; Giliberti, R.; Gigante, B.; Lombardi, A.; Volpe, M.; Lindpaintner, K. Altered structure, regulation and function of the gene encoding atrial natriuretic peptide in the stroke-prone spontaneously hypertensive rat. *Circ. Res.* **1999**, *85*, 900–905. [CrossRef] [PubMed]

12. Rubattu, S.; Giliberti, R.; De Paolis, P.; Stanzione, R.; Spinsanti, P.; Volpe, M. Effect of a regulatory mutation on the rat atrial natriuretic peptide gene transcription. *Peptides* **2002**, *23*, 555–560. [CrossRef]

13. De Paolis, P.; Nobili, V.; Lombardi, A.; Tarasi, D.; Barbato, D.; Ganten, U.; Brunetti, E.; Volpe, M.; Rubattu, S. Role of a molecular variant of the rat atrial natriuretic peptide gene on vascular remodeling. *Ann. Clin. Lab. Sci.* **2007**, *37*, 135–140. [PubMed]

14. Rubattu, S.; Giliberti, R.; Ganten, U.; Volpe, M. A differential brain ANP expression cosegregates with occurrence of early stroke in the stroke-prone phenotype of the spontaneously hypertensive rat. *J. Hypertens.* **1999**, *17*, 1849–1852. [CrossRef] [PubMed]

15. Gruchala, M.; Ciecwierz, D.; Wasag, B.; Targonski, R.; Dubaniewicz, W.; Nowak, A.; Sobiczewski, W.; Ochman, K.; Romanowski, P.; Limon, J.; et al. Association of the ScaI atrial natriuretic peptide gene polymorphism with nonfatal myocardial infarction and extent of coronary artery disease. *Am. Heart J.* **2003**, *145*, 125–131. [CrossRef] [PubMed]

16. Rubattu, S.; Stanzione, R.; Di Angelantonio, E.; Zanda, B.; Evangelista, A.; Tarasi, D.; Gigante, B.; Pirisi, A.; Brunetti, E.; Volpe, M. Atrial natriuretic peptide gene polymorphisms and the risk of ischemic stroke in humans. *Stroke* **2004**, *35*, 814–818. [CrossRef] [PubMed]

17. Zhang, L.; Cheng, L.; He, M.; Hu, B.; Wu, T. ANP T2238C, C-664G gene polymorphism and coronary heart disease in Chinese population. *J. Huazhong Univ. Sci. Technol. Med. Sci.* **2006**, *26*, 528–530. [CrossRef] [PubMed]

18. Barbato, E.; Bartunek, J.; Mangiacapra, F.; Sciarretta, S.; Stanzione, R.; Delrue, L.; Cotugno, M.; Marchitti, S.; Iaccarino, G.; Sirico, G.; et al. Influence of rs5065 atrial natriuretic peptide gene variant on coronary artery disease. *J. Am. Coll. Cardiol.* **2012**, *59*, 1763–1770. [CrossRef] [PubMed]

19. Rubattu, S.; De Giusti, M.; Farcomeni, A.; Abbolito, S.; Comito, F.; Cangianiello, S.; Squillace Greco, E.; Dito, E.; Pagliaro, B.; Cotugno, M.; et al. T2238C ANP gene variant and risk of recurrent acute coronary syndromes in an Italian cohort of ischemic heart disease patients. *J. Cardiovasc. Med.* **2016**, *17*, 601–607. [CrossRef] [PubMed]

20. Ziaee, S.; Kalayinia, S.; Boroumand, M.A.; Pourgholi, L.; Cheraghi, S.; Anvari, M.S.; Sheikhvatan, M. Association between the atrial natriuretic peptide rs5065 gene polymorphism and the presence and severity of coronary artery disease in an Iranian population. *Coron. Artery Dis.* **2014**, *25*, 242–246. [CrossRef] [PubMed]

21. Kim, J.H.; Jang, B.H.; Go, H.Y.; Park, S.; Shin, Y.C.; Kim, S.H.; Ko, S.G. Potential association between frequent nonsynonymous variant of NPPA and cardioembolic stroke. *DNA Cell Biol.* **2012**, *31*, 993–1000. [CrossRef] [PubMed]

22. Larifla, L.; Rambhojan, C.; Joannes, M.O.; Maimaitiming-Madani, S.; Donnet, J.P.; Marianne-Pepin, T.; Chout, R.; Roussel, R.; Foucan, L. Gene polymorphisms of FABP2, ADIPOQ and ANP and the risk of hypertriglyceridemia and methabolic syndrome in Afro-Caribbeans. *PLoS ONE* **2016**, *11*, e0163421. [CrossRef] [PubMed]

23. Roussel, R.; Tregouet, D.A.; Hadjadj, S.; Jeunemaitre, X.; Marre, M. Investigation of the human ANP gene in type 1 diabetic nephropathy: Case-control and follow-up studies. *Diabetes* **2004**, *53*, 1394–1398. [CrossRef] [PubMed]

24. Cannone, V.; Huntley, B.K.; Olson, T.M.; Heublein, D.M.; Scott, C.G.; Bailey, K.R.; Redfield, M.M.; Rodeheffer, R.J.; Burnett, J.C., Jr. Atrial natriuretic genetic variant rs5065 and risk for cardiovascular disease in the general community: A 9-year follow-up study. *Hypertension* **2013**, *62*, 860–865. [CrossRef] [PubMed]

25. Francia, P.; Ricotta, A.; Frattari, A.; Stanzione, R.; Modestino, A.; Mercanti, F.; Adduci, C.; Sensini, I.; Cotugno, M.; Balla, C.; et al. Atrial natriuretic peptide single nucleotide polymorphisms in patients with non-familial structural atrial fibrillation. *Clin. Med. Insights Cardiol.* **2013**, *7*, 153–159. [CrossRef] [PubMed]

26. Siebert, J.; Lewicki, Ł.; Myśliwska, J.; Młotkowska, M.; Rogowski, J. ScaI atrial natriuretic peptide gene polymorphisms and their possible association with postoperative atrial fibrillation—A preliminary report. *Arch. Med. Sci.* **2017**, *13*, 568–574. [CrossRef] [PubMed]

27. Niu, W. The relationship between natriuretic peptide precursor a gene T2238C polymorphism and hypertension: A meta-analysis. *Int. J. Hypertens.* **2011**, 653698. [CrossRef] [PubMed]

28. Lynch, A.I.; Boerwinkle, E.; Davis, B.R.; Ford, C.E.; Eckfeldt, J.H.; Leiendecker-Foster, C.; Arnett, D.K. Pharmacogenetic association of the NPPA T2238C genetic variant with cardiovascular disease outcomes in patients with hypertension. *JAMA* **2008**, *299*, 296–307. [CrossRef] [PubMed]

29. Li, T.Y.; Tse, M.Y.; Pang, S.C.; McLellan, C.S.; King, W.D.; Johri, A.M. Sex differences of the natriuretic peptide polymorphism associated with angiographic coronaty atherosclerosis. *Cardiol. Res.* **2017**, *8*, 1–6. [CrossRef] [PubMed]

30. Larifla, L.; Maimaitiming, S.; Velayoudom-Cephise, F.L.; Ferdinand, S.; Blanchet-Deverly, A.; Benabdallah, S.; Donnet, J.P.; Atallah, A.; Roussel, R.; Foucan, L. Association of 2238T>C polymorphism of the atrial natriuretic peptide gene with coronary artery disease in Afro-Caribbeans with type 2 diabetes. *Am. J. Hypertens.* **2012**, *25*, 524–527. [CrossRef] [PubMed]

31. Scarpino, S.; Marchitti, S.; Stanzione, R.; Evangelista, A.; Di Castro, S.; Savoia, C.; Quarta, G.; Sciarretta, S.; Ruco, L.; Volpe, M.; et al. ROS-mediated differential effects of the human atrial natriuretic peptide T2238C genetic variant on endothelial cells in vitro. *J. Hypertens.* **2009**, *27*, 1804–1813. [CrossRef] [PubMed]

32. Sciarretta, S.; Marchitti, S.; Bianchi, F.; Moyes, A.; Barbato, E.; Di Castro, S.; Stanzione, R.; Cotugno, M.; Castello, L.; Calvieri, C.; et al. The C2238 atrial natriuretic peptide molecular variant is associated with endothelial damage and dysfunction through natriuretic peptide receptor C signaling. *Circ. Res.* **2013**, *112*, 1355–1364. [CrossRef] [PubMed]

33. Rubattu, S.; Sciarretta, S.; Morriello, A.; Calvieri, C.; Battistoni, A.; Volpe, M. NPR-C: A component of the natriuretic peptide family with implications in human diseases. *J. Mol. Med.* **2010**, *88*, 889–897. [CrossRef] [PubMed]

34. Bellis, A.; Castaldo, D.; Trimarco, V.; Monti, M.G.; Chivasso, P.; Sadoshima, J.; Trimarco, B.; Morisco, C. Crosstalk between PKA and Akt protects endothelial cells from apoptosis in the late ischemic preconditioning. *Arterioscler. Thromb. Vasc. Biol.* **2009**, *29*, 1207–1212. [CrossRef] [PubMed]

35. Rubattu, S.; Marchitti, S.; Bianchi, F.; Di Castro, S.; Stanzione, R.; Cotugno, M.; Bozzao, C.; Sciarretta, S.; Volpe, M. The C2238/αANP variant is a negative modulator of both viability and function of coronary artery smooth muscle cells. *PLoS ONE* **2014**, *9*, e113108. [CrossRef] [PubMed]

36. Cheng, Y.; Zhang, C. MicroRNA-21 in cardiovascular disease. *J. Cardiovasc. Trans. Res.* **2010**, *3*, 251–255. [CrossRef] [PubMed]

37. Ji, R.; Cheng, Y.; Yue, J.; Yang, J.; Liu, X.; Chen, H.; Dean, D.B.; Zhang, C. MicroRNA expression signature and antisense-mediated depletion reveal an essential role of microRNA in vascular neointima formation. *Circ. Res.* **2007**, *100*, 1579–1588. [CrossRef] [PubMed]

38. Quintavalle, C.; Garofalo, M.; Croce, C.M.; Condorelli, G. ApoptomiRs in vascular cells: Their role in physiological and pathological angiogenesis. *Vasc. Pharmacol.* **2011**, *55*, 87–91. [CrossRef] [PubMed]

39. Song, J.T.; Hu, B.; Yan Qu, H.; Bi, C.I.; Huang, X.Z. Mechanical stretch modulates MicroRNA 21 expression, participating in proliferation and apoptosis in cultured human aortic smooth muscle cells. *PLoS ONE* **2012**, *7*, e47657. [CrossRef] [PubMed]

40. Kotlo, K.U.; Hesabi, B.; Danziger, R.S. Implication of microRNAs in atrial natriuretic peptide and nitric oxide signaling in vascular smooth muscle cells. *Am. J. Physiol. Cell Physiol.* **2011**, *301*, C929–C937. [CrossRef] [PubMed]

41. Roy, S.; Khanna, S.; Hussain, S.R.; Biswas, S.; Azad, A.; Rink, C.; Gnyawali, S.; Shilo, S.; Nuovo, G.J.; Sen, C.K. MicroRNA expression in response to murine myocardial infarction: miR-21 regulates fibroblast metalloprotease-2 via phosphatide and tensin homologue. *Cardiovasc. Res.* **2009**, *82*, 21–29. [CrossRef] [PubMed]

42. Oudit, G.Y.; Sun, H.; Kerfant, B.G.; Crackower, M.A.; Penninger, J.M.; Backx, P.H. The role of phosphoinositide-3 kinase and PTEN in cardiovascular physiology and disease. *J. Mol. Cell. Cardiol.* **2004**, *37*, 449–471. [CrossRef] [PubMed]

43. Lu, Y.; Zhang, Y.; Shan, H.; Pan, Z.; Li, X.; Li, B.; Xu, C.; Zhang, B.; Zhang, F.; Dong, D.; et al. MicroRNA-1 downregulation by propanolol in a rat model of myocardial infarction: A new mechanism for ischaemic cardioprotection. *Cardiovasc. Res.* **2009**, *84*, 434–441. [CrossRef] [PubMed]

44. Shu, M.; Zhou, Y.; Zhu, W.; Zhang, H.; Wu, S.; Chen, J.; Yan, G. MicroRNA 335 is required for differentiation of malignant glioma cells induced by activation of cAMP/Protein Kinase A pathway. *Mol. Pharmacol.* **2012**, *81*, 292–298. [CrossRef] [PubMed]

45. Stanzione, R.; Sciarretta, S.; Marchitti, S.; Bianchi, F.; Di Castro, S.; Scarpino, S.; Cotugno, M.; Frati, G.; Volpe, M.; Rubattu, S. C2238/αANP modulates Apolipoprotein E through Egr-1/miR199a in vascular smooth muscle cells in vitro. *Cell Death Dis.* **2015**, *6*, e2033. [CrossRef] [PubMed]

46. Mahley, R.W.; Rall, S.C., Jr. Apolipoprotein E: Far more than a lipid transport protein. *Annu. Rev. Genom. Hum. Genet.* **2000**, *1*, 507–537. [CrossRef] [PubMed]

47. Rodriguez-Vita, J.; Sanchez-Galan, E.; Santamaria, B.; Sanchez-Lopez, E.; Rodrigues-Diez, R.; Blanco-Colio, L.M.; Egido, J.; Ortiz, A.; Ruiz-Ortega, M. Essential role of TGF-β/Smad pathway on statin dependent vascular smooth muscle cell regulation. *PLoS ONE* **2008**, *3*, e3959. [CrossRef] [PubMed]

48. Zhang, T.; Dai, P.; Cheng, D.; Zhang, L.; Chen, Z.; Meng, X.; Zhang, F.; Han, X.; Liu, J.; Pan, J.; et al. Obesity occurring in apolipoprotein E-knockout mice has mild effects on fertility. *Reproduction* **2013**, *147*, 141–151. [CrossRef] [PubMed]

49. Beattie, J.H.; Duthie, S.J.; Kwun, I.; Ha, T.; Gordon, M.J. Rapid quantification of aortic lesions in ApoE$^{-/-}$ mice. *J. Vasc. Res.* **2009**, *46*, 347–352. [CrossRef] [PubMed]

50. Ward, H.; Mitrou, P.N.; Bowman, R.; Luben, R.; Wareham, N.J.; Khaw, K.T.; Bingham, S. APOE genotype, lipids, and coronary heart disease risk: A prospective population study. *Arch. Intern. Med.* **2009**, *169*, 1424–1429. [CrossRef] [PubMed]

51. Violi, F.; Pignatelli, P. Platelet oxidative stress and thrombosis. *Thromb. Res.* **2012**, *129*, 378–381. [CrossRef] [PubMed]

52. Loeb, A.L.; Gear, A.R. Potentiation of platelet aggregation by atrial natriuretic peptide. *Life Sci.* **1988**, *43*, 731–738. [CrossRef]

53. Carnevale, R.; Pignatelli, P.; Frati, G.; Nocella, C.; Stanzione, R.; Pastori, D.; Marchitti, S.; Valenti, V.; Santulli, M.; Barbato, E.; et al. C2238 ANP gene variant promotes increased platelet aggregation through the activation of Nox2 and the reduction of cAMP. *Sci. Rep.* **2017**, *7*, 3797. [CrossRef] [PubMed]

54. Pignatelli, P.; Sanguigni, V.; Lenti, L.; Ferro, D.; Finocchi, A.; Rossi, P.; Violi, F. Gp91phox-dependent expression of platelet CD40 ligand. *Circulation* **2004**, *110*, 1326–1329. [CrossRef] [PubMed]

55. Strisciuglio, T.; Barbato, E.; De Biase, C.; Di Gioia, G.; Cotugno, M.; Stanzione, R.; Trimarco, B.; Sciarretta, S.; Volpe, M.; Wijns, W.; et al. T2238C atrial natriuretic peptide gene variant and the response to antiplatelet therapy in stable ischemic heart disease patients. *J. Cardiovasc. Transl. Res.* **2017**. [CrossRef] [PubMed]

56. Schneider, D.J. Factors contributing to increased platelet reactivity in people with diabetes. *Diabetes Care* **2009**, *32*, 525–527. [CrossRef] [PubMed]

57. Rubattu, S.; Sciarretta, S.; Marchitti, S.; D'Agostino, M.; Battistoni, A.; Calvieri, C.; Volpe, M. NT-proANP/ANP is a determinant of vascular damage in humans. *High Blood Press Cardiovasc. Prev.* **2010**, *17*, 121–130. [CrossRef]

MDPI
St. Alban-Anlage 66
4052 Basel
Switzerland
Tel. +41 61 683 77 34
Fax +41 61 302 89 18
www.mdpi.com

International Journal of Molecular Sciences Editorial Office
E-mail: ijms@mdpi.com
www.mdpi.com/journal/ijms